Mängelexemplar

WP8   €27
'14

ex libris

# Springer Texts in Business and Economics

For further volumes:
www.springer.com/series/10099

Norman Schofield

# Mathematical Methods in Economics and Social Choice

Second Edition

Norman Schofield
Center in Political Economy
Washington University in Saint Louis
Saint Louis, MO, USA

ISSN 2192-4333  ISSN 2192-4341 (electronic)
ISBN 978-3-642-39817-9  ISBN 978-3-642-39818-6 (eBook)
DOI 10.1007/978-3-642-39818-6
Springer Heidelberg New York Dordrecht London

© Springer-Verlag Berlin Heidelberg 2004, 2014
This work is subject to copyright. All rights are reserved by the Publisher, whether the whole or part of the material is concerned, specifically the rights of translation, reprinting, reuse of illustrations, recitation, broadcasting, reproduction on microfilms or in any other physical way, and transmission or information storage and retrieval, electronic adaptation, computer software, or by similar or dissimilar methodology now known or hereafter developed. Exempted from this legal reservation are brief excerpts in connection with reviews or scholarly analysis or material supplied specifically for the purpose of being entered and executed on a computer system, for exclusive use by the purchaser of the work. Duplication of this publication or parts thereof is permitted only under the provisions of the Copyright Law of the Publisher's location, in its current version, and permission for use must always be obtained from Springer. Permissions for use may be obtained through RightsLink at the Copyright Clearance Center. Violations are liable to prosecution under the respective Copyright Law.
The use of general descriptive names, registered names, trademarks, service marks, etc. in this publication does not imply, even in the absence of a specific statement, that such names are exempt from the relevant protective laws and regulations and therefore free for general use.
While the advice and information in this book are believed to be true and accurate at the date of publication, neither the authors nor the editors nor the publisher can accept any legal responsibility for any errors or omissions that may be made. The publisher makes no warranty, express or implied, with respect to the material contained herein.

Printed on acid-free paper

Springer is part of Springer Science+Business Media (www.springer.com)

*Dedicated to the memory of Jeffrey Banks
and Richard McKelvey*

# Foreword

The use of mathematics in the social sciences is expanding both in breadth and depth at an increasing rate. It has made its way from economics into the other social sciences, often accompanied by the same controversy that raged in economics in the 1950s. And its use has deepened from calculus to topology and measure theory to the methods of differential topology and functional analysis.

The reasons for this expansion are several. First, and perhaps foremost, mathematics makes communication between researchers succinct and precise. Second, it helps make assumptions and models clear; this bypasses arguments in the field that are a result of different implicit assumptions. Third, proofs are rigorous, so mathematics helps avoid mistakes in the literature. Fourth, its use often provides more insights into the models. And finally, the models can be applied to different contexts without repeating the analysis, simply by renaming the symbols.

Of course, the formulation of social science questions must precede the construction of models and the distillation of these models down to mathematical problems, for otherwise the assumptions might be inappropriate.

A consequence of the pervasive use of mathematics in our research is a change in the level of mathematics training required of our graduate students. We need reference and graduate text books that address applications of advanced mathematics to a widening range of social sciences. This book fills that need.

Many years ago, Bill Riker introduced me to Norman Schofield's work and then to Norman. He is unique in his ability to span the social sciences and apply integrative mathematical reasoning to them all. The emphasis on his work and his book is on smooth models and techniques, while the motivating examples for presentation of the mathematics are drawn primarily from economics and political science. The reader is taken from basic set theory to the mathematics used to solve problems at the cutting edge of research. Students in every social science will find exposure to this mode of analysis useful; it elucidates the common threads in different fields. Speculations at the end of Chap. 5 provide students and researchers with many open research questions related to the content of the first four chapters. The answers are in these chapters. When the first edition appeared in 2004, I wrote in my Foreword that a goal of the reader should be to write Chap. 6. For the second edition of the book, Norman himself has accomplished this open task.

St. Louis, Missouri, USA  Marcus Berliant
2013

# Preface to the Second Edition

For the second edition, I have added a new chapter six. This chapter continues with the model presented in Chap. 3 by developing the idea of dynamical social choice. In particular the chapter considers the possibility of cycles enveloping the set of social alternatives.

A theorem of Saari (1997) shows that for any non-collegial set, $\mathbb{D}$, of decisive or winning coalitions, if the dimension of the policy space is sufficiently large, then the choice is empty under $\mathbb{D}$ for all smooth profiles in a residual subspace of $C^r(W, \Re^n)$. In other words the choice is generically empty.

However, we can define a social solution concept, known as the heart. When regarded as a correspondence, the heart is lower hemi-continuous. In general the heart is centrally located with respect to the distribution of voter preferences, and is guaranteed to be non-empty. Two examples are given to show how the heart is determined by the symmetry of the voter distribution.

Finally, to be able to use survey data of voter preferences, the chapter introduces the idea of stochastic social choice. In situations where voter choice is given by a probability vector, we can model the choice by assuming that candidates choose policies to maximise their vote shares. In general the equilibrium vote maximising positions can be shown to be at the electoral mean. The necessary and sufficient condition for this is given by the negative definiteness of the candidate vote Hessians. In an empirical example, a multinomial logit model of the 2008 Presidential election is presented, based on the American National Election Survey, and the parameters of this model used to calculate the Hessians of the vote functions for both candidates. According to this example both candidates should have converged to the electoral mean.

Saint Louis, Missouri, USA  Norman Schofield
June 13, 2013

# Preface to the First Edition

In recent years, the optimisation techniques, which have proved so useful in microeconomic theory, have been extended to incorporate more powerful topological and differential methods. These methods have led to new results on the qualitative behaviour of general economic and political systems. However, these developments have also led to an increase in the degree of formalism in published work. This formalism can often deter graduate students. My hope is that the progression of ideas presented in these lecture notes will familiarise the student with the geometric concepts underlying these topological methods, and, as a result, make mathematical economics, general equilibrium theory, and social choice theory more accessible.

The first chapter of the book introduces the general idea of mathematical structure and representation, while the second chapter analyses linear systems and the representation of transformations of linear systems by matrices. In the third chapter, topological ideas and continuity are introduced and used to solve convex optimisation problems. These techniques are also used to examine existence of a "social equilibrium." Chapter four then goes on to study calculus techniques using a linear approximation, the differential, of a function to study its "local" behaviour.

The book is not intended to cover the full extent of mathematical economics or general equilibrium theory. However, in the last sections of the third and fourth chapters I have introduced some of the standard tools of economic theory, namely the Kuhn Tucker Theorem, together with some elements of convex analysis and procedures using the Lagrangian. Chapter four provides examples of consumer and producer optimisation. The final section of the chapter also discusses, in a heuristic fashion, the smooth or critical Pareto set and the idea of a regular economy. The fifth and final chapter is somewhat more advanced, and extends the differential calculus of a real valued function to the analysis of a smooth function between "local" vector spaces, or manifolds. Modem singularity theory is the study and classification of all such smooth functions, and the purpose of the final chapter to use this perspective to obtain a generic or typical picture of the Pareto set and the set of Walrasian equilibria of an exchange economy.

Since the underlying mathematics of this final section are rather difficult, I have not attempted rigorous proofs, but rather have sought to lay out the natural path of development from elementary differential calculus to the powerful tools of singularity theory. In the text I have referred to work of Debreu, Balasko, Smale, and Saari, among others who, in the last few years, have used the tools of singularity theory to develop a deeper insight into the geometric structure of both the economy and the polity. These ideas are at the heart of recent notions of "chaos." Some speculations on this profound way of thinking about the world are offered in Sect. 5.6. Review exercises are provided at the end of the book.

I thank Annette Milford for typing the manuscript and Diana Ivanov for the preparation of the figures.

I am also indebted to my graduate students for the pertinent questions they asked during the courses on mathematical methods in economics and social choice, which I have given at Essex University, the California Institute of Technology, and Washington University in St. Louis.

In particular, while I was at the California Institute of Technology I had the privilege of working with Richard McKelvey and of discussing ideas in social choice theory with Jeff Banks. It is a great loss that they have both passed away. This book is dedicated to their memory.

Saint Louis, Missouri, USA                                          Norman Schofield

# Contents

| | | |
|---|---|---|
| **1** | **Sets, Relations, and Preferences** | 1 |
| 1.1 | Elements of Set Theory | 1 |
| | 1.1.1 A Set Theory | 2 |
| | 1.1.2 A Propositional Calculus | 4 |
| | 1.1.3 Partitions and Covers | 6 |
| | 1.1.4 The Universal and Existential Quantifiers | 7 |
| 1.2 | Relations, Functions and Operations | 7 |
| | 1.2.1 Relations | 7 |
| | 1.2.2 Mappings | 8 |
| | 1.2.3 Functions | 11 |
| 1.3 | Groups and Morphisms | 12 |
| 1.4 | Preferences and Choices | 24 |
| | 1.4.1 Preference Relations | 24 |
| | 1.4.2 Rationality | 25 |
| | 1.4.3 Choices | 27 |
| 1.5 | Social Choice and Arrow's Impossibility Theorem | 32 |
| | 1.5.1 Oligarchies and Filters | 33 |
| | 1.5.2 Acyclicity and the Collegium | 35 |
| | Further Reading | 38 |
| **2** | **Linear Spaces and Transformations** | 39 |
| 2.1 | Vector Spaces | 39 |
| 2.2 | Linear Transformations | 45 |
| | 2.2.1 Matrices | 46 |
| | 2.2.2 The Dimension Theorem | 49 |
| | 2.2.3 The General Linear Group | 53 |
| | 2.2.4 Change of Basis | 55 |
| | 2.2.5 Examples | 59 |
| 2.3 | Canonical Representation | 62 |
| | 2.3.1 Eigenvectors and Eigenvalues | 63 |
| | 2.3.2 Examples | 66 |
| | 2.3.3 Symmetric Matrices and Quadratic Forms | 67 |
| | 2.3.4 Examples | 71 |
| 2.4 | Geometric Interpretation of a Linear Transformation | 73 |

## 3 Topology and Convex Optimisation ... 77
- 3.1 A Topological Space ... 77
  - 3.1.1 Scalar Product and Norms ... 77
  - 3.1.2 A Topology on a Set ... 82
- 3.2 Continuity ... 88
- 3.3 Compactness ... 93
- 3.4 Convexity ... 99
  - 3.4.1 A Convex Set ... 99
  - 3.4.2 Examples ... 100
  - 3.4.3 Separation Properties of Convex Sets ... 104
- 3.5 Optimisation on Convex Sets ... 110
  - 3.5.1 Optimisation of a Convex Preference Correspondence ... 110
- 3.6 Kuhn-Tucker Theorem ... 115
- 3.7 Choice on Compact Sets ... 118
- 3.8 Political and Economic Choice ... 125
- Further Reading ... 132

## 4 Differential Calculus and Smooth Optimisation ... 135
- 4.1 Differential of a Function ... 135
- 4.2 $C^r$-Differentiable Functions ... 142
  - 4.2.1 The Hessian ... 142
  - 4.2.2 Taylor's Theorem ... 145
  - 4.2.3 Critical Points of a Function ... 149
- 4.3 Constrained Optimisation ... 155
  - 4.3.1 Concave and Quasi-concave Functions ... 155
  - 4.3.2 Economic Optimisation with Exogenous Prices ... 162
- 4.4 The Pareto Set and Price Equilibria ... 171
  - 4.4.1 The Welfare and Core Theorems ... 171
  - 4.4.2 Equilibria in an Exchange Economy ... 180
- Further Reading ... 187

## 5 Singularity Theory and General Equilibrium ... 189
- 5.1 Singularity Theory ... 189
  - 5.1.1 Regular Points: The Inverse and Implicit Function Theorem ... 189
  - 5.1.2 Singular Points and Morse Functions ... 196
- 5.2 Transversality ... 200
- 5.3 Generic Existence of Regular Economies ... 203
- 5.4 Economic Adjustment and Excess Demand ... 207
- 5.5 Structural Stability of a Vector Field ... 210
- 5.6 Speculations on Chaos ... 221
- Further Reading ... 227

## 6 Topology and Social Choice ... 231
- 6.1 Existence of a Choice ... 231
- 6.2 Dynamical Choice Functions ... 233

|  | 6.3 | Stochastic Choice | 238 |
|---|---|---|---|
|  |  | 6.3.1 The Model Without Activist Valence Functions | 244 |
|  | References | | 248 |
| **7** | **Review Exercises** | | 251 |
|  | 7.1 | Exercises to Chap. 1 | 251 |
|  | 7.2 | Exercises to Chap. 2 | 251 |
|  | 7.3 | Exercises to Chap. 3 | 253 |
|  | 7.4 | Exercises to Chap. 4 | 255 |
|  | 7.5 | Exercises to Chap. 5 | 255 |

**Subject Index** . . . . . . . . . . . . . . . . . . . . . . . . . . . . . . . 257

**Author Index** . . . . . . . . . . . . . . . . . . . . . . . . . . . . . . . 261

# Sets, Relations, and Preferences 1

In this chapter we introduce elementary set theory and the notation to be used throughout the book. We also define the notions of a binary relation, of a function, as well as the axioms of a group and field. Finally we discuss the idea of an individual and social preference relation, and mention some of the concepts of social choice and welfare economics.

## 1.1 Elements of Set Theory

Let $\mathcal{U}$ be a collection of objects, which we shall call the domain of discourse, the universal set, or universe. A set $B$ in this universe (namely a subset of $\mathcal{U}$) is a subcollection of objects from $\mathcal{U}$. $B$ may be defined either explicitly by enumerating the objects, for example by writing

$$B = \{\text{Tom, Dick, Harry}\}, \quad \text{or}$$
$$B = \{x_1, x_2, x_3, \ldots\}.$$

Alternatively $B$ may be defined implicitly by reference to some property $P(B)$, which characterises the elements of $B$, thus

$$B = \{x : x \text{ satisfies } P(B)\}.$$

For example:

$$B = \{x : x \text{ is an integer satisfying } 1 \leq x \leq 5\}$$

is a satisfactory definition of the set $B$, where the universal set could be the collection of all integers. If $B$ is a set, write $x \in B$ to mean that the element $x$ is a member of $B$. Write $\{x\}$ for the set which contains only one element, $x$.

If $A, B$ are two sets write $A \cap B$ for the *intersection*: that is the set which contains only those elements which are both in $A$ and $B$. Write $A \cup B$ for the *union*: that is the set whose elements are either in $A$ or $B$. The *null* set or *empty* set $\Phi$, is that subset of $\mathcal{U}$ which contains no elements in $\mathcal{U}$.

Finally if $A$ is a subset of $\mathcal{U}$, define the *negation* of $A$, or complement of $A$ in $\mathcal{U}$ to be the set $\mathcal{U} \backslash A = \overline{A} = \{x : x \text{ is in } \mathcal{U} \text{ but not in } A\}$.

### 1.1.1 A Set Theory

Now let $\Gamma$ be a family of subsets of $\mathcal{U}$, where $\Gamma$ includes both $\mathcal{U}$ and $\Phi$, i.e., $\Gamma = \{\mathcal{U}, \Phi, A, B, \ldots\}$.

If $A$ is a member of $\Gamma$, then write $A \in \Gamma$. Note that in this case $\Gamma$ is a collection or *family* of sets.

Suppose that $\Gamma$ satisfies the following properties:
1. for any $A \in \Gamma, \overline{A} \in \Gamma$,
2. for any $A, B$ in $\Gamma, A \cup B$ is in $\Gamma$,
3. for any $A, B$ in $\Gamma, A \cap B$ is in $\Gamma$.

Then we say that $\Gamma$ satisfies *closure* with respect to $(^-, \cup, \cap)$, and we call $\Gamma$ a *set theory*.

For example let $2^{\mathcal{U}}$ be the set of *all* subsets of $\mathcal{U}$, including both $\mathcal{U}$ and $\Phi$. Clearly $2^{\mathcal{U}}$ satisfies closure with respect to $(^-, \cup, \cap)$.

We shall call a set theory $\Gamma$ that satisfies the following axioms a *Boolean algebra*.

**Axioms**

| | | |
|---|---|---|
| S1. | Zero element | $A \cup \Phi = A, A \cap \Phi = \Phi$ |
| S2. | Identity element | $A \cup \mathcal{U} = \mathcal{U}, A \cap U = A$ |
| S3. | Idempotency | $A \cup A = A, A \cap A = A$ |
| S4. | Negativity | $A \cup \overline{A} = \mathcal{U}, A \cap \overline{A} = \Phi$ |
| | | $\overline{\overline{A}} = A$ |
| S5. | Commutativity | $A \cup B = B \cup A$ |
| | | $A \cap B = B \cap A$ |
| S6. | De Morgan Rule | $\overline{A \cup B} = \overline{A} \cap \overline{B}$ |
| | | $\overline{A \cap B} = \overline{A} \cup \overline{B}$ |
| S7. | Associativity | $A \cup (B \cup C) = (A \cup B) \cup C$ |
| | | $A \cap (B \cap C) = (A \cap B) \cap C$ |
| S8. | Distributivity | $A \cup (B \cap C) = (A \cup B) \cap (A \cup C)$ |
| | | $A \cap (B \cup C) = (A \cap B) \cup (A \cap C)$. |

We can illustrate each of the axioms by Venn diagrams in the following way.

Let the square on the page represent the universal set $\mathcal{U}$. A subset $B$ of points within $\mathcal{U}$ can then represent the set $B$. Given two subsets $A, B$ the union is the hatched area, while the intersection is the double hatched area. See Fig. 1.1.

We shall use $\subset$ to mean "included in". Thus "$A \subset B$" means that every element in $A$ is also an element of $B$. Thus:

## 1.1 Elements of Set Theory

**Fig. 1.1** Union

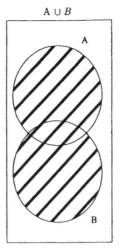
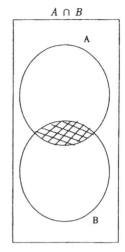

**Fig. 1.2** Inclusion

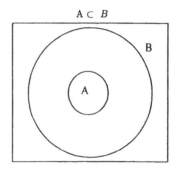

Suppose now that $P(A)$ is the property that characterizes $A$, or that

$$A = \{x : x \text{ satisfies } P(A)\}.$$

The symbol $\equiv$ means "identical to", so that $[x \in A] \equiv$ "$x$ satisfies $P(A)$".

Associated with any set theory is a *propositional calculus* which satisfies properties analogous with a Boolean algebra, except that we use $\wedge$ and $\vee$ instead of the symbols $\cap$ and $\cup$ for "and" and "or".

For example:

$$A \cup B = \{x : \text{"}x \text{ satisfies } P(A)\text{"} \vee \text{"}x \text{ satisfies } P(B)\text{"}\}$$
$$A \cap B = \{x : \text{"}x \text{ satisfies } P(A)\text{"} \wedge \text{"}x \text{ satisfies } P(B)\text{"}\}.$$

The analogue of "$\subset$" is "if...then" or "implies", which is written $\Rightarrow$.

Thus $A \subset B \equiv [\text{"}x \text{ satisfies } P(A)\text{"} \Rightarrow \text{"}x \text{ satisfies } P(B)\text{"}]$.

The analogue of "=" in set theory is the symbol "$\iff$" which means "if and only if", generally written "iff". For example,

$$[A = B] = [\text{``}x \text{ satisfies } P(A)\text{''} \iff \text{``}x \text{ satisfies } P(B)\text{''}].$$

Hence

$$[A = B] \equiv [\text{``}x \in A\text{''} \iff \text{``}x \in B\text{''}] \equiv [A \subset B \text{ and } B \subset A].$$

### 1.1.2 A Propositional Calculus

Let $\{\mathcal{U}, \Phi, P_1, \ldots, P_i, \ldots\}$ be a family of simple *propositions*. $\mathcal{U}$ is the *universal proposition* and always true, whereas $\Phi$ is the *null proposition* and always false. Two propositions $P_1, P_2$ can be combined to give a proposition $P_1 \wedge P_2$ (*i.e.*, $P_1$ *and* $P_2$) which is true iff both $P_1$ and $P_2$ are true, and a proposition $P_1 \vee P_2$ (*i.e.*, $P_1$ *or* $P_2$) which is true if either $P_1$ or $P_2$ is true. For a proposition $P$, the complement $\overline{P}$ in $\mathcal{U}$ is true iff $P$ is false, and is false iff $P$ is true.

Now extend the family of simple propositions to a family $\mathcal{P}$, by including in $\mathcal{P}$ any *propositional sentence* $S(P_1, \ldots, P_i, \ldots)$ which is made up of simple propositions combined under $^-, \vee, \wedge$. Then $\mathcal{P}$ satisfies *closure* with respect to $(^-, \vee, \wedge)$ and is called a *propositional calculus*.

Let $T$ be the truth function, which assigns to any simple proposition, $P_i$, the value 0 if $P_i$ is false and 1 if $P_i$ is true. Then $T$ extends to sentences in the obvious way, following the rules of logic, to give a truth function $T : \mathcal{P} \to \{0, 1\}$. If $T(S_1) = T(S_2)$ for all truth values of the constituent simple propositions of the sentences $S_1$ and $S_2$, then $S_1 = S_2$ (*i.e.*, $S_1$ and $S_2$ are identical propositions).

For example the truth values of the proposition $P_1 \vee P_2$ and $P_2 \vee P_1$ are given by the table:

| $T(P_1)$ | $T(P_2)$ | $T(P_1 \vee P_2)$ | $T(P_2 \vee P_1)$ |
|---|---|---|---|
| 0 | 0 | 0 | 0 |
| 0 | 1 | 1 | 1 |
| 1 | 0 | 1 | 1 |
| 1 | 1 | 1 | 1 |

Since $T(P_1 \vee P_2) = T(P_2 \vee P_1)$ for all truth values it must be the case that $P_1 \vee P_2 = P_2 \vee P_1$.

Similarly, the truth tables for $P_1 \wedge P_2$ and $P_2 \wedge P_1$ are:

| $T(P_1)$ | $T(P_2)$ | $T(P_1 \wedge P_2)$ | $T(P_2 \wedge P_1)$ |
|---|---|---|---|
| 0 | 0 | 0 | 0 |
| 0 | 1 | 0 | 0 |
| 1 | 0 | 0 | 0 |
| 1 | 1 | 1 | 1 |

Thus $P_1 \wedge P_2 = P_2 \wedge P_1$.

The propositional calculus satisfies commutativity of $\wedge$ and $\vee$. Using these truth tables the other properties of a Boolean algebra can be shown to be true.

## 1.1 Elements of Set Theory

For example:
(i) $P \vee \Phi = P$, $P \wedge \Phi = \Phi$.

| $T(P)$ | $T(\Phi)$ | $T(P \vee \Phi)$ | $T(P \wedge \Phi)$ |
|---|---|---|---|
| 0 | 0 | 0 | 0 |
| 1 | 0 | 1 | 0 |

(ii) $P \vee \mathcal{U} = \mathcal{U}$, $P \wedge \mathcal{U} = P$.

| $T(P)$ | $T(\mathcal{U})$ | $T(P \vee \mathcal{U})$ | $T(P \wedge \mathcal{U})$ |
|---|---|---|---|
| 0 | 1 | 1 | 0 |
| 1 | 1 | 1 | 1 |

(iii) Negation is given by reversing the truth value. Hence $\overline{\overline{P}} = P$.

| $T(P)$ | $T(\overline{P})$ | $T(\overline{\overline{P}})$ |
|---|---|---|
| 0 | 1 | 0 |
| 1 | 0 | 1 |

(iv) $P \vee \overline{P} = \mathcal{U}$, $P \wedge \overline{P} = \Phi$.

| $T(P)$ | $T(\overline{P})$ | $T(P \vee \overline{P})$ | $T(P \wedge \overline{P})$ |
|---|---|---|---|
| 0 | 1 | 1 | 0 |
| 1 | 0 | 1 | 0 |

*Example 1.1* Truth tables can be used to show that a propositional calculus $\mathcal{P} = (\mathcal{U}, \Phi, P_1, P_2, \ldots)$ with the operators ($^-, \vee, \wedge$) is a Boolean algebra.

Suppose now that $S_1(A_1, \ldots, A_n)$ is a compound set (or sentence) which is made up of the sets $A_1, \ldots, A_n$ together with the operators $\{\cup, \cap, ^-\}$.

For example suppose that

$$S_1(A_1, A_2, A_3) = A_1 \cup (A_2 \cap A_3),$$

and let $P(A_1), P(A_2), P(A_3)$ be the propositions that characterise $A_1, A_2, A_3$. Then

$$S_1(A_1, A_2, A_3) = \{x : x \text{ satisfies } "S_1\big(P(A_1), P(A_2), P(A_3)\big)"\}$$

$S_1(P(A_1), P(A_2), P(A_3))$ has precisely the same form as $S_1(A_1, A_2, A_3)$ except that $P(A_1)$ is substituted for $A_i$, and ($\wedge, \vee$) are substituted for ($\cap, \cup$).

In the example

$$S_1\big(P(A_1), P(A_2), P(A_3)\big) = P(A_1) \vee \big(P(A_2) \wedge P(A_3)\big).$$

Since $\mathcal{P}$ is a Boolean algebra, we know [by associativity] that $P(A_1) \vee (P(A_2) \wedge P(A_3)) = (P(A_1) \vee P(A_2)) \wedge (P(A_1) \vee P(A_3)) = S_2(P(A_1), P(A_2), P(A_3))$, say.

Hence the propositions $S_1(P(A_1), P(A_2), P(A_3))$ and $S_2(P(A_1), P(A_2), P(A_3))$, are identical, and the sentence

$$S_1(A_2, A_2, A_3) = \{x : x \text{ satisfies } P\big((A_1) \vee P(A_2)\big) \wedge \big(P(A_1) \vee P(A_3)\big)\}$$
$$= (A_1 \cup A_2) \cap (A_1 \cup A_3)$$
$$= S_2(A_1, A_2, A_3).$$

Consequently if $\Gamma = (\mathcal{U}, \Phi, A_1, A_2, \ldots)$ is a set theory, then by exactly this procedure $\Gamma$ can be shown to be a Boolean algebra.

Suppose now that $\Gamma$ is a set theory with universal set $\mathcal{U}$, and $X$ is a subset of $\mathcal{U}$. Let $\Gamma_X = (X, \Phi, A_1 \cap X, A_2 \cap X, \ldots)$. Since $\Gamma$ is a set theory on $\mathcal{U}$, $\Gamma_X$ must be a set theory on $X$, and thus there will exist a Boolean algebra in $\Gamma_X$.

To see this consider the following:
1. Since $A \in \Gamma$, then $\overline{A} \in \Gamma$. Now let $A_X = A \cap X$. To define the complement or negation (let us call it $\overline{A}_X$) of $A$ in $\Gamma_X$ we have $\overline{A}_X = \{x : x \text{ is in } X \text{ but not in } A\} = X \cap \overline{A}$. As we noted previously this is also often written $X - A$, or $X \backslash A$. But this must be the same as the complement or $A \cap X$ in $X$, i.e., $\overline{(A \cap X)} \cap X = (\overline{A} \cup \overline{X}) \cap X = (\overline{A} \cap X) \cup (\overline{X} \cap X) = \overline{A} \cap X$.
2. If $A, B \in \Gamma$ then $(A \cap B) \cap X = (A \cap X) \cap (B \cap X)$. (The reader should examine the behaviour of union.)

A notion that is very close to that of a set theory is that of a *topology*.

Say that a family $\Gamma = (\mathcal{U}, \Phi, A_1, A_2, \ldots)$ is a *topology* on $\mathcal{U}$ iff

**T1.** when $A_1, A_2 \in \Gamma$ then $A_1 \cap A_2 \in \Gamma$;

**T2.** If $A_j \in \Gamma$ for all $j$ belonging to some index set $J$ (possibly infinite) then $\bigcup_{j \in J} A_j \in \Gamma$.

**T3.** Both $\mathcal{U}$ and $\Phi$ belong to $\Gamma$.

Axioms **T1** and **T2** may be interpreted as saying that $\Gamma$ is closed under *finite intersection* and (infinite) *union*.

Let $X$ be any subset of $\mathcal{U}$. Then the *relative topology* $\Gamma_X$ induced from the topology $\Gamma$ on $\mathcal{U}$ is defined by

$$\Gamma_X = (X, \Phi, A_1 \cap X, \ldots)$$

where any set of the form $A \cap X$, for $A \in \Gamma$, belongs to $\Gamma_X$.

*Example 1.2* We can show that $\Gamma_X$ is a topology. If $U_1, U_2 \in \Gamma_X$ then there must exist sets $A_1, A_2 \in \Gamma$ such that $U_i = A_i \cap X$, for $i = 1, 2$. But then

$$U_1 \cap U_2 = (A_1 \cap X) \cap (A_2 \cap X)$$
$$= (A_1 \cap A_2) \cap X.$$

Since $\Gamma$ is a topology, $A_1 \cap A_2 \in \Gamma$. Thus $U_1 \cap U_2 \in \Gamma_X$. Union follows similarly.

### 1.1.3 Partitions and Covers

If $X$ is a set, a *cover* for $X$ is a family $\Gamma = (A_1, A_2, \ldots, A_j, \ldots)$ where $j$ belongs to an index set $J$ (possibly infinite) such that

$$X = \cup \{A_j : j \in J\}.$$

## 1.2 Relations, Functions and Operations

A *partition* for $X$ is a cover which is *disjoint*, *i.e.*, $A_j \cap A_k = \Phi$ for any distinct $j, k \in J$.

If $\Gamma_X$ is a cover for $X$, and $Y$ is a subset of $X$ then $\Gamma_Y = \{A_j \cap Y : j \in J\}$ is the induced cover on $Y$.

### 1.1.4 The Universal and Existential Quantifiers

Two operators which may be used to construct propositions are the universal and existential quantifiers.

For example, *"for all $x$* in $A$ it is the case that $x$ satisfies $P(A)$." The term "for all" is the universal quantifier, and generally written as $\forall$.

On the other hand we may say "there exists some $x$ in $A$ such that $x$ satisfies $P(A)$." The term "there exists" is the existential quantifier, generally written $\exists$.

Note that these have negations as follows:

$$\text{not}[\exists \, x \text{ s.t. } x \text{ satisfies } P] \equiv [\forall \, x : x \text{ does not satisfy} P]$$

$$\text{not}[\forall \, x : x \text{ satisfies} P] \equiv [\exists \, x \text{ s.t. } x \text{ does not satisfy } P].$$

We use *s.t.* to mean "such that".

## 1.2 Relations, Functions and Operations

### 1.2.1 Relations

If $X, Y$ are two sets, the *Cartesian product set* $X \times Y$ is the set of *ordered pairs* $(x, y)$ such that $x \in X$ and $y \in Y$.

For example if we let $\mathfrak{R}$ be the set of real numbers, then $\mathfrak{R} \times \mathfrak{R}$ or $\mathfrak{R}^2$ is the set

$$\{(x, y) : x \in \mathfrak{R}, y \in \mathfrak{R}\},$$

namely the plane. Similarly $\mathfrak{R}^n = \mathfrak{R} \times \cdots \times \mathfrak{R}$ ($n$ times) is the set of $n$-tuples of real numbers, defined by induction, *i.e.*, $\mathfrak{R}^n = \mathfrak{R} \times (\mathfrak{R} \times (\mathfrak{R} \times \cdots, \ldots))$.

A subset of the Cartesian product $Y \times X$ is called a *relation*, $P$, on $Y \times X$. If $(y, x) \in P$ then we sometimes write $yPx$ and say that $y$ stands in relation $P$ to $x$. If it is not the case that $(y, x) \in P$ then write $(y, x) \notin P$ or not $(yPx)$. $X$ is called the *domain* of $P$, and $Y$ is called the target or *codomain* of $P$.

If $V$ is a relation on $Y \times X$ and $W$ is a relation on $Z \times Y$, then define the relation $W \circ V$ to be the relation on $Z \times X$ given by $(z, x) \in W \circ V$ iff for some $y \in Y$, $(z, y) \in W$ and $(y, x) \in V$. The new relation $W \circ V$ on $Z \times X$ is called the composition of $W$ and $V$.

The *identity relation* (or diagonal) $e_X$ on $X \times X$ is

$$e_X = \{(x, x) : x \in X\}.$$

If $P$ is a relation on $Y \times X$, its *inverse*, $P^{-1}$, is the relation on $X \times Y$ defined by

$$P^{-1} = \{(x, y) \in X \times Y : (y, x) \in P\}.$$

Note that:

$$P^{-1} \circ P = \{(z, x) \in X \times X : \exists\, y \in Y \text{ s.t. } (z, y) \in P^{-1} \text{ and } (y, x) \in P\}.$$

Suppose that the *domain* of $P$ is $X$, i.e., for every $x \in X$ there is some $y \in Y$ s.t. $(y, x) \in P$. In this case for every $x \in X$, there exists $y \in Y$ such that $(x, y) \in P^{-1}$ and so $(x, x) \in P^{-1} \circ P$ for any $x \in X$. Hence $e_X \subset P^{-1} \circ P$. In the same way

$$P \circ P^{-1} = \{(t, y) \in Y \times Y : \exists x \in X \text{ s.t. } (t, x) \in P \text{ and } (x, y) \in P^{-1}\}$$

and so $e_Y \subset P \circ P^{-1}$.

## 1.2.2 Mappings

A relation $P$ on $Y \times X$ defines an assignment *or mapping* from $X$ to $Y$, which is called $\phi_P$ and is given by

$$\phi_P(x) = \{y : (y, x) \in P\}.$$

In general we write $\phi : X \to Y$ for a mapping which assigns to each element of $X$ the *set*, $\phi(x)$, of elements in $Y$. As above, the set $Y$ is called the *co-domain* of $\phi$.

The *domain* of a mapping, $\phi$, is the set $\{x \in X : \exists\, y \in Y \text{ s.t. } y \in \phi(x)\}$, and the *image* of $\phi$ is $\{y \in Y : \exists x \in X \text{ s.t. } y \in \phi(x)\}$.

Suppose now that $V, W$ are relations on $Y \times X, Z \times Y$ respectively. We have defined the composite relation $W \circ V$ on $Z \times X$. This defines a mapping $\phi_{W \circ V} : X \to Z$ by $z \in \phi_{W \circ V}(x)$ iff $\exists y \in Y$ such that $(y, x) \in V$ and $(z, y) \in W$. This in turn means that $y \in \phi_V(x)$ and $z \in \phi_W(y)$.

If $\phi : X \to Y$ and $\psi : Y \to Z$ are two mappings then define their *composition* $\psi \circ \phi : X \to Z$ by

$$(\psi \circ \phi)(x) = \psi[\phi(x)] = \cup\{\psi(y) : y \in \phi(x)\}.$$

Clearly $z \in \phi_{W \circ V}(x)$ iff $z \in \phi_W[\phi_V(x)]$.

Thus $\phi_{W \circ V}(x) = \phi_W[\phi_V(x)] = [(\phi_W \circ \phi_V)(x)]$, $\forall x \in X$. We therefore write $\phi_{W \circ V} = \phi_W \circ \phi_V$.

For example suppose $V$ and $W$ are given by

## 1.2 Relations, Functions and Operations

$$V : \{(2, 3), (3, 2), (1, 2), (4, 4), (4, 1)\} \quad \text{and}$$
$$W : \{(1, 4), (4, 4), (4, 1), (2, 1), (2, 3), (3, 2)\}$$

with mappings

$$\begin{array}{ccccc}
& \phi_V & & \phi_W & \\
1 & \rightarrow & 4 & \rightarrow & 1 \\
& \nearrow & & \searrow & \\
4 & & 1 & \rightarrow & 4 \\
& \nearrow & & \searrow & \\
2 & \rightarrow & 3 & \rightarrow & 2 \\
3 & \rightarrow & 2 & \rightarrow & 3
\end{array}$$

then the composite mapping $\phi_W \circ \phi_V = \phi_{W \circ V}$ is

$$\begin{array}{ccc}
1 & \rightarrow & 1 \\
\nearrow & \searrow & \\
4 & \rightarrow & 4 \\
& \searrow & \\
2 & \rightarrow & 2 \\
3 & \rightarrow & 3
\end{array}$$

with relation

$$W \circ V = \{(3, 3), (2, 2), (4, 2), (1, 4), (4, 4), (1, 1), (4, 1)\}.$$

Given a mapping $\phi : X \to Y$ then the reverse procedure to the above gives a relation, called the graph of $\phi$, or graph ($\phi$), where

$$\text{graph}(\phi) = \bigcup_{x \in X} (\phi(x) \times \{x\}) \subset Y \times X.$$

In the obvious way if $\phi : X \to Y$ and $\psi : Y \to Z$, are mappings, with composition $\psi \circ \phi : X \to Z$, then graph $(\psi \circ \phi) = \text{graph}(\psi) \circ \text{graph}(\phi)$.

Suppose now that $P$ is a relation on $Y \times X$, with inverse $P^{-1}$ on $X \times Y$, and let $\phi_P : X \to Y$ be the mapping defined by $P$. Then the mapping $\phi_{P^{-1}} : Y \to X$ is defined as follows:

$$\begin{aligned}
\phi_{P^{-1}}(y) &= \{x : (x, y) \in P^{-1}\} \\
&= \{x : (y, x) \in P\} \\
&= \{x : y \in \phi_P(x)\}.
\end{aligned}$$

More generally if $\phi : X \to Y$ is a mapping then the inverse mapping $\phi^{-1} : Y \to X$ is given by

$$\phi^{-1}(y) = \{x : y \in \phi(x)\}.$$

Thus

$$\phi_{P^{-1}} = (\phi_P)^{-1} : Y \to X.$$

For example let $\mathcal{Z}_4$ be the first four positive integers and let $P$ be the relation on $\mathcal{Z}_4 \times \mathcal{Z}_4$ given by

$$P = \{(2,3), (3,2), (1,2), (4,4), (4,1)\}.$$

Then the mapping $\phi_P$ and inverse $\phi_{P^{-1}}$ are given by:

$$\phi_P: \begin{array}{ccc} 1 & \to & 4 \\ & \nearrow & 1 \\ 4 & \nearrow & \\ 2 & \to & 3 \\ 3 & \to & 2 \end{array} \qquad \phi_{P^{-1}}: \begin{array}{ccc} 4 & \to & 1 \\ 1 & \searrow & \\ & \searrow & 4 \\ 3 & \to & 2 \\ 2 & \to & 3 \end{array}$$

If we compose $P^{-1}$ and $P$ as above then we obtain

$$P^{-1} \circ P = \{(1,1), (1,4), (4,1), (4,4), (2,2), (3,3)\},$$

with mapping

$$\begin{array}{ccc} & \phi_{P^{-1}} \circ \phi_P & \\ 1 & \to & 1 \\ & \nearrow \searrow & \\ 4 & \to & 4 \\ 2 & \to & 2 \\ 3 & \to & 3 \end{array}$$

Note that $P^{-1} \circ P$ contains the identity or diagonal relation $e = \{(1,1), (2,2), (3,3), (4,4)\}$ on $\mathcal{Z}_4 = \{1, 2, 3, 4\}$. Moreover $\phi_{P^{-1}} \circ \phi_P = \phi_{(P^{-1} \circ P)}$.

The mapping $id_X : X \to X$ defined by $id_X(x) = x$ is called the *identity* mapping on $X$. Clearly if $e_X$ is the identity relation, then $\phi_{e_X} = id_X$ and graph $(id_X) = e_X$.

If $\phi, \psi$ are two mappings $X \to Y$ then write $\psi \subset \phi$ iff for each $x \in X$, $\psi(x) \subset \phi(x)$.

As we have seen $e_X \subset P^{-1} \circ P$ and so

$$\phi_{e_X} = id_X \subset \phi_{(P^{-1} \circ P)} = \phi_{P^{-1}} \circ \phi_P = (\phi_P)^{-1} \circ \phi_P.$$

(This is only precisely true when $X$ is the domain of $P$, i.e., when for every $x \in X$ there exists some $y \in Y$ such that $(y, x) \in P$.)

## 1.2.3 Functions

If for all $x$ in the domain of $\phi$, there is exactly one $y$ such that $y \in \phi(x)$ then $\phi$ is called a *function*. In this case we generally write $f : X \to Y$, and sometimes $x \xrightarrow{f} y$ to indicate that $f(x) = y$. Consider the function $f$ and its inverse $f^{-1}$ given by

$$
\begin{array}{ccccccc}
& f & & f^{-1} & & f^{-1} \circ f & \\
1 & \to & 4 & \to & 1 & 1 & \to & 1 \\
& \nearrow & & \searrow & & & \nearrow \searrow & \\
4 & & 1 & & 4 & 4 & \to & 4 \\
2 & \to & 3 & \to & 2 & 2 & \to & 2 \\
3 & \to & 2 & \to & 3 & 3 & \to & 3
\end{array}
$$

Clearly $f^{-1}$ is not a function since it maps 4 to both 1 and 4, *i.e.*, the graph of $f^{-1}$ is $\{(1,4), (4,4), (2,3), (3,2)\}$. In this case $id_X$ is contained in $f^{-1} \circ f$ but is not identical to $f^{-1} \circ f$. Suppose that $f^{-1}$ is in fact a function. Then it is necessary that for each $y$ in the image there be at most one $x$ such that $f(x) = y$. Alternatively if $f(x_1) = f(x_2)$ then it must be the case that $x_1 = x_2$. In this case $f$ is called $1-1$ or *injective*. Then $f^{-1}$ is a function and

$$id_X = f^{-1} \circ f \quad \text{on the domain } X \text{ of } f$$

$$id_Y = f \circ f^{-1} \quad \text{on the image } Y \text{ of } f.$$

A mapping $\phi : X \to Y$ is said to be *surjective* (or called a surjection) iff every $y \in Y$ belongs to the image of $\phi$; that is, $\exists\, x \in X$ s.t. $y \in \phi(x)$.

A function $f : X \to Y$ which is both injective and surjective is said to be *bijective*.

*Example 1.3* Consider

$$
\begin{array}{ccccc}
& \pi & & \pi^{-1} & \\
1 & \to & 4 & \to & 1 \\
4 & \to & 2 & \to & 4 \\
2 & \to & 3 & \to & 2 \\
3 & \to & 1 & \to & 3.
\end{array}
$$

In this case the domain and image of $\pi$ coincide and $\pi$ is known as a *permutation*. Consider the possibilities where $\phi$ is a mapping $\Re \to \Re$, with graph $(\phi) \subset \Re^2$. (Remember $\Re$ is the set of real numbers.) There are three cases:

(i)  $\phi$ is a mapping:

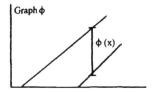

(ii)  $\phi$ is a non injective function:

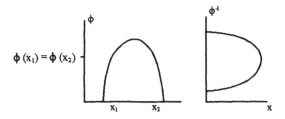

(iii)  $\phi$ is an injective function:

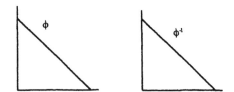

## 1.3 Groups and Morphisms

We earlier defined the composition of two mappings $\phi : X \to Y$ and $\psi : Y \to X$ to be $\psi \circ \phi : X \to Z$ given by $(\psi \circ \phi)(x) = \psi[\phi(x)] = \cup \{\psi(y) : y \in \phi(x)\}$. In the case of functions $f : X \to Y$ and $g : Y \to Z$ this translates to

$$(g \circ f)(x) = g[f(x)] = \{g(y) : y = f(x)\}.$$

Since both $f, g$ are functions the set on the right is a singleton set, and so $g \circ f$ is a function. Write $\mathcal{F}(A, B)$ for the set of functions from $A$ to $B$. Thus the composition operator, $\circ$, may be regarded as a function:

$$\circ : \quad \mathcal{F}(X, Y) \times \mathcal{F}(Y, Z) \quad \to \quad \mathcal{F}(X, Z)$$
$$(f, g) \quad\quad\quad\quad \to \quad g \circ f.$$

## 1.3 Groups and Morphisms

*Example 1.4* To illustrate consider the function (or matrix) $F$ given by

$$\begin{pmatrix} a & b \\ c & d \end{pmatrix} \begin{pmatrix} x_1 \\ x_2 \end{pmatrix} = \begin{pmatrix} ax_1 + bx_2 \\ cx_1 + dx_2 \end{pmatrix}.$$

This can be regarded as a function $F : \Re^2 \to \Re^2$ since it maps $(x_1, x_2) \to (ax_1 + bx_2, cx_1 + dx_2) \in \Re^2$.

Now let

$$F = \begin{pmatrix} a & b \\ c & d \end{pmatrix}, \quad H = \begin{pmatrix} e & f \\ g & h \end{pmatrix}.$$

$F \circ H$ is represented by

$$\begin{pmatrix} x_1 \\ x_2 \end{pmatrix} \xrightarrow{F} \begin{pmatrix} ax_1 + bx_2 \\ cx_1 + dx_2 \end{pmatrix} \xrightarrow{H} \begin{pmatrix} e(ax_1 + bx_2) + f(cx_1 + dx_2) \\ g(ax_1 + bx_2) + h(cx_1 + dx_2) \end{pmatrix}.$$

Thus

$$(H \circ F) \begin{pmatrix} x_1 \\ x_2 \end{pmatrix} = \begin{pmatrix} ea + fc \mid eb + fd \\ ga + hc \mid gb + hd \end{pmatrix} \begin{pmatrix} x_1 \\ x_2 \end{pmatrix}$$

or

$$(H \circ F) = \begin{pmatrix} e & f \\ g & h \end{pmatrix} \circ \begin{pmatrix} a & b \\ c & d \end{pmatrix} = \begin{pmatrix} ea + fc \mid eb + fd \\ ga + hc \mid gb + hd \end{pmatrix}.$$

The *identity* $E$ is the function

$$E \begin{pmatrix} x_1 \\ x_2 \end{pmatrix} = \begin{pmatrix} a & b \\ c & d \end{pmatrix} \begin{pmatrix} x_1 \\ x_2 \end{pmatrix} = \begin{pmatrix} x_1 \\ x_2 \end{pmatrix}.$$

Since this must be true for all $x_1, x_2$, it follows that $a = d = 1$ and $c = b = 0$.

Thus $E = \begin{pmatrix} 1 & 0 \\ 0 & 1 \end{pmatrix}$.

Suppose that the mapping $F^{-1} : \Re^2 \to \Re^2$ is actually a matrix. Then it is certainly a function, and by Sect. 1.2.3, $F^{-1} \circ F$ must be equal to the identity function on $\Re^2$, which here we call $E$. To determine $F^{-1}$, proceed as follows:

Let $F^{-1} = \begin{pmatrix} e & f \\ g & h \end{pmatrix}$. We know $F^{-1} \circ F = \begin{pmatrix} 1 & 0 \\ 0 & 1 \end{pmatrix}$.

Thus

$$ea + fc = 1 \quad | \quad eb + fd = 0$$
$$ga + hc = 0 \quad | \quad gb + hd = 1.$$

If $a \neq 0$ and $b \neq 0$ then $e = -\frac{fd}{b} = \frac{1-fc}{a}$.

Now let $|F| = (ad - bc)$, where $|F|$ is called the *determinant* of $F$. Clearly if $|F| \neq 0$, then $f = -b/|F|$. More generally, if $|F| \neq 0$ then we can solve the equations to obtain:

$$F^{-1} = \begin{pmatrix} e & f \\ g & h \end{pmatrix} = \frac{1}{|F|} \begin{pmatrix} d & -b \\ -c & a \end{pmatrix}.$$

If $|F| = 0$, then what we have called $F^{-1}$ is not defined. This suggests that when $|F| = 0$, the inverse $F^{-1}$ cannot be represented by a matrix, and in particular that $F^{-1}$ is not a function. In this case we shall call $F$ *singular*. When $|F| \neq 0$ then we shall call $F$ *non-singular*, and in this case $F^{-1}$ can be represented by a matrix, and thus a function. Let $M(2)$ stand for the set of $2 \times 2$ matrices, and let $M^*(2)$ be the subset of $M(2)$ consisting of non-singular matrices.

We have here defined a *composition operation*:

$$\begin{aligned} \circ : \quad M(2) \times M(2) &\to M(2) \\ (H, F) &\to H \circ F. \end{aligned}$$

Suppose we compose $E$ with $F$ then

$$E \circ F = \begin{pmatrix} 1 & 0 \\ 0 & 1 \end{pmatrix} \begin{pmatrix} a & b \\ c & d \end{pmatrix} = \begin{pmatrix} a & b \\ c & d \end{pmatrix} = F.$$

Finally for any $F \in M^*(2)$ it is the case that there exists a *unique* matrix $F^{-1} \in M(2)$ such that

$$F^{-1} \circ F = E.$$

Indeed if we compute the inverse $(F^{-1})^{-1}$ of $F^{-1}$ then we see that $(F^{-1})^{-1} = F$. Thus $F^{-1}$ itself belongs to $M^*(2)$.

$M^*(2)$ is an example of what is called a *group*.

More generally a *binary operation*, $\circ$, on a set $G$ is a function

$$\begin{aligned} \circ : \quad G \times G &\to G, \\ (x, y) &\to x \circ y. \end{aligned}$$

**Definition 1.1** A *group* $G$ is a set $G$ together with a binary operation, $\circ : G \times G \to G$ which
1. is associative: $(x \circ y) \circ z = x \circ (y \circ z)$ for all $x, y, z$ in $G$;
2. has an identity $e : e \circ x = x \circ e = x \ \forall x \in G$;
3. has for each $x \in G$ an inverse $x^{-1} \in G$ such that $x \circ x^{-1} = x^{-1} \circ x = e$.

When $G$ is a group with operation, $\circ$, write $(G, \circ)$ to signify this.

Associativity simply means that the order of composition in a sequence of compositions is irrelevant. For example consider the integers, $\mathcal{Z}$, under addition. Clearly $a + (b + c) = (a + b) + c$, where the left hand side means add $b$ to $c$, and then add

## 1.3 Groups and Morphisms

$a$ to this, while the right hand side is obtained by adding $a$ to $b$, and then adding $c$ to this. Under addition, the identity is that element $e \in \mathcal{Z}$ such that $a + e = a$. This is usually written 0. Finally the additive inverse of an integer $a \in \mathcal{Z}$ is $(-a)$ since $a + (-a) = 0$. Thus $(\mathcal{Z}, +)$ is a group.

However consider the integers under multiplication, which we shall write as "·". Again we have associativity since

$$a \cdot (b \cdot c) = (a \cdot b) \cdot c.$$

Clearly 1 is the identity since $1 \cdot a = a$. However the inverse of $a$ is that object $a^{-1}$ such that $a \cdot a^{-1} = 1$. Of course if $a = 0$, then no such inverse exists. For $a \neq 0$, $a^{-1}$ is more commonly written $\frac{1}{a}$. When $a$ is non-zero, and different from $\pm 1$, then $\frac{1}{a}$ is not an integer. Thus $(\mathcal{Z}, \cdot)$ is not a group. Consider the set $\mathcal{Q}$ of rationals, i.e., $a \in \mathcal{Q}$ iff $a = \frac{p}{q}$, where both $p$ and $q$ are integers. Clearly $1 \in \mathcal{Q}$. Moreover, if $a = \frac{p}{q}$ then $a^{-1} = \frac{q}{p}$ and so belongs to $\mathcal{Q}$. Although zero does not have an inverse, we can regard $(\mathcal{Q}\backslash\{0\}, \cdot)$ as a group.

**Lemma 1.1** *If $(G, \circ)$ is a group, then the identity $e$ is unique and for each $x \in G$ the inverse $x^{-1}$ is unique. By definition $e^{-1} = e$. Also $(x^{-1})^{-1} = x$ for any $x \in G$.*

*Proof*
1. Suppose there exist two distinct identities, $e$, $f$. Then $e \circ x = f \circ x$ for some $x$. Thus $(e \circ x) \circ x^{-1} = (f \circ x) \circ x^{-1}$. This is true because the composition operation

$$\big((e \circ x), x^{-1}\big) \to (e \circ x) \circ x^{-1}$$

   gives a unique answer.
   By associativity $(e \circ x) \circ x^{-1} = e \circ (x \circ x^{-1})$, etc.
   Thus $e \circ (x \circ x^{-1}) = f \circ (x \circ x^{-1})$. But $x \circ x^{-1} = e$, say.
   Since $e$ is an identity, $e \circ e = f \circ e$ and so $e = f$. Since $e \circ e = e$ it must be the case that $e^{-1} = e$.
2. In the same way suppose $x$ has two distinct inverses, $y$, $z$, so $x \circ y = x \circ z = e$. Then

$$y \circ (x \circ y) = y \circ (x \circ z)$$
$$(y \circ x) \circ y = (y \circ x) \circ z$$
$$e \circ y = e \circ z$$
$$y = z.$$

3. Finally consider the inverse of $x^{-1}$. Since $x \circ (x^{-1}) = e$ and by definition $(x^{-1})^{-1} \circ (x^{-1}) = e$ by part (2), it must be the case that $(x^{-1})^{-1} = x$. □

We can now construct some interesting groups.

**Lemma 1.2** *The set $M^*(2)$ of $2 \times 2$ non-singular matrices form a group under matrix composition, $\circ$.*

*Proof* We have already shown that there exists an identity matrix $E$ in $M^*(2)$. Clearly $|E| = 1$ and so $E$ has inverse $E$.

As we saw in Example 1.4, when we solved $H \circ F = E$ we found that

$$H = F^{-1} = \frac{1}{|F|} \begin{pmatrix} d & -b \\ -c & a \end{pmatrix}.$$

By Lemma 1.1, $(F^{-1})^{-1} = F$ and so $F^{-1}$ must have an inverse, i.e., $|F^{-1}| \neq 0$, and so $F^{-1}$ is non-singular. Suppose now that the two matrices $H, F$ belong to $M^*(2)$. Let

$$F = \begin{pmatrix} a & b \\ c & d \end{pmatrix}$$

and

$$H = \begin{pmatrix} e & f \\ g & h \end{pmatrix}.$$

As in Example 1.4,

$$|H \circ F| = \left| \begin{pmatrix} ea + fc & eb + fd \\ ga + hc & gb + hd \end{pmatrix} \right| = (ea + fc)(gb + hd) - (ga + hc)(eb + fd)$$
$$= (eh - gf)(ad - bc) = |H||F|.$$

Since both $H$ and $F$ are non-singular, $|H| \neq 0$ and $|F| \neq 0$ and so $|H \circ F| \neq 0$. Thus $H \circ F$ belongs to $M^*(2)$, and so matrix composition is a *binary operation* $M^*(2) \times M^*(2) \to M^*(2)$.

Finally the reader may like to verify that matrix composition on $M^*(2)$ is *associative*. That is to say if $F, G, H$ are non-singular $2 \times 2$ matrices then

$$H \circ (G \circ F) = (H \circ G) \circ F.$$

As a consequence $(M^*(2), \circ)$ is a group. □

*Example 1.5* For a second example consider the addition operation on $M(2)$ defined by

$$\begin{pmatrix} e & f \\ g & h \end{pmatrix} + \begin{pmatrix} a & b \\ c & d \end{pmatrix} = \begin{pmatrix} a+e & f+b \\ g+c & h+d \end{pmatrix}.$$

## 1.3 Groups and Morphisms

**Fig. 1.3** Rotation

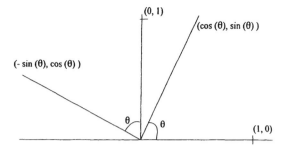

Clearly the identity matrix is $\begin{pmatrix} 0 & 0 \\ 0 & 0 \end{pmatrix}$ and the inverse of $F$ is

$$-F = -\begin{pmatrix} a & b \\ c & d \end{pmatrix} = \begin{pmatrix} -a & -b \\ -c & -d \end{pmatrix}.$$

Thus $(M(2), +)$ is a group.

Finally consider those matrices which represent *rotations* in $\Re^2$.

If we rotate the point $(1, 0)$ in the plane through an angle $\theta$ in the anticlockwise direction then the result is the point $(\cos\theta, \sin\theta)$, while the point $(0, 1)$ is transformed to $(-\sin\theta, \cos\theta)$. See Fig. 1.3. As we shall see later, this rotation can be represented by the matrix

$$\begin{pmatrix} \cos\theta & -\sin\theta \\ \sin\theta & \cos\theta \end{pmatrix}$$

which we will call $e^{i\theta}$.

Let $\Theta$ be the set of all matrices of this form, where $\theta$ can be any angle between 0 and 360°. If $e^{i\theta}$ and $e^{i\psi}$ are rotations by $\theta$, $\psi$ respectively, and we rotate by $\theta$ first and then by $\psi$, then the result should be identical to a rotation by $\psi + \theta$. To see this:

$$\begin{pmatrix} \cos\psi & -\sin\psi \\ \sin\psi & \cos\psi \end{pmatrix} \begin{pmatrix} \cos\theta & -\sin\theta \\ \sin\theta & \cos\theta \end{pmatrix}$$

$$= \begin{pmatrix} \cos\psi\cos\theta - \sin\psi\sin\theta & -\cos\psi\sin\theta - \sin\psi\cos\theta \\ \sin\psi\cos\theta + \cos\psi\sin\theta & -\sin\psi\sin\theta + \cos\psi\cos\theta \end{pmatrix}$$

$$= \begin{pmatrix} \cos(\psi+\theta) & -\sin(\psi+\theta) \\ \sin(\psi+\theta) & \cos(\psi+\theta) \end{pmatrix} = e^{i(\theta+\psi)}.$$

Note that $|e^{i\theta}| = \cos^2\theta + \sin^2\theta = 1$. Thus

$$\left(e^{i\theta}\right)^{-1} = \begin{pmatrix} \cos\theta & \sin\theta \\ -\sin\theta & \cos\theta \end{pmatrix} = \begin{pmatrix} \cos\theta & -\sin(-\theta) \\ \sin(-\theta) & \cos\theta \end{pmatrix} = e^{i(-\theta)}.$$

Hence the inverse to $e^{i\theta}$ is a rotation by $(-\theta)$, that is to say by $\theta$ but in the opposite direction. Clearly $E = e^{i0}$, a rotation through a zero angle. Thus $(\Theta, \circ)$ is a group. Moreover $\Theta$ is a subset of $M^*(2)$, since each rotation has a non-singular matrix. Thus $\Theta$ is a *subgroup* of $M^*(2)$.

A subset $\Theta$ of a group $(G, \circ)$ is a *subgroup* of $G$ iff the composition operation, $\circ$, restricted to $\Theta$ is "closed", and $\Theta$ is a group in its own right. That is to say (i) if $x, y \in \Theta$ then $x \circ y \in \Theta$, (ii) the identity $e$ belongs to $\Theta$ and (iii) for each $x$ in $\Theta$ the inverse, $x^{-1}$, also belongs to $\Theta$.

**Definition 1.2** Let $(X, \circ)$ and $(Y, \cdot)$ be two sets with binary operations, $\circ$, $\cdot$, respectively. A function $f : X \to Y$ is called a *morphism* (with respect to $(\circ, \cdot)$) iff $f(x \circ y) = f(x) \cdot f(y)$, for all $x, y \in X$. If moreover $f$ is bijective as a function, then it is called an *isomorphism*. If $(X, \circ), (Y, \cdot)$ are groups then $f$ is called a *homomorphism*.

A binary operation on a set $X$ is one form of mathematical structure that the set may possess. When an isomorphism exists between two sets $X$ and $Y$ then mathematically speaking their structures are identical.

For example let *Rot* be the set of all rotations in the plane. If $rot(\theta)$ and $rot(\psi)$ are rotations by $\theta$, $\psi$ respectively then we can combine them to give a rotation $rot(\psi + \theta)$, i.e.,

$$rot(\psi) \circ rot(\theta) = rot(\psi + \theta).$$

Here $\circ$ means do one rotation then the other. To the rotation, $rot(\theta)$ let $f$ assign the $2 \times 2$ matrix, called $e^{i\theta}$ as above. Thus

$$f : (rot, \circ) \to (\Theta, \circ),$$

where $f(rot(\theta)) = e^{i\theta}$.
Moreover

$$f\bigl(rot(\psi) \circ rot(\theta)\bigr) = f\bigl(rot(\psi + \theta)\bigr)$$

$$e^{i\psi} \circ e^{i\theta} = e^{i(\psi+\theta)}$$

Clearly the identity rotation is $rot(0)$ which corresponds to the zero matrix $e^{i0}$, while the inverse rotation to $rot(\theta)$ is $rot(-\theta)$ corresponding to $e^{-i\theta}$. Thus $f$ is a morphism.

Here we have a collection of geometric objects, called rotations, with their own structure and we have found another set of "mathematical" objects namely $2 \times 2$ matrices of a certain type, which has an identical structure.

**Lemma 1.3** *If $f : (X, \circ) \to (Y, \cdot)$ is a morphism between groups then*

## 1.3 Groups and Morphisms

(1) $f(e_X) = e_Y$ where $e_X, e_Y$ are the identities in $X, Y$.
(2) for each $x$ in $X$, $f(x^{-1}) = [f(x)]^{-1}$.

*Proof*
1. Since $f$ is a morphism $f(x \circ e_X) = f(x) \cdot f(e_X) = f(x)$. By Lemma 1.2, $e_Y$ is unique and so $f(e_X) = e_Y$.
2. $f(x \circ x^{-1}) = f(x) \cdot f(x^{-1}) = f(e_X) = e_Y$. By Lemma 1.2, $[f(x)]^{-1}$ is unique, and so $f(x^{-1}) = [f(x)]^{-1}$. $\square$

As an example, consider the determinant function $det : M(2) \to \Re$.

From the proof of Lemma 1.3, we know that for any $2 \times 2$ matrices, $H$ and $F$, it is the case that $|H \circ F| = |H| \, |F|$. Thus $det : (M(2), \circ) \to (\Re, \cdot)$ is a *morphism* with respect to matrix composition, $\circ$, in $M(2)$ and multiplication, $\cdot$, in $\Re$.

Note also that if $F$ is non-singular then $det(F) = |F| \neq 0$, and so $det : M^*(2) \to \Re \backslash \{0\}$.

It should be clear that $(\Re \backslash \{0\}, \cdot)$ is a group.

Hence $det : (M^*(2), \circ) \to (\Re \backslash \{0\}, \cdot)$ is a *homomorphism* between these two groups. This should indicate why those matrices in $M(2)$ which have zero determinant are those without an inverse in $M(2)$.

From Example 1.4, the identity in $M^*(2)$ is $E$, while the multiplicative identity in $\Re$ is 1. By Lemma 1.3, $det(E) = 1$.

Moreover $|F|^{-1} = \frac{1}{|F|}$ and so, by Lemma 1.3, $|F^{-1}| = \frac{1}{|F|}$. This is easy to check since

$$|F^{-1}| = \left| \frac{1}{|F|} \begin{pmatrix} d & -b \\ -c & a \end{pmatrix} \right| = \frac{da - bc}{|F|^2} = \frac{|F|}{|F|^2} = \frac{1}{|F|}.$$

However the determinant $det : M^*(2) \to \Re \backslash 0$ is not injective, since it is clearly possible to find two matrices, $H, F$ such that $|H| = |F|$ although $H$ and $F$ are different.

*Example 1.6* It is clear that the real numbers form a group $(\Re, +)$ under addition with identity 0, and inverse (to a) equal to $-a$. Similarly the reals form a group $(\Re \backslash \{0\}, \cdot)$ under multiplication, as long as we exclude 0.

Now let $\mathcal{Z}_2$ be the numbers $\{0, 1\}$ and define "addition modulo 2," written $+$, on $\mathcal{Z}_2$, by $0 + 0 = 0$, $0 + 1 = 1$, $1 + 0 = 1$, $1 + 1$, and "multiplication modulo 2," written $\cdot$, on $\mathcal{Z}_2$, by $0 \cdot 0 = 0$, $0 \cdot 1 = 1 \cdot 0 = 0$, $1 \cdot 1 = 1$.

Under "addition modulo 2," 0 is the identity, and 1 has inverse 1. Associativity is clearly satisfied, and so $(\mathcal{Z}_2, +)$ is a group. Under multiplication, 1 is the identity and inverse to itself, but 0 has no inverse. Thus $(\mathcal{Z}_2, \cdot)$ is not a group. Note that $(\mathcal{Z}_2 \backslash \{0\}, \cdot)$ is a group, namely the trivial group containing only one element. Let $\mathcal{Z}$ be the integers, and consider the function

$$f : \mathcal{Z} \to \mathcal{Z}_2,$$

defined by $f(x) = 0$ if $x$ is even, 1 if $x$ is odd.

We see that this is a morphism $f : (\mathcal{Z}, +) \to (\mathcal{Z}_2, +)$;
1. if $x$ and $y$ are both even then $f(x) = f(y) = 0$; since $x+y$ is even, $f(x+y) = 0$.
2. if $x$ is even and $y$ odd, $f(x) = 0$, $f(y) = 1$ and $f(x) + f(y) = 1$. But $x + y$ is odd, so $f(x+y) = 1$.
3. if $x$ and $y$ are both odd, then $f(x) = f(y) = 1$, and so $f(x) + f(y) = 0$. But $x + y$ is even, so $f(x+y) = 0$.

Since $(\mathcal{Z}, +)$ and $(\mathcal{Z}_2, +)$ are both groups, $f$ is a *homomorphism*. Thus $f(-a) = f(a)$.

On the other hand consider

$$f : (\mathcal{Z}, \cdot) \to (\mathcal{Z}_2, \cdot);$$

1. if $x$ and $y$ are both even then $f(x) = f(y) = 0$ and so $f(x) \cdot f(y) = 0 = f(xy)$.
2. if $x$ is even and $y$ odd, then $f(x) = 0$, $f(y) = 1$ and $f(x) \cdot f(y) = 0$. But $xy$ is even so $f(xy) = 0$.
3. if $x$ and $y$ are both odd, $f(x) = f(y) = 1$ and so $f(x)f(y) = 1$. But $xy$ is odd, and $f(xy) = 1$.

Hence $f$ is a *morphism*. However, neither $(\mathcal{Z}, \cdot)$ nor $(\mathcal{Z}_2, \cdot)$ is a group, and so $f$ is not a homomorphism.

A computer, since it is essentially a "finite" machine, is able to compute in binary arithmetic, using the two groups $(\mathcal{Z}_2, +), (\mathcal{Z}_2\backslash\{0\}, \cdot)$ rather than with the groups $(\mathfrak{R}, +), (\mathfrak{R}\backslash\{0\}, \cdot)$.

This is essentially because the additive and multiplicative groups based on $\mathcal{Z}_2$ form what is called a *field*.

**Definition 1.3**
1. A group $(G, \circ)$ is *commutative* or *abelian* iff for all $a, b \in G$, $a \circ b = b \circ a$.
2. A *field* $(\mathcal{F}, +, \cdot)$ is a set together with two operations called addition $(+)$ and multiplication $(\cdot)$ such that $(\mathcal{F}, +)$ is an abelian group with zero, or identity 0, and $(\mathcal{F}\backslash\{0\}, \cdot)$ is an abelian group with identity 1. For convenience the additive inverse of an element $a \in \mathcal{F}$ is written $(-a)$ and the multiplicative inverse of a non zero $a \in \mathcal{F}$ is written $a^{-1}$ or $\frac{1}{a}$.
Moreover, multiplication is *distributive* over addition, i.e., for all $a, b, c$ in $\mathcal{F}$, $a \cdot (b+c) = a \cdot b + a \cdot c$.

To give an indication of the notion of abelian group, consider $M^*(2)$ again. As we have seen

$$H \circ F = \begin{pmatrix} e & f \\ g & h \end{pmatrix} \circ \begin{pmatrix} a & b \\ c & d \end{pmatrix} = \begin{pmatrix} ea + fc & eb + fd \\ ga + hc & gb + hd \end{pmatrix}.$$

However,

$$F \circ H = \begin{pmatrix} a & b \\ c & d \end{pmatrix} \circ \begin{pmatrix} e & f \\ g & h \end{pmatrix} = \begin{pmatrix} ea + bg & af + bh \\ ce + dg & cf + dh \end{pmatrix}.$$

## 1.3 Groups and Morphisms

Thus $H \circ F \neq F \circ H$ in general and so $M^*(2)$ is non abelian. However, if we consider two rotations $e^{i\theta}, e^{i\psi}$ then $e^{i\psi} \circ e^{i\theta} = e^{i(\psi+\theta)} = e^{i\theta} \circ e^{i\psi}$. Thus the group $(\Theta, \circ)$ is abelian.

**Lemma 1.4** *Both $(\Re, +, \cdot)$ and $(\mathcal{Z}_2, +, \cdot)$ are fields.*

*Proof* Consider $(\mathcal{Z}_2, +, \cdot)$ first of all. As we have seen $(\mathcal{Z}_2, +)$ and $(\mathcal{Z}_2 \setminus \{0\}, \cdot)$ are groups. $(\mathcal{Z}_2, +)$ is obviously abelian since $0 + 1 = 1 + 0 = 1$, while $(\mathcal{Z}_2 \setminus \{0\}, \circ)$ is abelian since it has one element.

To check for distributivity, note that

$$1 \cdot (1+1) = 1 \cdot 0 = 0 = 1 \cdot 1 + 1 \cdot 1 = 1 + 1.$$

Finally to see that $(\Re, +, \cdot)$ is a field, we note that for any real numbers, $a, b, c$, $\Re$, $(b+c) = ab + ac$. $\square$

Given a field $(\mathcal{F}, +, \cdot)$ we define a new object called $\mathcal{F}^n$ where $n$ is a positive integer as follows. Any element $x \in \mathcal{F}^n$ is of the form

$$\begin{pmatrix} x_1 \\ \cdot \\ x_n \end{pmatrix}$$

where $x_1, \ldots, x_n$ all belong to $\mathcal{F}$.

**F1.** If $\alpha \in \mathcal{F}$, and $x \in \mathcal{F}^n$ define $\alpha x \in \mathcal{F}^n$ by

$$\alpha \begin{pmatrix} x_1 \\ \cdot \\ x_n \end{pmatrix} = \begin{pmatrix} \alpha x_1 \\ \cdot \\ \alpha x_n \end{pmatrix}.$$

**F2.** Define addition in $\mathcal{F}^n$ by

$$x + y = \begin{pmatrix} x_1 \\ \cdot \\ x_n \end{pmatrix} + \begin{pmatrix} y_1 \\ \cdot \\ y_n \end{pmatrix} = \begin{pmatrix} x_1 + y_1 \\ x_n + y_n \end{pmatrix}.$$

Since $\mathcal{F}$, by definition, is an abelian additive group, it follows that

$$x + y = \begin{pmatrix} x_1 + y_1 \\ \cdot \\ x_n + y_n \end{pmatrix} = \begin{pmatrix} y_1 + x_1 \\ \cdot \\ y_n + x_n \end{pmatrix} = y + x.$$

Now let
$$\underline{0} = \begin{pmatrix} 0 \\ \cdot \\ 0 \end{pmatrix}.$$

Clearly
$$x + \underline{0} = \begin{pmatrix} x_1 \\ \cdot \\ x_n \end{pmatrix} + \begin{pmatrix} 0 \\ \cdot \\ 0 \end{pmatrix} = x.$$

Hence $\underline{0}$ belongs to $\mathcal{F}^n$ and is an additive identity in $\mathcal{F}^n$.

Suppose we define
$$(-x) = -\begin{pmatrix} x_1 \\ \cdot \\ x_n \end{pmatrix} = \begin{pmatrix} -x_1 \\ \cdot \\ -x_n \end{pmatrix}.$$

Clearly
$$x + (-x) = \begin{pmatrix} x_1 \\ \cdot \\ x_n \end{pmatrix} + \begin{pmatrix} -x_1 \\ \cdot \\ -x_n \end{pmatrix} = \begin{pmatrix} x_1 - x_1 \\ x_n - x_n \end{pmatrix} = \underline{0}.$$

Thus for each $x \in \mathcal{F}^n$ there is an inverse, $(-x)$, in $\mathcal{F}^n$.

Finally, since F is an additive group
$$x + (y + z) = \begin{pmatrix} x_1 \\ \cdot \\ x_n \end{pmatrix} + \begin{pmatrix} y_1 + z_1 \\ \cdot \\ y_n + z_n \end{pmatrix} = \begin{pmatrix} x_1 + y_1 \\ \cdot \\ x_n + y_n \end{pmatrix} + \begin{pmatrix} z_1 \\ \cdot \\ z_n \end{pmatrix}$$
$$= (x + y) + z.$$

Thus $(\mathcal{F}^n, +)$ is an abelian group, with zero $\underline{0}$.

The fact that it is possible to multiply an element $x \in \mathcal{F}^n$ by a *scalar* $a \in \mathcal{F}$ endows $\mathcal{F}^n$ with further structure. To see this consider the example of $\mathfrak{R}^2$.

1.  If $a \in \mathfrak{R}$ and both $x, y$ belong to $\mathfrak{R}^2$, then
$$\alpha \left[ \begin{pmatrix} x_1 \\ x_2 \end{pmatrix} + \begin{pmatrix} y_1 \\ y_2 \end{pmatrix} \right] = \alpha \begin{pmatrix} x_1 + y_1 \\ x_2 + y_2 \end{pmatrix} = \begin{pmatrix} \alpha x_1 + \alpha y_1 \\ \alpha x_2 + \alpha y_2 \end{pmatrix}$$

by distribution, and
$$= \begin{pmatrix} \alpha x_1 \\ \alpha x_2 \end{pmatrix} + \begin{pmatrix} \alpha y_1 \\ \alpha y_2 \end{pmatrix} = \alpha \begin{pmatrix} x_1 \\ x_2 \end{pmatrix} + \alpha \begin{pmatrix} y_1 \\ y_2 \end{pmatrix}$$

by **F1**. Thus $\alpha(x + y) = \alpha x + \alpha y$.

## 1.3 Groups and Morphisms

2.
$$(\alpha + \beta)\begin{pmatrix} x_1 \\ x_2 \end{pmatrix} = \begin{pmatrix} (\alpha + \beta)x_1 \\ (\alpha + \beta)x_2 \end{pmatrix}$$

by **F1**

$$= \begin{pmatrix} \alpha x_1 + \beta x_1 \\ \alpha x_2 + \beta x_2 \end{pmatrix} = \begin{pmatrix} \alpha x_1 \\ \alpha x_2 \end{pmatrix} + \begin{pmatrix} \beta x_1 \\ \beta x_2 \end{pmatrix}$$

by **F2**

$$= \alpha \begin{pmatrix} x_1 \\ x_2 \end{pmatrix} + \beta \begin{pmatrix} x_1 \\ x_2 \end{pmatrix}$$

by **F1**. Therefore, $(\alpha + \beta)x = \alpha x + \beta x$.

3.
$$(\alpha \beta) \begin{pmatrix} x_1 \\ x_2 \end{pmatrix} = \begin{pmatrix} (\alpha\beta)x_1 \\ (\alpha\beta)x_2 \end{pmatrix}$$

by **F1** $= \alpha \begin{pmatrix} \beta x_1 \\ \beta x_2 \end{pmatrix}$ by associativity and **F1**, and $= \alpha(\beta x)$ by **F1**.
Thus $(\alpha\beta)x = \alpha(\beta x)$.

4.
$$\begin{pmatrix} x_1 \\ x_2 \end{pmatrix} = \begin{pmatrix} 1 \cdot x_1 \\ 1 \cdot x_2 \end{pmatrix} = \begin{pmatrix} x_1 \\ x_2 \end{pmatrix}.$$

Therefore $1(x) = x$.

These four properties characterise what is know as a *vector space*.

Finally, consider the operation of a matrix $F$ on the set of elements in $\Re^2$. By definition

$$F(x+y) = \begin{pmatrix} a & b \\ c & d \end{pmatrix} \left[ \begin{pmatrix} x_1 \\ x_2 \end{pmatrix} + \begin{pmatrix} y_1 \\ y_2 \end{pmatrix} \right] = \begin{pmatrix} a & b \\ c & d \end{pmatrix} \begin{pmatrix} x_1 + y_1 \\ x_2 + y_2 \end{pmatrix}$$

$$= \begin{pmatrix} a(x_1 + y_1) + b(x_2 + y_2) \\ c(x_1 + y_1) + d(x_2 + y_2) \end{pmatrix} = \begin{pmatrix} ax_1 + bx_2 \\ cx_1 + dx_2 \end{pmatrix} + \begin{pmatrix} ay_1 + by_2 \\ cy_1 + dy_2 \end{pmatrix}$$

by **F2**

$$= \begin{pmatrix} a & b \\ c & d \end{pmatrix} \begin{pmatrix} x_1 \\ x_2 \end{pmatrix} + \begin{pmatrix} a & b \\ c & d \end{pmatrix} \begin{pmatrix} y_1 \\ y_2 \end{pmatrix} = F(x) + F(y).$$

Hence $F : (\Re^2, +) \to (\Re^2, +)$ is a *morphism* from the abelian group $(\Re^2, +)$ into itself.

By Lemma 1.3, we know that $F(\underline{0}) = \underline{0}$, and for any element $x \in \Re^2$,

$$F(-x) = F(-1(x)) = -F(x) = -1F(x).$$

**Fig. 1.4** Indifference and strict preference

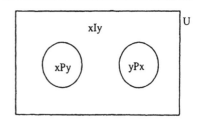

A morphism between vector spaces is called a *linear transformation*. Vector spaces and linear transformations are discussed in Chap. 2.

## 1.4 Preferences and Choices

### 1.4.1 Preference Relations

A *binary relation* $P$ on $X$ is a subset of $X \times X$; more simply $P$ is called a relation on $X$. For example let $X \equiv \Re$ (the real line) and let $P$ be ">" meaning strictly greater than. The relation ">" clearly satisfies the following properties:
1. it is never the case that $x > x$
2. it is never the case that $x > y$ *and* $y > x$
3. it is always the case that $x > y$ and $y > z$ implies $x > z$.

These properties can be considered more abstractly. A relation $P$ on $X$ is:
1. *symmetric* iff $xPy \Rightarrow yPx$
   *asymmetric* iff $xPy \Rightarrow$ not $(yPx)$
   *antisymmetric* iff $xPy$ and $yPx \Rightarrow x = y$
2. *reflexive* iff $(xPx) \; \forall \; x \in X$
   *irreflexive* iff not $(xPx) \; \forall x \in X$
3. *transitive* iff $xPy$ and $yPz \Rightarrow xPz$
4. *connected* iff for any $x, y \in X$ either $xPy$ or $yPx$.

By analogy with the relation ">" a relation $P$, which is both *irreflexive* and *asymmetric* is called a *strict preference relation*.

Given a strict preference relation $P$ on $X$, we can define two new relations called $I$, for *indifference*, and $R$ for *weak preference* as follows.
1. $xIy$ iff not $(xPy)$ and not $(yPx)$
2. $xRy$ iff $xPy$ or $xIy$.

By de Morgan's rule $xIy$ iff not $(xPy \vee yPx)$. Thus for any $x, y \in X$ either $xIy$ or $xPy$ or $yPx$. Since $P$ is asymmetric it cannot be the case that both $xPy$ and $yPx$ are true. Thus the propositions "$xPy$", "$yPx$", "$xIy$" are disjoint, and hence form a partition of the universal proposition, $U$.

Note that $(xPy \vee xIy) \equiv$ not $(yPx)$ since these three propositions form a (disjoint) partition. Thus $(xRy)$ iff not $(yPx)$.

In the case that $P$ is the strict preference relation ">", it should be clear that indifference is identical to "=" and weak preference to "> or =" usually written "$\geq$".

### 1.4 Preferences and Choices

**Lemma 1.5** *If P is a strict preference relation then indifference (I) is reflexive and symmetric, while weak preference (R) is reflexive and connected. Moreover if $xRy$ and $yRx$ then $xIy$.*

*Proof*
1. Since not $(xPx)$ this must imply $xIx$, so $I$ is reflexive
2. 
$$xIy \iff \text{not } (xPy) \wedge \text{not } (yPx)$$
$$\iff \text{not } (yPx) \wedge \text{not } (xPy)$$
$$\iff yIx.$$

   Hence $I$ is symmetric.
3. $xRy \iff xPy$ or $xIy$. Thus $xIx \Rightarrow xRx$, so $R$ is reflexive.
4. $xRy \iff xPy$ or $yIx$ and $yRx \iff yPx$ or $yIx$.
   ' Not $(xRy \vee yRx) \iff$ not $(xPy \vee yPx \vee xIy)$. But $xPy \vee yPx \vee xIy$ is always true since these three propositions form a partition of the universal set. Thus not $(xRy \vee yRx)$ is always false, and so $xRy \vee yRx$ is always true. Thus $R$ is connected.
5. Clearly

$$xRy \text{ and } yRx \iff (xPy \wedge yPx) \vee xIy$$
$$\iff xIy$$

since $xPy \wedge yPx$ is always false by asymmetry. $\square$

In the case that $P$ corresponds to ">" then $x \geq y$ and $y \geq x \Rightarrow x = y$, so "$\geq$" is antisymmetric.

Suppose that $P$ is a strict preference relation on $X$, and there exists a function $u : X \to \Re$, called a *utility function*, such that $xPy$ iff $u(x) > u(y)$. Therefore

$$xRy \quad \text{iff } u(x) \geq u(y)$$
$$xIy \quad \text{iff } u(x) = u(y).$$

The order relation ">" on the real line is transitive (since $x > y > z \Rightarrow x > z$). Therefore $P$ must be transitive when it is representable by a utility function.

We therefore have reason to consider "rationality" properties, such as transitivity, of a strict preference relation.

#### 1.4.2 Rationality

**Lemma 1.6** *If P is irreflexive and transitive on X then it is asymmetric.*

*Proof* To show that $A \wedge B \Rightarrow C$ we need only show that $B \wedge$ not $(C) \Rightarrow$ not $(A)$.

Therefore suppose that $P$ is transitive but fails asymmetry. By the latter assumption there exists $x, y \in X$ such that $xPy$ and $yPx$. By transitivity this gives $xPx$, which violates irreflexivity. □

Call a strict preference relation, $P$, on $X$ *negatively transitive* iff it is the case that, for all $x, y, z \in X$, not $(xPy) \wedge$ not $(yPz) \Rightarrow$ not $(xPz)$. Note that $xRy \Longleftrightarrow$ not $(yPx)$. Thus the negative transitivity of $P$ is equivalent to the property

$$yRx \wedge zRy \Rightarrow zRx.$$

Hence $R$ must be *transitive*.

**Lemma 1.7** *If $P$ is a strict preference relation that is negatively transitive then $P$, $I$, $R$ are all transitive.*

*Proof*
1. By the previous observation, $R$ is transitive.
2. To prove $P$ is transitive, suppose otherwise, *i.e.*, that there exist $x, y, z$ such that $xPy$, $yPz$ but not $(xPz)$. By definition not $(xPz) \Longleftrightarrow zRx$. Moreover $yPz$ or $yIz \Longleftrightarrow yRz$. Thus $yPz \Longrightarrow yRz$. By transitivity of $R$, $zRx$ and $yRz$ gives $yRx$, or not $(xPy)$. But we assumed $xPy$. By contradiction we must have $xPz$.
3. To show $I$ is transitive, suppose $xIy$, $yIz$ but not $(xIz)$. Suppose $xPz$, say. But then $xRz$. Because of the two indifferences we may write $zRy$ and $yRx$. By transitivity of $R$, $zRx$. But $zRx$ and $xRz$ imply $xIz$, a contradiction. In the same way if $zPx$, then $zRx$, and again $xIz$. Thus $I$ must be transitive. □

Note that this lemma also implies that $P$, $I$ and $R$ combine transitively. For example, if $xRy$ and $yPz$ then $xPz$.

To show this, suppose, in contradiction, that *not* $(xPz)$.

This is equivalent to $zRx$. If $xRy$, then by transitivity of $R$, we obtain $zRy$ and so *not* $(yPz)$. Thus $xRy$ and *not* $(xPz) \Rightarrow$ *not* $(yPz)$. But $yPz$ and *not* $(yPz)$ cannot both hold. Thus $xRy$ and $yPz \Rightarrow xPz$. Clearly we also obtain $xIy$ and $yPz \Rightarrow xPz$ for example.

When $P$ is a negatively transitive strict preference relation on $X$, then we call it a *weak order* on $X$. Let $O(X)$ be the set of weak orders on $X$. If $P$ is a transitive strict preference relation on $X$, then we call it a *strict partial order*. Let $T(X)$ be the set of strict partial orders on $X$. By Lemma 1.7, $O(X) \subset T(X)$.

Finally call a preference relation *acyclic* if it is the case that for any finite sequence $x_1, \ldots, x_r$ of points in $X$ if $x_j P x_{j+1}$ for $j = 1, \ldots, r-1$ then it cannot be the case that $x_r P x_1$.

Let $A(X)$ be the set of acyclic strict preference relations on $X$. To see that $T(x) \subset A(X)$, suppose that $P$ is transitive, but cyclic, *i.e.*, that there exists a finite cycle $x_1 P x_2 \ldots P x_r P x_1$. By transitivity $x_{r-1} P x_r P x_1$ gives $x_{r-1} P x_1$, and by repetition we obtain $x_2 P x_1$. But we also have $x_1 P x_2$, which violates asymmetry.

### 1.4.3 Choices

As we noted previously, if $P$ is a strict preference relation on a set $X$, then a maximal element, or *choice*, on $X$ is an element $x$ such that for no $y \in X$ is it the case that $yPx$. We can express this another way. Since $P \subset X \times X$, there is a mapping

$$\phi_P : X \to X \quad \text{where } \phi_P(x) = \{y : yPx\}.$$

We shall call $\phi_P$ the *preference correspondence* of $P$. The *choice* of $P$ on $X$ is the set $C_p(X) = \{x : \phi_P(X) = \Phi\}$. Suppose now that $P$ is a strict preference relation on $X$. For each subset $Y$ of $X$, let

$$C_P(Y) = \{x \in Y : \phi_P(X) \cap Y = \Phi\}.$$

This defines a *choice correspondence* $C_P : 2^X \to 2^X$ from $2^X$, the set of all subsets of $X$, into itself.

An important question in social choice and welfare economics concerns the existence of a "social" choice correspondence, $C_P$, which guarantees the non-emptiness of the social choice $C_P(Y)$ for each *feasible set*, $Y$, in $X$, and an appropriate social preference, $P$.

**Lemma 1.8** *If $P$ is an acyclic strict preference relation on a finite set $X$, then $C_P(Y)$ is non-empty for each subset $Y$ of $X$.*

*Proof* Suppose that $X = \{x_1, \ldots, x_r\}$. If all elements in $X$ are indifferent then clearly $C_P(X) = X$.

So we can assume that if the cardinality $|Y|$ of $Y$ is at least 2, then $x_2 P x_1$ for some $x_2, x_1$. We proceed by induction on the cardinality of $Y$.

If $Y = \{x_1\}$ then obviously $x_1 = C_P(Y)$.

If $Y = \{x_1, x_2\}$ then either $x_1 P x_2, x_2 P x_1$, or $x_1 I x_2$ in which case $C_P(Y) = \{x_1\}, \{x_2\}$ or $\{x_1, x_2\}$ respectively. Suppose $C_P(Y) \neq \Phi$ whenever the cardinality $|Y|$ of $Y$ is 2, and consider $Y' = \{x_1, x_2, x_3\}$.

Without loss of generality suppose that $x_2 \in C_P(\{x_1, x_2\})$, but that neither $x_1$ nor $x_2 \in C_P(Y')$. There are two possibilities (i) If $x_2 P x_1$ then by asymmetry of $P$, not $(x_1 P x_2)$. Since $x_2 \notin C_P(Y')$ then $x_3 P x_2$, so not $(x_2 P x_3)$. Suppose that $C_P(Y') = \Phi$. Then $x_1 P x_3$, and we obtain a cycle $x_1 P x_3 P x_2 P x_1$. This contradicts acyclicity, so $x_3 \in C_P(Y')$. (ii) If $x_2 I x_1$ and $x_3 \notin C_P(Y')$ then either $x_1 P x_3$ or $x_2 P x_3$. But neither $x_1$ nor $x_2 \in C_P(Y')$ so $x_3 P x_1$ and $x_3 P x_2$. This contradicts asymmetry of $P$. Consequently $x_3 \in C_P(Y')$.

It is clear that this argument can be generalised to the case when $|Y| = k$ and $Y'$ is a superset of $Y$ (i.e., $Y \subset Y'$ with $|Y'| = k + 1$.) So suppose $C_P(Y) \neq \Phi$. To show $C_P(Y') \neq \Phi$ when $Y' = Y \cup \{x_{k+1}\}$, suppose $[C_P(Y) \cup \{x_{k+1}\}] \cap C_P(Y') \neq \Phi$. Then there must exist some $x \in Y$ such that $x P x_{k+1}$.

If $x \in C_P(Y)$ then $x_{k+1} P x$, since $x \notin C_P(Y')$ and $zPx$ for no $z \in Y$. Hence we obtain the asymmetry $x P x_{k+1} P x$. On the other hand if $x \in Y \setminus C_P(Y)$, then there must exist a chain $x_r P x_{r-1} P \ldots x_1 P x$ with $r < k$, such that $x_r \in C_P(Y)$. Since $x_r \notin$

$C_P(Y')$ it must be the case that $x_{k+1} P x_r$. This gives a cycle $x P x_{k+1} P x_r P \ldots x$. By contradiction, $C_P(Y') \neq \Phi$.

By induction if $C_P(Y) \neq \Phi$ then $C_P(Y') \neq \Phi$ for any superset $Y'$ of $Y$. Since $C_P(Y) \neq \Phi$ whenever $|Y| = 2$, it is evident that $C_P(Y) \neq \Phi$ for any finite subset $Y$ of $X$. □

If $P$ is a strict preference relation on $X$ and $P$ is representable by a utility function $u : X \to \Re$ then it must be the case that all of $P, I, R$ are transitive. To see this, we note the following:
1.  $xRy$ and $yRz$ iff $u(x) \geq u(y) \geq u(z)$. Since "$\geq$" on $\Re$ is transitive it follows that $u(x) \geq u(z)$ and so $xRz$.
2.  $xIy$ and $yIz$ iff $u(x) = u(y) = u(z)$, and thus $xIz$.

In this case indifference, $I$, is reflexive, symmetric and transitive. Such a relation on $X$ is called an *equivalence relation*.

For any point $x$ in $X$, let $[x]$ be the equivalence class of $x$ in $X$, *i.e.*, $[x] = \{y : yIx\}$.

Every point in $X$ belongs to exactly *one* equivalence class. To see this suppose that $x \in [y]$ and $x \in [z]$, then $xIy$ and $xIz$. By symmetry $zIx$, and by transitivity $zIy$. Thus $[y] = [z]$.

The set of equivalence classes in $X$ under an equivalence relation, $I$, is written $X/I$. Clearly if $u : X \to \Re$ is a utility function then an equivalence class $[x]$ is of the form

$$[x] = \{y \in X : u(x) = u(y)\},$$

which we may also write as $u^{-1}[u(x)]$.

If $X$ is a finite set, and $P$ is representable by a utility function then

$$C_P(X) = \{x \in X : u(x) = s\}$$

where $s$ is max $[u(y) : y \in X]$, the maximum value of $u$ on $X$.

Social choice theory is concerned with the existence of a *choice* under a social preference relation $P$ which in some sense aggregates individual preferences for all members of a society $M = \{1, \ldots, i, \ldots, m\}$. Typically the social preference relation cannot be representable by a "social" utility function. For example suppose a society consists of $n$ individuals, each one of whom has a preference relation $P_i$ on the feasible set $X$.

Define a social preference relation $P$ on $X$ by $xPy$ iff $xP_iy$ for all $i \in M$ ($P$ is called the *strict Pareto rule*).

It is clear that if each $P_i$ is transitive, then so must be $P$. As a result, $P$ must be acyclic. If $X$ is a finite set, then by Lemma 1.8, there exists a choice $C_P(X)$ on $X$.

The same conclusion follows if we define $xQy$ iff $xR_jy \; \forall \; j \in M$, and $xP_iy$ for some $i \in M$.

If we assume that each individual has negatively transitive preferences, then $Q$ will be transitive, and will again have a *choice*. $Q$ is called the *weak Pareto rule*. Note that a point $x$ belongs to $C_Q(X)$ iff it is impossible to move to another point $y$

## 1.4 Preferences and Choices

**Fig. 1.5** The preference relation $Q$

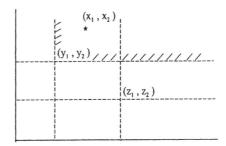

which makes nobody "worse off", but makes some members of the society "better off". The set $C_Q(X)$ is called the *Pareto set*. Although the social preference relation $Q$ has a choice, there is no social utility function which represents $Q$. To see this suppose the society consists of two individuals 1, 2 with transitive preferences $xP_1yP_1z$ and $zP_2xP_2y$.

By the definition $xQy$, since both individuals prefer $x$ to $y$. However conflict of preference between $y$ and $z$, and between $x$ and $z$ gives $yIz$ and $xIz$, where $I$ is the social indifference rule associated with $Q$. Consequently $I$ is not transitive and there is no "social utility function" which represents $Q$. Moreover the elements of $X$ cannot be partitioned into disjoint indifference equivalence classes.

To see the same phenomenon geometrically define a preference relation $P$ on $\Re^2$ by

$$(x_1, x_2) P(y_1, y_2) \iff x_1 > y_1 \wedge x_2 > y_2.$$

From Fig. 1.5 $(x_1, x_2)$ $P$ $(y_1, y_2)$. However $(x_1, x_2)$ $I$ $(z_1, z_2)$ and $(y_1, y_2)$ $I$ $(z_1, z_2)$. Again there is no social utility function representing the preference relation $Q$. Intuitively it should be clear that when the feasible set is "bounded" in some way in $\Re^2$, then the preference relation $Q$ has a *choice*. We shall show this more generally in a later chapter. (See Lemma 3.9 below.)

In Fig. 1.5, we have represented the preference $Q$ in $\Re^2$ by drawing the set preferred to the point $(y_1, y_2)$, say, as a subset of $\Re^2$.

An alternative way to describe the preference is by the graph of $\phi_P$. For example, suppose $X$ is the unit interval $[0, 1]$ in $\Re$, and let the horizontal axis be the *domain* of $\phi_P$, and the vertical axis be the *co-domain* of $\phi_P$. In Fig. 1.6, the preference $P$ is identical to the relation $>$ on the interval (so $yPx$ iff $y > x$). The graph of $\phi_P$ is then the shaded set in the figure. Note that $yPx$ iff $xP^{-1}y$. Because $P$ is irreflexive, the diagonal $e_X = \{(x, x) : x \in X\}$ cannot belong to graph $(\phi_P)$. To find graph $(\phi_P^{-1})$ we simply "reflect" graph $(\phi_P)$ in the diagonal.

The shaded set in Fig. 1.7 represents graph $(\phi_P^{-1})$. Because $P$ is asymmetric, it is impossible for both $yPx$ and $xPy$ to be true. This means that graph$(\phi_P) \cap$ graph$(\phi_P^{-1}) = \Phi$ (the empty set). This can be seen by superimposing Figs. 1.6 and 1.7. A preference of the kind illustrated in Fig. 1.6 is often called *monotonic* since increasing values of $x$ are preferred.

**Fig. 1.6** The graph of the preference $P$

**Fig. 1.7** The graph of the inverse preference

**Fig. 1.8** Representation of the preference $P$

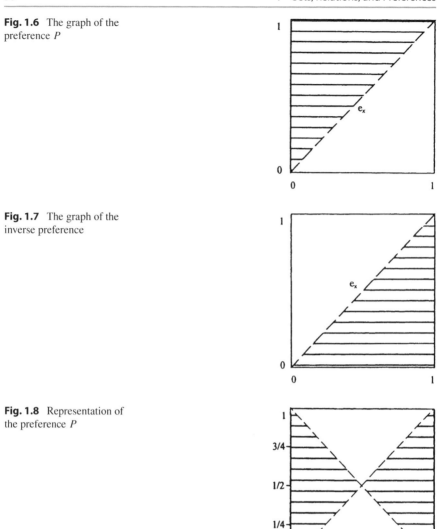

To illustrate a transitive, but non-monotonic strict preference, consider Fig. 1.8 which represents the preference $yPx$ iff $x < y < 1-x$, for $x \leq \frac{1}{2}$, or $1-x < y < x$, for $x > \frac{1}{2}$. For example if $x = \frac{1}{4}$, then $\phi_P(x)$ is the interval $(\frac{1}{4}, \frac{3}{4})$, namely all points between $\frac{1}{4}$, and $\frac{3}{4}$, excluding the end points.

It is obvious that $P$ represents a utility function

$$u(x) = x \quad \text{if } x \leq \frac{1}{2}$$

## 1.4 Preferences and Choices

**Fig. 1.9** A Euclidean preference

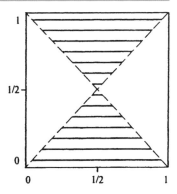

$$u(x) = 1 - x \quad \text{if } \frac{1}{2} < x \leq 1.$$

Clearly the choice of $P$ is $C_P(x) = \frac{1}{2}$. Such a most preferred point is often called a "bliss point" for the preference. Indeed a preference of this kind is usually called "Euclidean", since the preference is induced from the distance from the bliss point. In other words, $yPx$ iff $|y - \frac{1}{2}| < |x - \frac{1}{2}|$. Note again that this preference is transitive and of course acyclic. The fact that the $P$ is asymmetric can be seen from noting that the shaded set in Fig. 1.8 (graph $\phi_P$) and the shaded set in Fig. 1.9 (graph $\phi_P^{-1}$) do not intersect.

Figure 1.10 represents a more complicated asymmetric preference. Here

$$\phi_P(x) = \left(x, x + \frac{1}{2}\right) \quad \text{if } x \leq \frac{1}{2}$$

$$= \left(\frac{1}{2}, x\right) \cup \left(0, x - \frac{1}{2}\right) \quad \text{if } x > \frac{1}{2}.$$

Clearly there is a cycle, say $\frac{1}{4} P \frac{1}{8} P \frac{11}{16} P \frac{1}{4}$. Moreover the choice $C_P(X)$ is empty. This example illustrates that when acyclicity fails, then it is possible for the choice to be empty.

To give an example where $P$ is both acyclic on the interval, yet no choice exists, consider Fig. 1.11. Define $\phi_P(x) = (x, x + \frac{1}{2})$ if $x \leq \frac{1}{2}$ and $\phi_P(x) = (\frac{1}{2}, x)$ if $x > \frac{1}{2}$.

$P$ is still asymmetric, but we cannot construct a cycle. For example, if $x = \frac{1}{4}$ then $yPx$ for $y \in (\frac{1}{4}, \frac{3}{4})$ but if $zPy$ then $z > \frac{1}{4}$. Note however that $\phi_P(\frac{1}{2}) = (\frac{1}{2}, 1)$ so $C_P(x) = \Phi$.

This example shows that Lemma 1.8 cannot be extended directly to the case that $X$ is the interval. In Chap. 3 below we show that we have to impose "continuity" on $P$ to obtain an analogous result to Lemma 1.8.

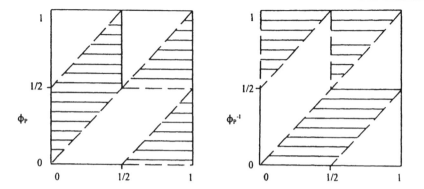

**Fig. 1.10** An empty Choice Set

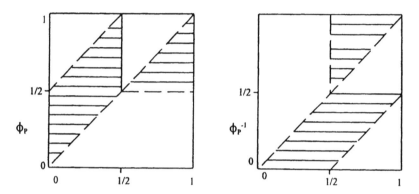

**Fig. 1.11** An acyclic preference with empty choice

## 1.5 Social Choice and Arrow's Impossibility Theorem

The discussion following Lemma 1.8 showed that even the weak Pareto rule, $Q$, did not give rise to transitive indifference, and thus could not be represented by a "social utility function". However $Q$ does give rise to transitive strict preference. We shall show that any rule that gives transitive strict preference must be "oligarchic" in the same way that $Q$ is oligarchic. In other words any rule that gives transitive strict preference must be based on the Pareto (or unanimity) choice of some subset, say $\Theta$ of the society, $M$. Arrow's Theorem (Arrow 1951) shows that if it is desired that indifference be transitive, then the rule must be "dictatorial", in the sense that it obeys the preference of a single individual.

## 1.5 Social Choice and Arrow's Impossibility Theorem

### 1.5.1 Oligarchies and Filters

The literature on social choice theory is very extensive and technical, and this section will not attempt to address its many subtleties. The general idea to examine the possible rationality properties of a "social choice rule"

$$\sigma : S(X)^M \longrightarrow S(X).$$

Here $S(X)$ stands for the set of strict preference relations on the set $X$ and $M = \{1, \ldots, i, \ldots\}$ is a society. Usually $M$ is a finite set of cardinality $m$. Sometimes however $M$ will be identified with the set of integers $\mathcal{Z}$. $S(X)^M$ is the set of strict preference profiles for this society. For example if $|M| = m$, then a profile $\pi = \{P_1, \ldots, P_m\}$ is a list of preferences for the members of the society. We use $A(X)^M, T(X)^M, O(X)^M$ for profiles whose individual preferences are acyclic, strict partial orders or weak orders, respectively. Social choice theory is based on *binary* comparisons. This means that if two profiles $\pi^1$ and $\pi^2$ agree on a pair of alternatives $\{x, y\}$, say, then the social preferences $\sigma(\pi^1)$ and $\sigma(\pi^2)$ also agree on $\{x, y\}$. A key idea is that of a *decisive coalition*. Say a subset $A \subset M$ is *decisive* under the rule $\sigma$ iff for any profile $\pi = (P_1, \ldots, P_m)$ such that $x P_i y$ for all $i \in A$ then $x(\sigma(\pi))y$. That is to say whenever $A$ is decisive, and its members agree that $x$ is preferred to $y$ then the social preference chooses $x$ over $y$. The set of decisive coalitions under the rule is written $\mathcal{D}_\sigma$, or more simply, $\mathcal{D}$. To illustrate this idea, suppose $M = \{1, 2, 3\}$ and $\mathcal{D}_\sigma$ comprises any coalition with at least two members. It is easy to construct a profile $\pi \in A(x)^M$ such that $\sigma$ is not even acyclic. For example, choose a profile $\pi$ on the alternatives $\{x, y, z\}$ such that

$$x P_1 y P_1 z, \qquad y P_2 z P_2 x, \qquad z P_3 x P_3 y.$$

Since both 1 and 2 prefer $y$ to $z$ we must have $y\sigma(\pi)z$. But in the same way we find that $z\sigma(\pi)x$ and $x\sigma(\pi)y$, giving a social preference cycle on $\{x, y, z\}$. In general restricting the image of $\sigma$ so that it lies in $A(X), T(X)$, and $O(X)$ imposes constraints on $\mathcal{D}_\sigma$. We now examine these constraints.

**Lemma 1.9** *If $\sigma : T(X)^M \longrightarrow T(X)$, and $M, A, B$ all belong to $\mathcal{D}_\sigma$, then $A \cap B \in \mathcal{D}_\sigma$.*

*Outline of Proof* Partition $M$ into the four sets $V_1 = A \cap B$, $V_2 = A \backslash B, V_3 = B \backslash A, V_4 = M \backslash (A \cup B)$ and suppose that each individual has preferences on the alternatives $\{x, y, z\}$ as follows:

$i \in V_1$: $\quad z P_i x P_i y$

$i \in V_2$: $\quad x P_i y$, with preferences for $z$ unspecified

$i \in V_3$: $\quad z P_i x$, with preferences for $y$ unspecified

$i \in V_4$: $\quad$ completely unspecified.

Now $A\setminus B = \{i \in A,\text{ but } i \notin B\}$, so $V_1 \cup V_2 = A$. Since $A$ is decisive and every individual in $A$ prefers $x$ to $y$ we obtain $x\sigma(\pi)y$. In the same way $V_1 \cup V_3 = B$, and $B$ is decisive, so $z\sigma(\pi)x$. Since we require $\sigma(\pi)$ to be transitive, it is necessary that $z\sigma(\pi)y$. Since individual preferences are assumed to belong to $T(X)$, we require that $zP_i y$ for all $i \in V_1$. We have not however specified the preferences for the rest of the society. Thus $V_1 = A \cap B$ must be decisive for $\{x, z\}$ in the sense that $V_1$ can choose between $x$ and $z$, independently of the rest of the society. But this must be true for every pair of alternatives. Thus $A \cap B \in \mathcal{D}_\sigma$. $\square$

In general it could be possible for $\mathcal{D}_\sigma$ to be empty. However it is usual to assume that $\sigma$ satisfies the *strict Pareto rule*. That is to say for any $x$, $y$ if $xP_i y$, for all $i \in M$ then $x\sigma(\pi)y$. This simply means that $M \in \mathcal{D}_\sigma$. Moreover, this implies that $\Phi \notin \mathcal{D}_\sigma$. To see this, suppose that $\Phi \in \mathcal{D}_\sigma$ and consider a profile with $xP_i y\ \forall\ i \in M$. Since nobody prefers $y$ to $x$, and the empty set is decisive, we obtain $y\sigma(\pi)x$. But by the Pareto rule, we have $x\sigma(\pi)y$. We assume however that $\sigma(\pi)$ is always a strict preference relation, and so $y\sigma(\pi)x$ cannot occur. Finally if $A \in \mathcal{D}_\sigma$ then any set $B$ which contains $A$ must also be decisive. Thus Lemma 1.9 can be interpreted in the following way.

**Lemma 1.10** *If $\sigma : T(X)^M \longrightarrow T(X)$ and $\sigma$ satisfies the strict Pareto rule, then $\mathcal{D}_\sigma$ satisfies the following conditions*:
**D1.** (monotonicity) $A \subset B$ and $A \in \mathcal{D}_\sigma$ implies $B \in \mathcal{D}_\sigma$.
**D2.** (identity) $A \in \mathcal{D}_\sigma$ and $\Phi \notin \mathcal{D}_\sigma$.
**D3.** (closed under intersection) $A, B \in \mathcal{D}_\sigma$ implies $A \cap B \in \mathcal{D}_\sigma$.

A collection $\mathcal{D}$ of subsets of $M$ which satisfy **D1**, **D2**, and **D3** is called a *filter*.

Note also that when $M$ is finite, then $\mathcal{D}_\sigma$ must also be finite. By repeating Lemma 1.9 for each pair of coalitions, we find that $\Theta = \cap\{A_i : A_i \in \mathcal{D}_\sigma\}$ must be non-empty and also decisive. This set $\Theta$ is usually called the *oligarchy*. In the case that $\sigma$ is simply the strict Pareto rule, then the oligarchy is the whole society, $M$.

However, any rule that gives a transitive strict preference relation must be equivalent to the Pareto rule based on some oligarchy, possibly a strict subset of $M$.

For example, majority rule for the society $M = \{1, 2, 3\}$ defines the decisive coalitions $\{1, 2\}, \{1, 3\}, \{2, 3\}$. These three coalitions contain no oligarchy. We can immediately infer that this rule cannot be transitive. In fact, as we have seen, it is not even acyclic. Below, we explore this further. If we require the rule always to satisfy the condition of negative transitivity then the oligarchy will consist of a single individual, when $M$ is finite.

**Lemma 1.11** *If $\sigma : T(X)^M \longrightarrow O(X)$ and $M \in \mathcal{D}_\sigma$ and $A \subset M$ and $A \notin \mathcal{D}_\sigma$, then $M\setminus A \in \mathcal{D}_\sigma$.*

*Proof* Since $A \notin \mathcal{D}_\sigma$, we can find a profile $\pi$ and a pair $\{x, y\}$ such that $yP_i x$ for all $i \in A$, yet not $(y\sigma(\pi)x)$. Let us write this latter condition as $xRy$, where $R$ stands for weak social preference.

## 1.5 Social Choice and Arrow's Impossibility Theorem

Suppose now there is an alternative $z$ such that $xP_iz$ for all $i \in M \backslash A$, and that $yP_iz$ for all $i \in M$. By the Pareto condition ($M \in \mathcal{D}_\sigma$) we obtain $yPz$ (where $P$ stands for $\sigma(\pi)$). By negative transitivity of $P$, $xRy$ and $yPz$ implies $xPz$ (see Lemma 1.7). However we have not specified the preferences of $A$ on $\{x, z\}$. But we have shown that if the members of $M \backslash A$ prefer $x$ to $z$, then so does the society. It then follows that $M \backslash A$ must be decisive. $\square$

It follows from this lemma that if $\sigma : T(X)^M \longrightarrow O(X)$ and $M \in \mathcal{D}_\sigma$, then whenever $A \in \mathcal{D}_\sigma$, there is some proper subset $B$ (such that $B \subset A$ yet $B \neq A$) with $B \in \mathcal{D}_\sigma$. To see this consider any proper subset $C$ of $A$ with $C \notin \mathcal{D}_\sigma$. By Lemma 1.11, $M \backslash C \in \mathcal{D}_\sigma$. But since $O(X)$ belongs to $T(X)$, we can use the property **D3** of Lemma 1.10 to infer that $A \cap (M \backslash C) \in \mathcal{D}_\sigma$. But $A \cap (M \backslash C) = A \backslash C$, and since $C$ is a proper subset of $A$, $A \backslash C \neq \Phi$. Hence $A \backslash C \in \mathcal{D}_\sigma$. In the case $M$ has finite cardinality, we can repeat this argument to show that there must be some individual $i$ such that $\{i\} \in \mathcal{D}_\sigma$. But then $i$ is a dictator in the sense that, for any $x$, $y$ if $\pi$ is a profile with $xP_iy$ then $x\sigma(\pi)y$.

**Arrow's Impossibility Theorem** *If $\sigma : O(X)^M \longrightarrow O(X)$ and $M \in \mathcal{D}_\sigma$, with $|M|$ finite, then there is a dictator $\{i\}$, say, such that $\{i\} \in \mathcal{D}_\sigma$.*

It is obvious that if $\mathcal{D}_\sigma$ is non-empty, then all the coalitions in $\mathcal{D}_\sigma$ must contain the dictator $\{i\}$. In particular $\{i\} = \cap \{M_i : i \in \mathcal{D}_\sigma\}$, and $\{i\} \in \mathcal{D}_\sigma$.

A somewhat similar result holds when $M$ is a "space" rather than a finite set. In this case there need not be a dictator in $M$. However in this case, the filter $\mathcal{D}_\sigma$ defines an "invisible dictator". That is to say, we can imagine the coalitions in $\mathcal{D}_\sigma$ becoming smaller and smaller, so that they define the invisible dictator in the limit. See Kirman and Sondermann (1972).

### 1.5.2 Acyclicity and the Collegium

As Lemma 1.8 demonstrated, if $P$ is an acyclic preference relation on a finite set, then the choice for $P$ will be non-empty. Given Arrow's Theorem, it is therefore useful to examine the properties of a social choice rule that are compatible with acyclicity. To this end we introduce the notion of the *Nakamura number* for a rule, $\sigma$ (Nakamura 1979).

**Definition 1.4** Let $\mathcal{D}$ be a family of subsets of the finite set $M$. The collegium $K(\mathcal{D})$ is the intersection

$$\cap \{A_i : A_i \in \mathcal{D}\}.$$

That is to say $K(\mathcal{D})$ is the largest set in $M$ such that $K(\mathcal{D}) \subset A$ for all $A \in \mathcal{D}$.

If $K(\mathcal{D})$ is empty then $\mathcal{D}$ is said to be *non-collegial*. Otherwise $\mathcal{D}$ is *collegial*.

If $\sigma$ is a social choice rule, and $\mathcal{D}_\sigma$ its family of decisive coalitions, then $K(\sigma) = K(\mathcal{D}_\sigma)$ is the collegium for $\sigma$. Again $\sigma$ is called collegial or non-collegial depending on whether $K(\sigma)$ is non-empty or empty. The Nakamura number of a non-collegial family $\mathcal{D}$ is written $k(\mathcal{D})$ and is the cardinality of the smallest non-collegial subfamily of $\mathcal{D}$. That is, there exists some subfamily $\mathcal{D}'$ of $\mathcal{D}$ with $|\mathcal{D}'| = k(\mathcal{D})$ such that $K(\mathcal{D}') = \Phi$. Moreover if $\mathcal{D}''$ is a subfamily of $\mathcal{D}$ with $|\mathcal{D}''| \leq k(\mathcal{D}) - 1$ then $K(\mathcal{D}'') \neq \Phi$.

In the case $\mathcal{D}$ is collegial define $k(\mathcal{D}) = \infty$. For a social choice rule define $k(\sigma) = k(\mathcal{D}_\sigma)$, where $\mathcal{D}_\sigma$ is the family of decisive coalitions for $\sigma$.

*Example 1.7*
(i) To illustrate this definition, suppose $\mathcal{D}$ consists of the four coalitions, $\{A_1, A_2, A_3, A_4\}$ where $A_1 = \{2, 3, 4\}$, $A_2 = \{1, 3, 4\}$, $A_3 = \{1, 2, 4, 5\}$ and $A_4 = \{1, 2, 3, 5\}$. Of course $\mathcal{D}$ will be monotonic, so supersets of these coalitions will be decisive. It is evident that $A_1 \cap A_2 \cap A_3 = \{4\}$ and so if $\mathcal{D}' = \{A_1, A_2, A_3\}$ then $K(\mathcal{D}') \neq \Phi$. However $K(\mathcal{D}') \cap A_4 = \Phi$ and so $K(\mathcal{D}) = \Phi$. Thus $k(\mathcal{D}) = 4$.

(ii) An especially interesting case is of a $q$-majority rule where each individual has one vote, and any coalition with at least $q$ voters (out of $m$) is decisive. In this case it is easy to show then that $k(\sigma) = 2 + [\frac{q}{m-q}]$ where $[\frac{q}{m-q}]$ is the greatest integer strictly less than $\frac{q}{m-q}$. In the case that $m = 4$ and $q = 3$, then we find that $[\frac{3}{1}] = 2$, so $k(\sigma) = 4$.

On the other hand for all other simple majority rules where $m = 2s + 1$ or $2s$ and $q = s + 1$ (and $s$ is integer) then

$$\left[\frac{q}{m-q}\right] = \left[\frac{s+1}{s}\right] \text{ or } \left[\frac{s+1}{s-1}\right]$$

depending on whether $m$ is odd or even. In both cases $[\frac{q}{m-q}] = 1$. Thus $k(\sigma) = 3$ for any simple majority rule with $m \neq 4$.

(iii) Finally, observe that for any simple majoritarian rule, if $M_1, M_2$ both belong to $\mathcal{D}$, then $A_1 \cap A_2 \neq \Phi$. So in general, any non-collegial subfamily of $\mathcal{D}$ must include at least three coalitions. Consequently any majoritarian rule, $\sigma$, has $k(\sigma) \geq 3$.

The Nakamura number allows us to construct social preference cycles.

**Nakamura Lemma** *Suppose that $\sigma$ is a non-collegial voting rule, with Nakamura number $k(\sigma) = k$. Then there exists an acyclic profile $\pi = (P_1, \ldots, P_n)$ for the society $M$, on a set $X = \{x_1, \ldots, x_k\}$ of cardinality $k$, such that $\sigma(\pi)$ is cyclic on $W$, and the choice of $\sigma(\pi)$ is empty.*

*Proof* We wish to construct a cycle by considering $k$ different decisive coalitions, $A_1, \ldots, A_k$ and assigning preferences to the members of each coalition such that

## 1.5 Social Choice and Arrow's Impossibility Theorem

$$x_i P_i x_2 \quad \text{for all } i \in A_1$$
$$\vdots$$
$$x_{k-1} P_i x_k \quad \text{for all } i \in A_{k-1}$$
$$x_k P_i x_1 \quad \text{for all } i \in A_k.$$

We now construct such a profile, $\pi$. Let $\mathcal{D}_k = \{A_1, \ldots, A_{k-1}\}$. By the definition of the Nakamura number, this must be collegial. Hence there exists some individual $\{k\}$, say, with $k \in A_1 \cap \cdots \cap A_{k-1}$. We can assign $k$ the acyclic preference profile $x_1 P_k x_2 P_k \ldots P_k x_k$.

In the same way, for each subfamily $\mathcal{D}_j = \{A, \ldots, A_{j-1}, A_{j+1}, \ldots, A_k\}$ there exists a collegium containing individual $j$, to whom we assign the preference

$$x_{j+1} P_j x_{j+2} \ldots x_k P_j x_1 \ldots x_j P_j x_j.$$

We may continue this process to assign acyclic preferences to each member of the various collegia of subfamilies of $\mathcal{D}$, so as to give the required cyclic social preference. $\square$

**Lemma 1.12** *A necessary condition for a social choice rule $\sigma$ to be acyclic on the finite set $X$ of cardinality at least $m = |M|$, for each acyclic profile $\pi$ on $X$, is that $\sigma$ be collegial.*

*Proof* Suppose $\sigma$ is not collegial. It is easy to show that the Nakamura number $k(\sigma)$ will not exceed $m$. By the Nakamura Theorem there is an acyclic profile $\pi$ on a set $X$ of cardinality $m$, such that $\sigma(\pi)$ is cyclic on $X$. Thus acyclicity implies that $\sigma$ must be collegial. $\square$

Note that this lemma emphasizes the size of the set of alternatives. It is worth observing here that in the previous proofs of Arrow's Theorem, the cardinality of the set of alternatives was implicitly assumed to be at least 3.

These techniques using the Nakamura number can be used to show that a simple social rule will be acyclic whenever it is collegial. Say a social choice rule is *simple* iff whenever $x\sigma(\pi)y$ for the profile $\pi$, then $x P_i y$ for all $i$ in some coalition that is decisive for $\sigma$.

Note that a social choice rule need not, in general, be simple. If $\sigma$ is simple then all the information necessary to analyse the rule is contained in $\mathcal{D}_\sigma$.

**Lemma 1.13** *Let $\sigma$ be a simple choice rule on a finite set $X$:*
 (i) *If $\sigma$ is dictatorial, then $\sigma(\pi) \subset O(X)$ for all $\pi \in O(X)^M$*
 (ii) *If $\sigma$ is oligarchic, then $\sigma(\pi) \subset T(X)$ for all $\pi \in T(X)^M$*
 (iii) *If $\sigma$ is collegial, then $\sigma(\pi) \subset A(X)$ for all $\pi \in A(X)^M$.*

*Proof*

(i) If $i$ is a dictator, then $x I_i y$ implies that $x$ and $y$ are socially indifferent. Because $P_i$ belongs to $O(X)$ so must $\sigma(\pi)$.
(ii) In the same way, if $\Theta$ is the oligarchy and $x\sigma(\pi)y$ then $xP_i y$ for all $i$ in $\Theta$. Thus $\sigma(\pi)$ must be transitive.
(iii) If there is a cycle $x_1\sigma(\pi)\ldots,x_k\sigma(\pi)x_1$ then each of these social preferences must be supported by a decisive coalition. Since the collegium is non-empty, there is some individual, $i$, say, who has such a cyclic preference. This contradicts the assumption that $\pi \in A(X)^M$.

□

Another way of expressing part (iii) of this lemma is to introduce the idea of a prefilter. Say $\mathcal{D}$ is a *prefilter* if and only if it satisfies **D1** (monotonicity) and **D2** (identity) introduced earlier, and also non-empty intersection (so $K(\mathcal{D}) \neq \Phi$). Clearly if $\sigma$ is simple, and $\mathcal{D}_\sigma$ is a prefilter, then $\sigma$ is acyclic and consistent with the Pareto rule.

In Chap. 3 we shall further develop the notion of social choice using the notion of the Nakamura number, in the situation where $X$ has a geometric structure.

## Further Reading

The first version of the impossibility theorem can be found in

Arrow, K. J. (1951). *Social choice and individual values*. New York: Wiley.

The ideas of the filter and Nakamura number are in

Kirman, A. P., & Sondermann, D. (1972). Arrow's Impossibility Theorem, many agents and invisible dictators. *Journal of Economic Theory, 5*, 267–278.
Nakamura, K. (1979). The vetoers in a simple game with ordinal preferences. *International Journal of Game Theory, 8*, 55–61.

# Linear Spaces and Transformations

## 2.1 Vector Spaces

We showed in Sect. 1.3 that when $\mathcal{F}$ was a field, the $n$-fold product set $\mathcal{F}^n$ had an additional operation defined on it, which was induced from addition in $\mathcal{F}$, so that $(\mathcal{F}^n, +)$ became an abelian group with zero $\underline{0}$. Moreover we were able to define a product $\cdot : \mathcal{F} \times \mathcal{F}^n \to \mathcal{F}^n$ which takes $(\alpha, x)$ to a new element of $\mathcal{F}^n$ called $(\alpha x)$. Elements of $\mathcal{F}^n$ are known as *vectors*, and elements of $\mathcal{F}$ as *scalars*. The properties that we discovered in $\mathcal{F}^n$ characterise a *vector space*. A vector space is also known as a linear space.

**Definition 2.1** A *vector space* $(V, +)$ is an abelian additive group with zero $\underline{0}$, together with a field $(\mathcal{F}, +, \cdot)$ with zero 0 and identity 1. An element of $V$ is called a *vector* and an element of $\mathcal{F}$ a *scalar*. Moreover for any $\alpha \in \mathcal{F}, v \in V$ there is a scalar multiplication $(\alpha, v) \to \alpha v \in V$ which satisfies the following properties:

**V1.** $\alpha(v_1 + v_2) = \alpha v_1 + \alpha v_2$, for any $\alpha \in \mathcal{F}, v_1, v_2 \in V$.
**V2.** $(\alpha + \beta)v = \alpha v + \beta v$, for any $\alpha, \beta \in \mathcal{F}, v \in V$.
**V3.** $(\alpha \beta)v = \alpha(\beta v)$, for any $\alpha, \beta \in \mathcal{F}, v \in V$.
**V4.** $1 \cdot v = v$, for $1 \in \mathcal{F}$, and for any $v \in V$.

Call $V$ a vector space over the field $\mathcal{F}$. From the previous discussion the set $\Re^n$ becomes an abelian group $(\Re^n, +)$ under addition. We shall frequently write

$$x = \begin{pmatrix} x_1 \\ \vdots \\ x_n \end{pmatrix}$$

**Fig. 2.1** Vector addition

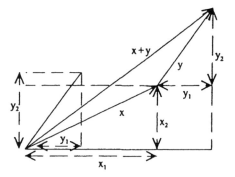

for a vector in $\Re^n$, where $x_1, \ldots, x_n$ are called the coordinates of $x$. Vector addition as in Fig. 2.1 is then defined by

$$x + y = \begin{pmatrix} x_1 \\ \vdots \\ x_n \end{pmatrix} + \begin{pmatrix} y_1 \\ \vdots \\ y_n \end{pmatrix} = \begin{pmatrix} x_1 + y_1 \\ \vdots \\ x_n + y_n \end{pmatrix}.$$

A vector space over $\Re$ is called a *real vector space*.

For example $(\mathcal{Z}_2, +, \cdot)$ is a field and so $(\mathcal{Z}_2)^n$ is a vector space over the field $\mathcal{Z}_2$. It may not be possible to represent each vector in a vector space by a list of coordinates. For example, consider the set of all functions with domain $X$ and image in $\Re$. Call this set $\Re^X$. If $f, g \in \Re^X$, define $f + g$ to be that function which maps $x \in X$ to $f(x) + g(x)$. Clearly there is a zero function $\underline{0}$ defined by $\underline{0}(x) = \underline{0}$, and each $f$ has an inverse $(-f)$ defined by $(-f)(x) = -(f(x))$. Finally for $\alpha \in \Re$, $f \in \Re^X$, define $\alpha f : X \to \Re$ by $(\alpha f)(x) = \alpha(f(x))$. Thus $\Re^X$ is a vector space over $\Re$.

**Definition 2.2** Let $(V, +)$ be a vector space over a field, $\mathcal{F}$. A subset $V'$ of $V$ is called a *vector subspace* of $V$ if and only if
1. $v_1, v_2 \in V' \Rightarrow v_1 + v_2 \in V'$, and
2. if $\alpha \in F$ and $v \in V'$ then $\alpha v' \in V'$.

**Lemma 2.1** *If $(V, +)$ is a vector space with zero $\underline{0}$ and $V'$ is a vector subspace, then, for each $v \in V'$, the inverse $(-v) \in V'$, and $\underline{0} \in V'$, so $(V', +)$ is a subgroup of $(V, +)$.*

*Proof* Suppose $v \in V'$. Since $\mathcal{F}$ is a field, there is an identity 1, with additive inverse $-1$. But by **V2**, $(1 - 1)v = 1 \cdot v + (-1)v = 0 \cdot v$, since $1 - 1 = 0$. Now $(1 + 0)v = 1 \cdot v + 0 \cdot v$, and so $0 \cdot v = \underline{0}$. Thus $(-1)v = (-v)$. Since $V'$ is a vector subspace, $(-1)v \in V'$, and so $(-v) \in V'$. But then $v + (-v) = \underline{0}$, and so $\underline{0} \in V'$. □

From now on we shall simply write $V$ for a vector space, and refer to the field only on occasion.

## 2.1 Vector Spaces

**Definition 2.3** Let $V' = \{v_1, \ldots, v_r\}$ be a set of vectors in the vector space $V$. A vector $v$ is called a *linear combination* of the set $V'$ iff $v$ can be written in the form

$$v = \sum_{i=1}^{r} \lambda_i v_i$$

where each $\lambda_i$, $i = 1, \ldots, r$ belongs to the field $\mathcal{F}$. The *span* of $V'$, written $\text{Span}(V')$ is the set of vectors which are linear combinations of the set $V'$. If $V'' = \text{Span}(V')$, then $V'$ is said *to span $V''$*.

For example, suppose

$$V' = \left\{ \begin{pmatrix} 1 \\ 2 \end{pmatrix} \begin{pmatrix} 2 \\ 1 \end{pmatrix} \right\}.$$

Since we can solve the equation

$$\begin{pmatrix} x \\ y \end{pmatrix} = \alpha \begin{pmatrix} 1 \\ 2 \end{pmatrix} + \beta \begin{pmatrix} 2 \\ 1 \end{pmatrix}$$

for any $(x, y) \in \mathfrak{R}^2$, by setting $\alpha = \frac{1}{3}(2y - x)$ and $\beta = \frac{1}{3}(2x - y)$, it is clear that $V'$ is a span for $\mathfrak{R}^2$.

**Lemma 2.2** *If $V'$ is a finite set of vectors in the vector space, $V$, then $\text{Span}(V')$ is a vector subspace of $V$.*

*Proof* We seek to show that for any $\alpha, \beta \in \mathcal{F}$ and any $u, w \in \text{Span}(V')$, then $\alpha u + \beta w \in \text{Span}(V')$. By definition, if $V' = \{v_1, \ldots, v_r\}$, then $u = \sum_{i=1}^{r} \eta_i v_i$ and $w = \sum_{i=1}^{r} \mu_i v_i$, where $\eta_i, \mu_i \in \mathcal{F}$ for $i = 1, \ldots, r$. But then $\alpha u + \beta w = \alpha \sum_{i=1}^{r} \eta_i v_i + \beta \sum_{i=1}^{r} \mu_i v_i = \sum_{i=1}^{r} \lambda_i v_i$, where $\lambda_i = \alpha \eta_i + \beta \mu_i \in \mathcal{F}$, for $i = 1, \ldots, r$. Thus $\alpha u + \beta w \in \text{Span}(V')$.

Note that, by this lemma, the zero vector $\underline{0}$ belongs to $\text{Span}(V')$. □

**Definition 2.4** Let $V' = \{v_1, \ldots, v_r\}$ be a set of vectors in $V$. $V'$ is called a *frame* iff $\sum_{i=1}^{r} \alpha_i v_i = \underline{0}$ implies that $\alpha_i = 0$ for $i = 1, \ldots, r$. (Here each $\alpha_i$ belongs to the field $\mathcal{F}$.) In this case the set $V'$ is called a *linearly independent set*. If $V'$ is not a frame, the vectors in $V'$ are said to be linearly dependent. Say a vector is *linearly dependent* on $V' = \{v_1, \ldots, v_r\}$ iff $v \in \text{Span}(V')$.

Note that if $V'$ is a frame, then
1. $\underline{0} \notin V'$ since $\alpha \underline{0} = \underline{0}$ for every non-zero $\alpha \in F$.
2. If $v \in V'$ then $(-v) \notin V'$, otherwise $1 \cdot v + 1(-v) = \underline{0}$ would belong to $V'$, contradicting (1).

**Lemma 2.3**
1. $V'$ is not a frame iff there is some vector $v \in V'$ which is linearly dependent on $V' \setminus \{v\}$.
2. If $V'$ is a frame, then any subset of $V'$ is a frame.
3. If $V'$ spans $V''$, but $V'$ is not a frame, then there exists some vector $v \in V'$ such that $V''' = V' \setminus \{v\}$ spans $V''$.

*Proof* Let $V' = \{v_1, \ldots, v_r\}$ be the set of vectors in the vector space $V$.
1. Suppose $V'$ is not a frame. Then there exists an equation $\sum_{j=1}^{r} \alpha_j v_j = 0$, where, for at least one $k$, it is the case that $\alpha_k \neq 0$. But then $v_k = -\frac{1}{\alpha_k}(\sum_{j \neq k} \alpha_j v_j)$. Let $v_k = v$. Then $v$ is linearly dependent on $V' \setminus \{v\}$. On the other hand suppose that $v_1$, say, is linearly dependent on $\{v_2, \ldots, v_r\}$. Then $v_1 = \sum_{j=2}^{r} \alpha_j v_j$, and so $\underline{0} = -v_1 + \sum_{j=2}^{r} \alpha_j v_j = \sum_{j=1}^{r} \alpha_j v_j$ where $\alpha_1 = -1$. Since $\alpha_1 \neq 0$, $V'$ cannot be a frame.
2. Suppose $V''$ is a subset of $V'$, but that $V''$ is not a frame. For convenience let $V'' = \{v_1, \ldots, v_k\}$ where $k \leq r$. Then there is a non-zero solution

$$\underline{0} \neq \sum_{j=1}^{k} \alpha_j v_j.$$

Since $V''$ is a subset of $V'$, this implies that $V'$ cannot be a frame. Thus if $V'$ is a frame, so is any subset $V''$.
3. Suppose that $V'$ is not a frame, but that it spans $V''$. By part (1), there exists a vector $v_1$, say, in $V'$ such that $v_1$ belongs to $\mathrm{Span}(V' \setminus \{v_1\})$. Thus $v_1 = \sum_{j=2}^{r} \alpha_j v_j$. Since $V'$ is a span for $V''$, any vector $v$ in $V''$ can be written

$$v = \sum_{j=1}^{r} \beta_j v_j$$

$$= \beta_1 \left( \sum_{j=2}^{r} \alpha_j v_j \right) + \sum_{j=2}^{r} \beta_j v_j.$$

Thus $v$ is a linear combination of $V' \setminus \{v_1\}$ and so $V'' = \mathrm{Span}(V' \setminus \{v_1\})$. Let $V''' = V' \setminus \{v_1\}$ to complete the proof. $\square$

**Definition 2.5** A *basis* for a vector space $V$ is a frame $V'$ which spans $V$.

For example, we previously considered

$$V' = \left\{ \begin{pmatrix} 1 \\ 2 \end{pmatrix}, \begin{pmatrix} 2 \\ 1 \end{pmatrix} \right\}$$

## 2.1 Vector Spaces

and showed that any vector in $\Re^2$ could be written as

$$\begin{pmatrix} x \\ y \end{pmatrix} = \left(\frac{2y-x}{3}\right)\begin{pmatrix} 1 \\ 2 \end{pmatrix} + \left(\frac{2x-y}{3}\right)\begin{pmatrix} 2 \\ 1 \end{pmatrix} = \lambda_1 \begin{pmatrix} 1 \\ 2 \end{pmatrix} + \lambda_2 \begin{pmatrix} 2 \\ 1 \end{pmatrix}.$$

Thus $V'$ is a span for $\Re^2$. Moreover if $(x, y) = (0, 0)$ then $\lambda_1 = \lambda_2 = 0$ and so $V'$ is a frame. Hence $V'$ is a basis for $\Re^2$. If $V' = \{v_1, \ldots, v_n\}$ is a basis for a vector space $V$ then any vector $v \in V$ can be written

$$v = \sum_{j=1}^{n} \alpha_j v_j$$

and the elements $(\alpha_1, \ldots, \alpha_n)$ are known as the *coordinates* of the vector $v$, *with respect to* the basis $V'$.

For example the *natural basis* for $\Re^n$ is the set $V' = \{e_1, \ldots, e_n\}$ where $e_i = (0, \ldots, 1, \ldots, 0)$ with a 1 in the $i$th position.

**Lemma 2.4** $\{e_1, \ldots, e_n\}$ *is a basis for* $\Re^n$.

*Proof* We can write any vector $x$ in $\Re^n$ as $\{x_1, \ldots, x_n\}$. Clearly

$$x = \begin{pmatrix} x_1 \\ \vdots \\ x_n \end{pmatrix} = x_1 \begin{pmatrix} 1 \\ 0 \\ \cdot \end{pmatrix} + \cdots x_n \begin{pmatrix} 0 \\ \vdots \\ 1 \end{pmatrix}.$$

If $x = 0$ then $x_1 = \cdots = x_n = 0$ and so $\{e_1, \ldots, e_n\}$ is a frame, as well as a span, and thus a basis for $\Re^n$. □

However a single vector $x$ will have different coordinates depending on the basis chosen. For example the vector $(x, y)$ has coordinates $(x, y)$ in the basis $\{e_1, e_2\}$ but coordinates $(\frac{2y-x}{3}, \frac{2x-y}{3})$ with respect to the basis $\{\binom{1}{2}, \binom{2}{1}\}$.

Once the basis is chosen, the coordinates of any vector with respect to that basis are unique. □

**Lemma 2.5** *Suppose* $V' = \{v_1, \ldots, v_n\}$ *is a basis for* $V$. *Let* $v = \sum_{i=1}^{n} \alpha_i v_i$. *Then the coordinates* $(\alpha_1, \ldots, \alpha_n)$, *with respect to the basis, are unique.*

*Proof* If the coordinates were not unique then it would be possible to write $v = \sum_{i=1}^{n} \beta_i v_i = \sum_{i=1}^{n} \alpha_i v_i$ with $\beta_i \neq \alpha_i$ for some $i$.

But $0 = v - v = \sum_{i=1}^{n} \alpha_i v_i - \sum_{i=1}^{n} \beta_i v_i = \sum_{i=1}^{n} (\alpha_i - \beta_i) v_i$.

Since $V'$ is a frame, $\alpha_i - \beta_i = 0$ for $i = 1, \ldots, n$. Thus $\alpha_i = \beta_i$ for all $i$, and so the coordinates are unique. □

Note in particular that with respect to *any* basis $\{v_1, \ldots, v_n\}$ for $V$, the unique zero vector $\underline{0}$ always has coordinates $(0, \ldots, 0)$.

**Definition 2.6** A space $V$ is *finitely generated* iff there exists a span $V'$, for $V$, which has a finite number of elements.

**Lemma 2.6** *If $V$ is a finitely generated vector space, then it has a basis with a finite number of elements.*

*Proof* Since $V$ is finitely generated, there is a finite set $V_1 = \{v_1, \ldots, v_n\}$ which spans $V$. If $V_1$ is a frame, then it is a basis. If $V_1$ is linearly dependent, then by Lemma 2.3(3) there is a vector $v \in V_1$, such that $\text{Span}(V_2) = V$, where $V_2 = V_1 \setminus \{v\}$. Again if $V_2$ is a frame, then it is a basis. If there were no subset $V_r = \{v_1, \ldots, v_{n-r+1}\}$ of $V_1$ which was a frame, then $V_1$ would have to be the empty set, implying that $V$ was an empty set. But this contradicts $\underline{0} \in V$. □

**Lemma 2.7** *If $V$ is a finitely generated vector space, and $V_1$ is a frame, then there is a basis $V_2$ for $V$ which includes $V_1$.*

*Proof* Let $V_1 = \{v_1, \ldots, v_r\}$. If $\text{Span}(V_1) = V$ then $V_1$ is a basis. So suppose that $\text{Span}(V_1) \neq V$. Then there exists an element $v_{r+1} \in V$ which does not belong to $\text{Span}(V_1)$. We seek to show that $V_2 = V_1 \cup \{v_{r+1}\}$ is a frame. Consider $\underline{0} = \alpha_{r+1} v_{r+1} + \sum_{i=1}^{r} \alpha_i v_i$.

If $\alpha_{r+1} = 0$, then the linear independence of $V_1$ implies that $\alpha_i = 0$, for $i = 1, \ldots, r$. Thus $V_2$ is a frame. If $\alpha_{r+1} \neq 0$, then

$$v_{r+1} = -\frac{1}{\alpha_{r+1}} \left( \sum_{i=1}^{r} \alpha_i v_i \right).$$

But this implies that $v_{r+1}$ belongs to $\text{Span}(V_1)$ and therefore that $V = \text{Span}(V_1)$. Thus $V_2$ is a frame. If $V_2$ is a span for $V$, then it is a basis. If $V_2$ is not a span, reiterate this process. Since $V$ is finitely generated, there must be some frame $V_{n-r+1} = \{v_1, \ldots, v_r, v_{r+1}, \ldots, v_n\}$ which is a span, and thus a basis for $V$. □

These two lemmas show that if $V$ is a finitely generated vector space, and $\{v_1, \ldots, v_m\}$ is a span then some subset $\{v_1, \ldots, v_n\}$, with $n \leq m$, is a basis. A basis is a *minimal* span.

On the other hand if $X = \{v_1, \ldots, v_r\}$ is a frame, but not a span, then elements may be added to $X$ in such a way as to preserve linear independence, until this "superset" of $X$ becomes a basis. Consequently a basis is a *maximal* frame. These two results can be combined into one theorem.

**Exchange Theorem** *Suppose that $V$ is a finitely generated vector space. Let $X = \{x_1, \ldots, x_n\}$ be a frame and $Y = \{y_1, \ldots, y_n\}$ a span. Then there is some subset $Y'$ of $Y$ such that $X \cup Y'$ is a basis for $V$.*

*Proof* By induction, let $X_s = \{x_1, \ldots, x_s\}$, for each $s = 1, \ldots, n$, and let $X_0 = \phi$. □

## 2.2 Linear Transformations

We know already from Lemma 2.6 that there is some subset $Y_0$ of $Y$ such that $X_0 \cup Y_0$ is a basis for $V$. Suppose for some $s < m$, there is a subset $Y_s$ of $Y$ such that $X_s \cup Y_s$ is a basis.

Let $Y_s = \{y_1, \ldots, y_t\}$. Now $x_{s+1} \notin \operatorname{Span}(X_s \cup Y_s)$ since $X_s \cup Y_s$ is a basis. Thus $x_{s+1} = \sum_1^s \alpha_1 x_1 + \sum_1^t \beta_i y_i$. But $X_{s+1} = \{x_1, \ldots, x_{s+1}\}$ is a frame, since it is a subset of $X$.

Thus at least one $\beta_j \neq 0$. Let $Y_{s+1} = Y_s \backslash \{y_j\}$, so $Y_j \notin \operatorname{Span}(X_{s+1} \cup Y_{s+1})$ and so $X_{s+1} \cup Y_{s+1} = \{x_1, \ldots, x_{s+1}\} \cup \{y_1, \ldots, y_{j-1}, y_{j+1}, \ldots, y_t\}$ is a basis for $V$.

Thus if there is some subset $Y_s$ of $Y$ such that $X_s \cup Y_s$ is a basis, there is a subset $Y_{s+1}$ of $Y$ such that $X_{s+1} \cup Y_{s+1}$ is a basis.

By induction, there is a subset $Y_m = Y'$ of $Y$ such that $X_m \cup Y_m = X \cup Y'$ is a basis. □

**Corollary 2.8** *If $X = \{x_1, \ldots, x_m\}$ is a frame in a vector space $V$, and $Y = \{y_1, \ldots, y_n\}$ is a span for $V$, then $m \leq n$.*

**Lemma 2.9** *If $V$ is a finitely generated vector space, then any two bases have the same number of vectors, where this number is called the dimension of $V$, and written $\dim(V)$.*

*Proof* Let $X, Y$ be two bases with $m, n$ number of elements. Consider $X$ as a frame and $Y$ as a span. Thus $m \leq n$. However $Y$ is also a frame and $X$ a span. Thus $n \leq m$. Hence $m = n$. □

If $V'$ is a vector subspace of a finitely generated vector space $V$, then any basis for $V'$ can be extended to give a basis for $V$. To see this, there must exist some finite set $V'' = \{v_1, \ldots, v_r\}$ of vectors all belonging to $V'$ such that $\operatorname{Span}(V'') = V'$. Otherwise $V$ could not be finitely generated. As before eliminate members of $V''$ until a frame is obtained. This gives a basis for $V'$. Clearly $\dim(V') \leq \dim(V)$. Moreover if $V'$ has a basis $V''' = \{v_1, \ldots, v_r\}$ then further linear independent vectors belonging to $V \backslash V'$ can be added to $V'''$ to give a basis for $V$.

As we showed in Lemma 2.3, the vector space $\Re^n$ has a basis $\{e_1, \ldots, e_n\}$ consisting of $n$ elements. Thus $\dim(\Re^n) = n$.

If $V^m$ is a vector subspace of $\Re^n$ of dimension $m$, where of course $m \leq n$, then in a certain sense $V^m$ is identical to a copy of $\Re^m$ through the origin $\underline{0}$. We make this more explicit below.

## 2.2 Linear Transformations

In Chap. 1 we considered a morphism from the abelian group $(\Re^2, +)$ to itself. A morphism between vector spaces is called a *linear transformation*.

**Definition 2.7** Let $V, U$ be two vector spaces of dimension $n, m$ respectively, over the same field $\mathcal{F}$. Then a *linear transformation* $T : V \to U$ is a function from $V$ to $U$ with domain $V$, such that

1. for any $a \in \mathcal{F}$, any $v \in V$, $T(\alpha v) = \alpha(T(v))$
2. for any $v_1, v_2 \in V$, $T(v_1 + v_2) = T(v_1) + T(v_2)$.

Note that a linear transformation is simply a morphism between $(V, +)$ and $(U, +)$ which respects the operation of the field $\mathcal{F}$. We shall show that any linear transformation $T$ can be represented by an array of the form

$$M(T) = \begin{pmatrix} a_{11} & & a_{1n} \\ \vdots & \vdots & \vdots \\ a_{m1} & & a_{mn} \end{pmatrix}$$

consisting of $n \times m$ elements in $F$. An array such as this is called an $n$ by $m$ (or $n \times m$) *matrix*. The set of $n \times m$ matrices we shall write as $M(n, m)$.

### 2.2.1 Matrices

For convenience we shall consider finitely generated vector spaces over $\Re$, so that we restrict attention to linear transformations between $\Re^n$ and $\Re^m$, for any integers $n$ and $m$. Now let $V = \{v_1, \ldots, v_n\}$ be a basis for $\Re^n$ and $U = \{u_1, \ldots, u_m\}$ a basis for $\Re^m$.

Since $V$ is a basis for $\Re^n$, any vector $x \in \Re^n$ can be written as $x = \sum_{j=1}^{n} x_j v_j$, with coordinates $(x_1, \ldots, x_n)$.

If $T$ is a linear transformation, then $T(\alpha v_1 + \beta v_2) = T(\alpha v_1) + T(\beta v_2) = \alpha T(v_1) + \beta T(v_2)$. Therefore

$$T(x) = T\left(\sum_{j=1}^{n} x_j v_j\right) = \sum_{j=1}^{n} x_j T(v_j).$$

Since each $T(v_j)$ lies in $\Re^m$ we can write $T(v_j) = \sum_{i=1}^{m} a_{ij} u_i$, where $(a_{1j}, a_{2j}, \ldots, a_{mj})$ are the coordinates of $T(v_j)$ with respect to the basis $U$ for $\Re^m$.

Thus

$$T(x) = \sum_{j=1}^{n} x_j \sum_{i=1}^{m} a_{ij} u_i = \sum_{i=1}^{m} y_i u_i$$

where the $i$th coordinate, $y_i$, of $T(x)$ is equal to $\sum_{j=1}^{n} a_{ij} x_j$.

We obtain a set of linear equations:

$$y_1 = a_{11} x_1 + a_{12} x_2 + \cdots a_{1j} x_j + \cdots a_{1n} x_n$$
$$\vdots \qquad \vdots$$
$$y_i = a_{i1} x_1 + a_{i2} x_2 + \cdots a_{ij} x_j + \cdots a_{in} x_n$$
$$\vdots \qquad \vdots$$
$$y_m = a_{m1} x_1 + a_{m2} x_2 + \cdots a_{mj} x_j + \cdots a_{mn} x_n.$$

## 2.2 Linear Transformations

This set of equations is more conveniently written

$$\text{row } i \begin{pmatrix} a_{11} & \cdots & a_{ij} & \cdots & a_{1n} \\ \vdots & & & & \\ a_{i1} & & a_{ij} & & a_{in} \\ \vdots & & & & \\ a_{m1} & \cdots & a_{mj} & \cdots & a_{mn} \end{pmatrix} \begin{pmatrix} x_1 \\ \vdots \\ x_j \\ \vdots \\ x_n \end{pmatrix} = \begin{pmatrix} y_1 \\ \vdots \\ y_i \\ \vdots \\ y_m \end{pmatrix}.$$

$j$th column

or as $M(T)x = y$, where $M(T)$ is the $n \times m$ array whose $i$th row is $(a_{i1}, \ldots, a_{in})$ and whose $j$th column is $(a_{1j}, \ldots, a_{mj})$. This matrix is commonly written as $(a_{ij})$ where it is understood that $i = 1, \ldots, m$ and $j = 1, \ldots, n$.

Note that the operation of $M(T)$ on $x$ is as follows: to obtain the $i$th coordinate, $y_i$, take the $i$th row vector $(a_{i1}, \ldots, a_{in})$ and form the *scalar product* of this with the column vector $(x_1, \ldots, x_n)$, where this scalar product is defined to be $\sum_{j=1}^{n} a_{ij} x_j$.

The coefficients of $T(v_j)$ with respect to the basis $(u_1, \ldots, u_m)$ are $(a_{1j}, \ldots, a_{mj})$ and these turn up as the $j$th column of the matrix. Thus we could write the matrix as

$$M(T) = \begin{pmatrix} T(v_1) \ldots T(v_j) \ldots T(v_n) \end{pmatrix}$$

where $T(v_j)$ is the column of coordinates in $\Re^m$. Suppose now that $W = \{w_1, \ldots, w_p\}$ is a basis for $\Re^p$ and $S : \Re^m \to \Re^p$ is a linear transformation. Then to represent $S$ as a matrix with respect to the two sets of bases, $U$ and $W$, for each $i = 1, \ldots, m$, we need to know

$$S(u_i) = \sum_{k=1}^{p} b_{ki} w_k.$$

Then as before $S$ is represented by the matrix

$$M(S) = \begin{pmatrix} b_{11} & \cdots & b_{1i} & \cdots & b_{1m} \\ \vdots & & b_{ki} & & \vdots \\ b_{p1} & \cdots & b_{pi} & \cdots & b_{pm} \end{pmatrix}$$

where the $i$th column is the column of coordinates of $S(u_i)$ in $\Re^p$.

We can compute the composition

$$(S \circ T) : \Re^n \xrightarrow{T} \Re^m \xrightarrow{S} \Re^p.$$

The question is how should we compose the two matrices $M(S)$ and $M(T)$ so that the result "corresponds" to the matrix $M(S \circ T)$ which represents $S \circ T$.

First of all we show that $S \circ T : \Re^n \to \Re^p$ is a linear transformation, so that we know that it can be represented by an $(n \times p)$ matrix.

**Lemma 2.10** *If $T : \Re^n \to \Re^m$ and $S : \Re^m \to \Re^p$ are linear transformations, then $S \circ T : \Re^n \to \Re^p$ is a linear transformation.*

*Proof* Consider $\alpha, \beta \in \Re$, $v_1, v_2 \in \Re^n$. Then

$$\begin{aligned}(S \circ T)(\alpha v_1 + \beta v_2) &= S[T(\alpha v_1 + \beta v_2)] \\ &= S(\alpha T(v_1) + \beta T(v_2)) \quad \text{since } T \text{ is linear} \\ &= \alpha S(T(v_1)) + \beta S(T(v_2)) \quad \text{since } S \text{ is linear} \\ &= \alpha (S \circ T)(v_1) + \beta (S \circ T)(v_2).\end{aligned}$$

Thus $S \circ T$ is linear.

By the previous analysis, $(S \circ T)$ can be represented by an $(n \times p)$ matrix whose $j$th column is $(S \circ T)(v_j)$. Thus

$$\begin{aligned}(S \circ T)(v_j) &= S\left(\sum_{i=1}^{m} a_{ij} u_i\right) \\ &= \sum_{i=1}^{m} a_{ij} S(u_i) \\ &= \sum_{i=1}^{m} a_{ij} \sum_{k=1}^{p} b_{ki} w_k \\ &= \sum_{k=1}^{p} \left(\sum_{i=1}^{m} a_{ij} b_{ki}\right) w_k.\end{aligned}$$

Thus the $k$th entry in the $j$th column of $M(S \circ T)$ is $\sum_{i=1}^{m} b_{ki} a_{ij}$.
Thus $(S \circ T)$ can be represented by the matrix

$$M(S \circ T) = k\text{th row} \begin{pmatrix} & \overset{\longleftarrow \; n \; \longrightarrow}{} & \\ \cdots & \sum_{i=1}^{m} b_{ki} a_{ij} & \cdots \\ & j\text{th column} & \end{pmatrix} p$$

The $j$th column in this matrix can be obtained more simply by operating the matrix $M(S)$ on the $j$th column vector $T(v_j)$ in the matrix $M(T)$.
Thus $M(S \circ T) = (M(S)(T(v_1)) \ldots M(S)(T(v_n))) = M(S) \circ M(T)$.

## 2.2 Linear Transformations

$$\text{kth row of } \begin{pmatrix} b_{k1} \ldots & b_{ki} \ldots & b_{km} \\ \leftarrow & m \text{ columns} & \rightarrow \end{pmatrix} \begin{pmatrix} \leftarrow n \rightarrow \\ a_{ij} \\ \vdots \\ a_{ij} \\ \vdots \\ a_{mj} \\ j\text{th column} \end{pmatrix} m \text{ rows}$$

$$= M(S) \circ M(T).\qquad\square$$

Thus the "natural" method of matrix composition corresponds to the composition of linear transformations.

Now let $L(\Re^n, \Re^n)$ stand for the set of linear transformations from $\Re^n$ to $\Re^n$. As we have shown, if $S, T$ belong to this set then $S \circ T$ is also a linear transformation from $\Re^n$ to $\Re^n$. Thus composition of functions ($\circ$) is a binary operation $L(\Re^n, \Re^n) \times L(\Re^n, \Re^n) \to L(\Re^n, \Re^n)$.

Let $M : L(\Re^n, \Re^n) \to M(n, n)$ be the mapping which assigns to any linear transformation $T : \Re^n \to \Re^n$ the matrix $M(T)$ as above. Note that $M$ is dependent on the choice of bases $\{v_1, \ldots, v_n\}$ and $\{u_1, \ldots, u_n\}$ for the domain and codomain, $\Re^n$. There is in general no reason why these two bases should be the same.

Now let $\circ$ be the method of matrix composition which we have just defined. Thus the mapping $M$ satisfies

$$M(S \circ T) = M(S) \circ M(T)$$

for any two linear transformations, $S$ and $T$. Suppose now that we are given a linear transformation, $T \in L(\Re^n, \Re^n)$. Clearly the matrix $M(T)$ which represents $T$ with respect to the two bases is unique, and so $M$ is a function.

On the other hand suppose that $T, S$ are both represented by the same matrix $A = (a_{ij})$.

By definition $T(v_j) = S(v_j) = \sum_{i=1}^{m} a_{ij} u_i$ for each $j = 1, \ldots, n$.

But then $T(x) = S(x)$ for any $x \in \Re^n$, and so $T = S$. Thus $M$ is injective.

Moreover if $A$ is any matrix, then it represents a linear transformation, and so $M$ is surjective. Thus we have a bijective morphism

$$M : \big(L(\Re^n, \Re^n), \circ\big) \to \big(M(n, n), \circ\big).$$

As we saw in the case of $2 \times 2$ matrices, the subset of non-singular matrices in $M(n, n)$ forms a group. We repeat the procedure for the more general case.

### 2.2.2 The Dimension Theorem

Let $T : V \to U$ be a linear transformation between the vector spaces $V, U$ of dimension $n, m$ respectively over a field $\mathcal{F}$. The transformation is characterised by two subspaces, of $V$ and $U$.

**Definition 2.8**
1. the *kernel* of a transformation $T : V \to U$ is the set $\text{Ker}(T) = \{x \in V : T(x) = \underline{0}\}$ in $V$.
2. The *image* of the transformation is the set $\text{Im}(T) = \{y \in U : \exists\, x \in V \text{ s.t. } T(x) = y\}$.

Both these sets are vector subspaces of $U, V$ respectively. To see this suppose $v_1, v_2 \in \text{Ker}(T)$, and $\alpha, \beta \in F$. Then $T(\alpha v_1 + \beta v_2) = \alpha T(v_1) + \beta T(v_2) = \underline{0} + \underline{0} = \underline{0}$. Hence $\alpha v_1 + \beta v_2 \in \text{Ker}(T)$.

If $u_1, u_2 \in \text{Im}(T)$ then there exists $v_1, v_2 \in V$ such that $T(v_1) = u_1, T(v_2) = u_2$. But then

$$\alpha \mu_1 + \beta \mu_2 = \alpha T(v_1) + \beta T(v_2)$$
$$= T(\alpha v_1 + \beta v_2).$$

Since $V$ is a vector space, $\alpha v_1 + \beta v_2 \in V$ and so $\alpha u_1 + \beta u_2 \in \text{Im}(T)$.

By the exchange theorem there exists a basis $k_1, \ldots, k_p$ for $\text{Ker}(T)$, where $p = \dim \text{Ker}(T)$ and a basis $u_1, \ldots, u_s$ for $\text{Im}(T)$ where $s = \dim(\text{Im}(T))$. Here $p$ is called the *kernel rank* of $T$, often written $kr(T)$, and $s$ is the *rank* of $T$, or $rk(T)$.

**The Dimension Theorem** *If $T : V \to U$ is a linear transformation between vector spaces over a field $\mathcal{F}$, where dimension $(V) \neq n$, then the dimension of the kernel and image of $T$ satisfy the relation*

$$\dim(\text{Im}(T)) + \dim(\text{Ker}(T)) = n.$$

*Proof* Let $\{u_1, \ldots, u_s\}$ be a basis for $\text{Im}(T)$ and for each $i = 1, \ldots, s$, let $v_i$ be the vector in $V^n$ such that $T(v_i) = u_i$.

Let $v$ be any vector in $V$. Then

$$T(v) = \sum_{i=1}^{s} \alpha_i u_i, \quad \text{for } T(v) \in \text{Im}(T).$$

So

$$T(v) = \sum_{i=1}^{s} \alpha_i T(v_i)$$
$$= T\left(\sum_{i=1}^{s} \alpha_i v_i\right), \quad \text{and}$$
$$T\left(v - \sum_{i=1}^{s} \alpha_i v_i\right) = \underline{0},$$

the zero vector in $U$, i.e., $v - \sum_{i=1}^{s} \alpha_i v_i \in \text{kernel } T$. Let $\{k_1, \ldots, k_p\}$ be the basis for $\text{Ker}(T)$.

## 2.2 Linear Transformations

Then $v - \sum_{i=1}^{s} \alpha_i v_i = \sum_{j=1}^{p} \beta_j k_j$, or $v = \sum_{i=1}^{s} \alpha_i v_i + \sum_{j=1}^{p} \beta_j k_j$. Thus $(v_1, \ldots, v_s, k_1, \ldots, k_p)$ is a span for $V$.

Suppose we consider

$$\sum_{i=1}^{s} \alpha_i v_i + \sum_{j=1}^{p} \beta_j k_j = \underline{0}. \tag{*}$$

Then, since $T(k_j) = 0$ for $j = 1, \ldots, p$,

$$T\left(\sum_{i=1}^{s} \alpha_i v_i + \sum_{j=1}^{p} \beta_j k_j\right) = \sum_{i=1}^{s} \alpha_i T(v_i) + \sum_{j=1}^{p} \beta_j T(k_j)$$

$$= \sum_{i=1}^{s} \alpha_i T(v_i) = \sum_{i=1}^{s} \alpha_i u_i = \underline{0}.$$

Now $\{u_1, \ldots, u_s\}$ is a basis for $\text{Im}(T)$, and hence these vectors are linearly independent. So $\alpha_i = 0$, $i = 1, \ldots, s$. Therefore (*) gives $\sum_{j=1}^{p} \beta_j k_j = \underline{0}$.

However $\{k_1, \ldots, k_p\}$ is a basis for $\text{Ker}(T)$ and therefore a frame, so $\beta_j = 0$ for $j = 1, \ldots, p$. Hence $\{v_1, \ldots, v_s, k_1, \ldots, k_p\}$ is a frame, and therefore a basis for $V$. By the exchange theorem the dimension of $V$ is the unique number of vectors in a basis. Therefore $s + p = n$. □

Note that this theorem is true for general vector spaces. We specialise now to vector spaces $\Re^n$ and $\Re^m$.

Suppose $\{v_1, \ldots, v_n\}$ is a basis for $\Re^n$. The coordinates of $v_j$ with respect to this basis are $(0, \ldots, 1, \ldots, 0)$ with 1 in the $j$th place. As we have noted the image of $v_j$ under the transformation $T$ can be represented by the $j$th column $(a_{ij}, \ldots, a_{mj})$ in the matrix $M(T)$, with respect to the original basis $(e_1, \ldots, e_m)$, say, for $\Re^m$. Call the $n$ different column vectors of this matrix $a_1, \ldots, a_j, \ldots, a_n$.

Then the equation $M(T)(x) = y$ is identical to the equation $\sum_{j=1}^{n} x_j a_j = y$ where $x = (x_1, \ldots, x_n)$.

Clearly any vector $y$ in the image of $M(T)$ can be written as a linear combination of the columns $A = \{a_1, \ldots, a_n\}$. Thus $\text{Span}(A) = \text{Im}(M(T))$. Suppose now that $A$ is not a frame. In this case $a_n$, say, can be written as a linear combination of $\{a_1, \ldots, a_{n-1}\}$, i.e., $\sum_{j=1}^{n} k_{1j} a_j = \underline{0}$ and $k_{1n} \neq 0$. Then the vector $k_1 = (k_{11}, \ldots, k_{1n})$ satisfies $M(T)(k_1) = \underline{0}$. Thus $k_1$ belongs to $\text{Ker}(M(T))$.

Eliminate $a_n$, say, and proceed in this way. After $p$ iterations we will have obtained $p$ kernel vectors $\{k_1, \ldots, k_p\}$ and the remaining column vectors $\{a_1, \ldots, a_{n-p}\}$ will form a frame, and thus a basis for the image of $M(T)$.

Consequently $\dim(\text{Im}(M(T))) = n - p = n - \dim(\text{Ker}(M(T)))$. The number of linearly independent columns in the matrix $M(T)$ is called the *rank* of $M(T)$, and is clearly the dimension of the image of $M(T)$. In particular if $M_1(T)$ and $M_2(T)$ are two matrix representations with respect to different bases, of the linear transformation $T$, then $\text{rank } M_1(T) = \text{rank } M_2(T) = \text{rank}(T)$.

Thus rank($T$) is an invariant, in the sense of being independent of the particular bases chosen for $\Re^n$ and $\Re^m$.

In the same way the kernel rank of $T$ is an invariant; that is, for any matrix representation $M(T)$ of $T$ we have $\ker \operatorname{rank}(M(T)) = \ker \operatorname{rank}(T)$.

In general if $y \in \operatorname{Im}(T)$, $x_0$ satisfies $T(x_0) = y$, and $k$ belongs to the kernel, then

$$T(x_0 + k) = T(x_0) + T(k) = y + \underline{0} = y.$$

Thus if $x_0$ is a solution to the equation $T(x_0) = y$, the point $x_0 + k$ is also a solution. More generally $x_0 + \operatorname{Ker}(T) = \{x_0 + k : k \in \operatorname{Ker}(T)\}$ will also be the set of solutions. Thus for a particular $y \in \operatorname{Im}(T)$, $T^{-1}(y) = \{x : T(x) = y\} = x_0 + \operatorname{Ker}(T)$.

By the dimension theorem $\dim \operatorname{Ker}(T) = n - \operatorname{rank}(T)$. Thus $T^{-1}(y)$ is a geometric object of "dimension" $\dim \operatorname{Ker}(T) = n - \operatorname{rank}(T)$.

We defined $T$ to be *injective* iff $T(x_0) = T(x)$ implies $x_0 = x$. Thus $T$ is injective iff $\operatorname{Ker}(T) = \{\underline{0}\}$. In this case, if there is a solution to the equation $T(x_0) = y$, then this solution is unique.

Suppose that $n \leq m$, and that the $n$ different column vectors of the matrix are linearly independent. In this case $\operatorname{rank}(T) = n$ and so $\dim \operatorname{Ker}(T) = 0$. Thus $T$ must be *injective*. In particular if $n < m$ then not every $y \in \Re^m$ belongs to the image of $T$, and so not every equation $T(x) = y$ has a solution. Suppose on the other hand that $n > m$. In this case the maximum possible rank is $m$ (since $n$ vectors cannot be linearly independent in $\Re^m$ when $n > m$). If $\operatorname{rank}(T) = m$, then there must exist a kernel of dimension $(n - m)$.

Moreover $\operatorname{Im}(T) = \Re^m$, and so for every $y \in \Re^m$ there exists a solution to this equation $T(x) = y$. Thus $T$ is *surjective*. However the solution is not unique, since $T^{-1}(y) = x + \operatorname{Ker}(T)$ is of dimension $(n - m)$ as before.

Suppose now that $n = m$, and that $T : \Re^n \to \Re^n$ has maximal rank $n$. Then $T$ is both injective and surjective and thus an *isomorphism*. Indeed $T$ will have an inverse function $T^{-1} : \Re^n \to \Re^n$. Moreover $T^{-1}$ is linear. To see this note that if $x_1 = T^{-1}(y_1)$ and $x_2 = T^{-1}(y_2)$ then $T(x_1) = y_1$ and $T(x_2) = y_2$ so $T(x_1 + x_2) = y_1 + y_2$. Thus $T^{-1}(y_1 + y_2) = x_1 + x_2 = T^{-1}(y_1) + T^{-1}(y_2)$. Moreover if $x = T^{-1}(\alpha y)$ then $T(x) = \alpha y$. If $\alpha \neq 0$, then $\frac{1}{\alpha}T(x) = T(\frac{1}{\alpha}x) = y$ or $\frac{1}{\alpha}x = T^{-1}(y)$. Hence $x = \alpha T^{-1}(y)$. Thus $T^{-1}(\alpha y) = \alpha T^{-1}(y)$. Since $T^{-1}$ is linear it can be represented by a matrix $M(T^{-1})$. As we know $M : (L(\Re^n, \Re^n), \circ) \to (M(n,n), \circ)$ is a bijective morphism, so $M$ maps the identity linear transformation, Id, to the identity *matrix*

$$M(\operatorname{Id}) = I = \begin{pmatrix} 1 & \cdots & 0 \\ \vdots & & \vdots \\ 0 & \cdots & 1 \end{pmatrix}.$$

When $T$ is an isomorphism with inverse $T^{-1}$, then the representation $M(T^{-1})$ of $T^{-1}$ is $[M(T)]^{-1}$. We now show how to compute the *inverse matrix* $[M(T)]^{-1}$ of an isomorphism.

### 2.2.3 The General Linear Group

To compute the inverse of an $n \times n$ matrix $A$, we define, by induction, the determinant of $A$. For a $1 \times 1$ matrix $(a_{11})$ define $\det(A_{11}) = a_{11}$, and for a $2 \times 2$ matrix $A = \begin{pmatrix} a_{11} & a_{12} \\ a_{21} & a_{22} \end{pmatrix}$ define $\det A = a_{11}a_{22} - a_{21}a_{12}$.

For an $n \times n$ matrix $A$ define the $(i, j)$th *cofactor* to be the determinant of the $(n-1) \times (n-1)$ matrix $A(i, j)$ obtained from $A$ by removing the $i$th row and $j$th column, then multiplying by $(-1)^{i+j}$. Write this cofactor as $A_{ij}$. For example in the $3 \times 3$ matrix, the cofactor in the $(1, 1)$ position is

$$A_{11} = \det \begin{pmatrix} a_{22} & a_{23} \\ a_{32} & a_{33} \end{pmatrix} = a_{22}a_{33} - a_{32}a_{23}.$$

The $n \times n$ matrix $(A_{ij})$ is called the *cofactor matrix*.

The determinant of the $n \times n$ matrix $A$ is then $\sum_{j=1}^{n} a_{1j} A_{1j}$. The determinant is also often written as $|A|$.

This procedure allows us to define the determinant of an $n \times n$ matrix. For example if $A = (a_{ij})$ is a $3 \times 3$ matrix, then

$$|A| = a_{11}\begin{vmatrix} a_{22} & a_{23} \\ a_{32} & a_{33} \end{vmatrix} - a_{12}\begin{vmatrix} a_{21} & a_{23} \\ a_{31} & a_{33} \end{vmatrix} + a_{13}\begin{vmatrix} a_{21} & a_{22} \\ a_{31} & a_{32} \end{vmatrix}$$
$$= a_{11}(a_{22}a_{33} - a_{32}a_{23}) - a_{12}(a_{21}a_{33} - a_{31}a_{23}) + a_{13}(a_{21}a_{32} - a_{31}a_{22}).$$

An alternative way of defining the determinant is as follows. A permutation of $n$ is a bijection $s : \{1, \ldots, n\} \to \{1, \ldots, n\}$, with degree $d(s)$ the number of exchanges needed to give the permutation.

Then $|A| = \sum_s (-1)^{d(s)} \Pi_{i=1}^n a_{is(i)} = a_{11}a_{22}a_{33}\cdots + \cdots$ where the summation is over all permutations. The two definitions are equivalent, and it can be shown that

$$|A| = \sum_{j=1}^{n} a_{ij} A_{ij} \quad \text{(for any } i = 1, \ldots, n\text{)}$$
$$= \sum_{i=1}^{n} a_{ij} A_{ij} \quad \text{(for any } j = 1, \ldots, n\text{)}$$

while

$$0 = \sum_{i=1}^{n} a_{ij} A_{ik} \quad \text{if } j \neq k$$
$$= \sum_{j=1}^{n} a_{ij} A_{kj} \quad \text{if } i \neq k.$$

Thus

$$(a_{ij})(A_{jk})^t = \left(\sum_{j=1}^{n} a_{ij} A_{kj}\right)$$

$$= \begin{pmatrix} |A| & \cdots & 0 \\ \vdots & & \vdots \\ 0 & \cdots & |A| \end{pmatrix} = |A|I.$$

Here $(A_{jk})^t$ is the $n \times n$ matrix obtained by transposing the rows and columns of $(A_{jk})$. Now the matrix $A^{-1}$ satisfies $A \circ A^{-1} = I$, and if $A^{-1}$ exists then it is unique. Thus $A^{-1} = \frac{1}{|A|}(A_{ij})^t$.

Suppose that the matrix $A$ is non-singular, so $|A| \neq 0$. Then we can construct an inverse matrix $A^{-1}$.

Moreover if $A(x) = y$ then $y = A^{-1}(x)$ which implies that $A$ is both injective and surjective. Thus rank$(A) = n$ and the column vectors of $A$ must be linearly independent.

As we have noted, however, if $A$ is not injective, with Ker$(A) \neq \{0\}$, then rank$(A) < n$, and the column vectors of $A$ must be linearly dependent. In this case the inverse $A^{-1}$ is not a function and cannot therefore be represented by a matrix and so we would expect $|A|$ to be zero.

**Lemma 2.11** *If $A$ is an $n \times n$ matrix with* rank$(A) < n$ *then* $|A| = 0$.

*Proof* Let $A'$ be the matrix obtained from $A$ by adding a multiple $(\alpha)$ of the $k$th column of $A$ to the $j$th column of $A$. The $j$th column of $A'$ is therefore $a_j + \alpha a_k$. This operation leaves the $j$th column of the cofactor matrix unchanged. Thus

$$|A'| = \sum_{i=1}^{n} a'_{ij} A_{ij}$$

$$= \sum_{i=1}^{n} (a_{ij} + \alpha a_{ik}) A_{ij}$$

$$= \sum_{i=1}^{n} a_{ij} A_{ij} + \alpha \sum_{i=1}^{n} a_{ik} A_{ij}$$

$$= |A| + 0 = |A|.$$

Suppose now that the columns of $A$ are linearly dependent, and that $a_j = \sum_{k \neq j} \alpha_k a_k$ for example. Let $A'$ be the matrix obtained from $A$ by substituting $a'_j = 0 = a_j - \sum_{k \neq j} \alpha_k a_k$ for the $j$th column.

By the above $|A'| = \sum_{i=1}^{n} a'_{ij} A_{ij} = 0 = |A|$. □

## 2.2 Linear Transformations

Suppose now that $A$, $B$ are two non-singular matrices $(a_{ij})$, $(b_{ki})$. The composition is then $B \circ A = (\sum_{i=1}^{m} b_{ki} a_{ij})$ with determinant

$$|B \circ A| = \sum_{s} (-1)^{d(s)} \prod_{k=1}^{n} \circ \left( \sum_{i=1}^{m} b_{ki} a_{is(k)} \right).$$

This expression can be shown to be equal to

$$\sum_{s} (-1)^{d(s)} \prod_{i=1}^{n} a_{is(i)} \sum_{s} (-1)^{d(s)} \prod_{i=1}^{n} b_{ks(k)} = |B||A| \neq 0.$$

Hence the composition $(B \circ A)$ has an inverse $(B \circ A)^{-1}$ given by $A^{-1} \circ B^{-1}$.

Now let $(GL(\Re^n, \Re^n), \circ)$ be the set of invertible linear transformations, with $\circ$ composition of functions, and let $M^*(n, n)$ be the set of non-singular $n \times n$ matrices. Choice of bases $\{v_1, \ldots, v_n\}, \{u_1, \ldots, u_n\}$ for the domain and codomain defines a morphism

$$M : (GL(\Re^n, \Re^n), \circ) \to (M^*(n, n), \circ).$$

Suppose now that $T$ belongs to $GL(\Re^n, \Re^n)$. As we have seen this is equivalent to $|M(T)| \neq 0$, so the image of $M$ is precisely $M^*(n, n)$. Moreover if $|M(T)| \neq 0$ then $|M(T^{-1})| = \frac{1}{|M(T)|}$ and $M(T^{-1})$ belongs to $M^*(n, n)$. On the other hand if $S, T \in GL(\Re^n, \Re^n)$ then $S \circ T$ also has rank $n$, and has inverse $T^{-1} \circ S^{-1}$ with rank $n$.

The matrix $M(S \circ T)$ representing $T \circ S$ has inverse

$$M(T^{-1} \circ S^{-1}) = M(T^{-1}) \circ M(S^{-1})$$
$$= [M(T)]^{-1} \circ [M(S)]^{-1}.$$

Thus $M$ is an *isomorphism* between the two groups $(GL(\Re^n, \Re^n), \circ))$ and $(M^*(n, n), \circ)$.

The group of invertible linear transformations is also called the *general linear group*.

### 2.2.4 Change of Basis

Let $L(\Re^n, \Re^m)$ stand for the set of linear transformations from $\Re^n$ to $\Re^m$, and let $M(n, m)$ stand for the set of $n \times m$ matrices. We have seen that the choice of bases for $\Re^n, \Re^m$ defines a *function*.

$$M : L(\Re^n, \Re^m) \to M(n, m)$$

which take a linear transformation $T$ to its representation $M(T)$. We now examine the relationship between two representations $M_1(T)$, $M_2(T)$ of a single linear transformation.

**Basis Change Theorem** *Let $\{v_1, \ldots, v_n\}$ and $\{u_1, \ldots, u_m\}$ be bases for $\Re^n, \Re^m$ respectively.*

*Let $T$ be a linear transformation which is represented by a matrix $A = (a_{ij})$ with respect to these bases. If $V' = \{v'_1, \ldots, v'_n\}$, $U' = \{u'_1, \ldots, u'_m\}$ are new bases for $\Re^n, \Re^m$ then $T$ is represented by the matrix $B = Q^{-1} \circ A \circ P$, where $P, Q$ are respectively $(n \times n)$ and $(m \times m)$ invertible matrices.*

*Proof* For each $v'_k \in V' = \{v'_1, \ldots, v'_n\}$ let $v'_k = \sum_{i=1}^n b_{ik} v_i$ and $b_k = (b'_{1k}, \ldots, b_{nk})$.

Let $P = (b_1, \ldots, b_n)$ where the $k$th column of $P$ is the column of coordinates of $b_k$. With respect to the new basis $V'$, $v'_k$ has coordinates $e_k = (0, \ldots, 1, \ldots, 0)$ with a 1 in the $k$th place.

But then $P(e_k) = b_k$ the coordinates of $v'_k$ with respect to $V$.

Thus $P$ is the matrix that transforms coordinates with respect to $V'$ into coordinates with respect to $V$. Since $V$ is a basis, the columns of $P$ are linearly independent, and so rank $P = n$, and $P$ is invertible.

In the same way let $u'_k = \sum_{i=1}^m c_{ik} u_i$, $c_k = (c_{1k}, \ldots, c_{mk})$ and $Q = (c_1, \ldots, c_m)$ the matrix with columns of these coordinates.

Hence $Q$ represents change of basis from $U'$ to $U$. Since $Q$ is an invertible $m \times m$ matrix it has inverse $Q^{-1}$ which represents change of basis from $U$ to $U'$.

Thus we have the diagram

$$\begin{array}{ccc} \{v_1, \ldots, v_n\} & \xrightarrow{A} & \{u_1, \ldots, u_m\} \\ P \uparrow & & Q^{-1} \downarrow \uparrow Q \\ \{v'_1, \ldots, v'_n\} & \xrightarrow{B} & \{u'_1, \ldots, u'_m\} \end{array}$$

from which we see that the matrix $B$, representing the linear transformation $T : \Re^n \to \Re^m$ with respect to the new bases is given by $B = Q^{-1} \circ A \circ P$. □

**Isomorphism Theorem** *Any linear transformation $T : \Re^n \to \Re^m$ of rank $r$ can be represented, by suitable choice of bases for $\Re^n$ and $\Re^m$, by an $n \times m$ matrix*

$$\begin{pmatrix} I_r & 0 \\ 0 & 0 \end{pmatrix} \quad \text{where } I_r = \begin{pmatrix} 1 & 0 \\ 0 & 1 \end{pmatrix} \text{ is the } (r \times r) \text{ identity matrix.}$$

*In particular*
1. *if $n < m$ and $T$ is injective then there is an isomorphism $S : \Re^m \to \Re^m$ such that $S \circ T(x_1, \ldots, x_n) = (x_1, \ldots, x_n, 0, \ldots, 0)$ with $(n - m)$ zero entries, for any vector $(x_1, \ldots, x_n)$ in $\Re^n$*

## 2.2 Linear Transformations

2. if $n \geq m$ and $T$ is surjective then there are isomorphisms $\Re : \Re^n \to \Re^n$, $S : \Re^m \to \Re^m$ such that $S \circ T \circ R(x_1, \ldots, x_n) = (x_1, \ldots, x_m)$. If $n = m$, then $S \circ T \circ R$ is the identity isomorphism.

*Proof* Of necessity $\mathrm{rank}(T) = r \leq \min(n, m)$. If $r < n$, let $p = n - r$ and choose a basis $k_1, \ldots, k_p$ for $\mathrm{Ker}(T)$. Let $V = \{v_1, \ldots, v_n\}$ be the original basis for $\Re^n$. By the exchange theorem there exists $r = (n - p)$ different members $\{v_1, \ldots, v_r\}$ say of $V$ such that $V' = \{v_1, \ldots, v_r, k_1, \ldots, k_p\}$ is a basis for $\Re^n$.

Choose $V'$ as the new basis for $\Re^n$, and let $P$ be the basis change matrix whose columns are the column vectors in $V'$. As in the proof of the dimension theorem the image of the vectors $v_1, \ldots, v_{n-p}$ under $T$ provide a basis for the image of $T$. Let $U = \{u_1, \ldots, u_m\}$ be the original basis of $\Re^m$. By the exchange theorem there exists some subset $U' = \{u_1, \ldots, u_{m-r}\}$ of $U$ such that $U'' = \{T(v_1), \ldots, T(v_r), u_1, \ldots, u_{m-r}\}$ form a basis for $\Re^m$. Note that $T(v_1), \ldots, T(v_r)$ are represented by the $r$ linearly independent columns of the original matrix $A$ representing $T$. Now let $Q$ be the matrix whose columns are the members of $U''$. By the basis change theorem, $B = Q^{-1} \circ A \circ P$, where $B$ is the matrix representing $T$ with respect to these new bases. Thus we obtain

$$\{v_1, \ldots, v_n\} \xrightarrow{A} \{u_1, \ldots, u_m\}$$

$$P \uparrow \qquad\qquad Q^{-1} \downarrow$$

$$\{v_1, \ldots, v_r, k_1 \ldots\} \xrightarrow{B} \{T(v_1) \ldots T(v_r), u_1, \ldots, u_{m-r}\}.$$

With respect to these new bases, the matrix $B$ representing $T$ has the required form: $\begin{pmatrix} I_r & 0 \\ 0 & 0 \end{pmatrix}$.

1. If $n < m$ and $T$ is injective then $r = n$. Hence $P$ is the identity matrix, and so $B = Q^{-1} \circ A$.

   But $Q^{-1}$ is an $m \times m$ invertible matrix, and thus represents an isomorphism $\Re^n \longrightarrow \Re^n$, while

$$B \begin{pmatrix} x_1 \\ x_n \end{pmatrix} = \begin{pmatrix} I_n \\ 0 \end{pmatrix} \begin{pmatrix} x_1 \\ x_n \end{pmatrix} = \begin{pmatrix} x_1 \\ \vdots \\ x_n \\ \vdots \\ 0 \end{pmatrix}.$$

   Write a vector $x = \sum_{i=1}^{n} x_i v_i$ as $(x_1, \ldots, x_n)$, and let $S$ be the linear transformation $\Re^m \to \Re^m$ represented by the matrix $Q^{-1}$. Then $S \circ T(x_1, \ldots, x_n) = (x_1, \ldots, x_n, 0, \ldots, 0)$.

2. If $n \geq m$ and $T$ is surjective then $\mathrm{rank}(T) = m$, and $\dim \mathrm{Ker}(T) = n - m$. Thus $B = (I_m \ 0) = Q^{-1} \circ A \circ P$. Let $S, R$ be the linear transformations represented by $Q^{-1}$ and $P$ respectively.

Then $S \circ T \circ R(x_1, \ldots, x_n) = (x_1, \ldots, x_m)$. If $n = m$ then $S \circ T \circ R$ is the identity transformation. □

Suppose now that $V, U$ are the two bases for $\Re^n, \Re^m$ as in the basis theorem. A linear transformation $T : \Re^n \longrightarrow \Re^m$ is represented by a matrix $M_1(T)$ with respect to these bases. If $V', U'$ are two new bases, then $T$ will be represented by the matrix $M_2(T)$, and by the basis theorem

$$M_2(T) = Q^{-1} \circ M_1(T) \circ P$$

where $Q, P$ are non-singular ($m \times m$) and ($n \times n$) matrices respectively. Since $M_1(T)$ and $M_2(T)$ represent the same linear transformation, they are in some sense equivalent. We show this more formally.

Say the two matrices $A, B \in M(n, m)$ are *similar* iff there exist non singular square matrices $P \in M^*(n, n)$ and $Q \in M^*(m, m)$ such that $B = Q^{-1} \circ A \circ P$, and in this case write $B \sim A$.

**Lemma 2.12** *The similarity relation* ($\sim$) *on $M(n, m)$ is an equivalence relation.*

*Proof*
1. To show that $\sim$ is reflexive note that $A = I_m^{-1} \circ A \circ I_n$ where $I_m, I_n$ are respectively the ($m \times m$) and ($n \times n$) identity matrices.
2. To show that $\sim$ is symmetric we need to show that $B \sim A$ implies that $A \sim B$. Suppose therefore that $B = Q^{-1} \circ A \circ P$.
Since $Q \in M^*(m, m)$ it has inverse $Q^{-1} \in M^*(m, m)$.
Moreover $(Q^{-1})^{-1} \circ Q^{-1} = I_m$, and thus $Q = (Q^{-1})^{-1}$. Thus

$$Q \circ B \circ P^{-1} = (Q \circ Q^{-1}) \circ A \circ (P \circ P^{-1})$$
$$= A$$
$$= (Q^{-1})^{-1} \circ B \circ (P^{-1}).$$

Thus $A \sim B$.
3. To show $\sim$ is transitive, we seek to show that $C \sim B \sim A$ implies $C \sim A$. Suppose therefore that $C = R^{-1} \circ B \circ S$ and $B = Q^{-1} \circ A \circ P$, where $R, Q \in M^*(m, m)$ and $S, P \in M^*(n, n)$. Then

$$C = (R^{-1} \circ Q^{-1}) \circ A \circ P \circ S$$
$$= (Q \circ R)^{-1} \circ A \circ (P \circ S).$$

Now $(M^*(m, m), \circ), (M^*(n, n), \circ)$ are both groups and so $Q \circ R \in M^*(m, m)$, $P \circ S \in M^*(n, n)$. Thus $C \sim A$. □

## 2.2 Linear Transformations

The isomorphism theorem shows that if there is a linear transformation $T : \mathfrak{R}^n \longrightarrow \mathfrak{R}^m$ of rank $r$, then the $(n \times m)$ matrix $M_1(T)$ which represents $T$, with respect to some pair of the bases, is similar to an $n \times m$ matrix

$$B = \begin{pmatrix} I_r & 0 \\ 0 & 0 \end{pmatrix} \quad \text{i.e., } M(T) \sim B.$$

If $S$ is a second linear transformation of rank $r$ then $M_1(S) \sim B$.
By Lemma 2.12, $M_1(S) \sim M_1(T)$.

Suppose now that $U', V'$ are a second pair of bases for $\mathfrak{R}^n, \mathfrak{R}^m$ and let $M_2(S), M_2(T)$ represent $S$ and $T$. Clearly $M_2(S) \sim M_2(T)$.

Thus if $S, T$ are linear transformations $\mathfrak{R}^m \longrightarrow \mathfrak{R}^n$ we may say that $S, T$ are equivalent iff for any choice of bases the matrices $M(S), M(T)$ which represent $S, T$ are similar.

For any linear transformation $T \in L(\mathfrak{R}^n, \mathfrak{R}^m)$ let $[T]$ be the equivalence class $\{S \in L(\mathfrak{R}^n, \mathfrak{R}^m) : S \sim T\}$. Alternatively a linear transformation $S$ belongs to $[T]$ iff $\text{rank}(S) = \text{rank}(T)$. Consequently the equivalence relation partitions $L(\mathfrak{R}^n, \mathfrak{R}^m)$ into a finite number of distinct equivalence classes where each class is classified by its rank, and the rank runs from 0 to $\min(n, m)$.

### 2.2.5 Examples

*Example 2.1* To illustrate the use of these procedures in the solution of linear equations, consider the case with $n < m$ and the equation $A(x) = y$ where

$$A = \begin{pmatrix} 1 & -1 & 2 \\ 5 & 0 & 3 \\ -1 & -4 & 5 \\ 3 & 2 & -1 \end{pmatrix} \quad \text{and} \quad y_1 = \begin{pmatrix} -1 \\ 1 \\ -1 \\ 1 \end{pmatrix}, \quad y_2 = \begin{pmatrix} 0 \\ 5 \\ -5 \\ 5 \end{pmatrix}.$$

To find $\text{Im}(A)$, we first of all find $\text{Ker}(A)$. The equation $A(x) = \underline{0}$ gives four equations

$$x_1 - x_2 + 2x_3 = 0$$
$$5x_1 + 0 + 3x_3 = 0$$
$$-x_1 - 4x_2 + 5x_3 = 0$$
$$3x_1 + 2x_2 - x_3 = 0$$

with solution $k = (x_1, x_2, x_3) = (-3, 7, 5)$.

Thus $\text{Ker}(A) \supset \{\lambda k \in \mathfrak{R}^3 : \lambda \in \mathfrak{R}\}$. Hence $\dim \text{Im}(A) \leq 2$. Clearly the first two columns $(a_1, a_2)$ of $A$ are linearly independent and so $\dim \text{Im}(A) = 2$. However $y_2 = a_1 + a_2$. Thus a particular solution to the equation $A(x) = y_2$ is $x_0 = (1, 1, 0)$.

The full set of solutions to the equation is

$$x_0 + \operatorname{Ker}(A) = \{(1, 1, 0) + \lambda(-3, 7, 5) : \lambda \in \Re\}.$$

To see whether $y_1 \in \operatorname{Im}(A)$ we need only attempt to solve the equation $y_1 = \alpha a_1 + \beta a_2$. This gives

$$-1 = \alpha - \beta$$
$$1 = 5\alpha$$
$$-1 = -\alpha - 4\beta$$
$$1 = 3\alpha + 2\beta.$$

From the first two equations $\alpha = \frac{1}{5}, \beta = \frac{6}{5}$, which is incompatible with the fourth equation. Thus $y_1$ cannot belong to $\operatorname{Im}(A)$.

*Example 2.2* Consider now an example of the case $n > m$, where

$$A = \begin{pmatrix} 2 & 1 & 1 & 1 & 1 \\ 1 & 2 & -1 & 1 & 1 \end{pmatrix} : \Re^5 \longrightarrow \Re^2.$$

Obviously the first two columns are linearly independent and so $\dim \operatorname{Im}(A) \geq 2$. Let $\{a_i : i = 1, \ldots, 5\}$ be the five column vectors of the matrix and consider the equation

$$\begin{pmatrix} 2 \\ 1 \end{pmatrix} - \begin{pmatrix} 1 \\ 2 \end{pmatrix} - \begin{pmatrix} 1 \\ -1 \end{pmatrix} = \begin{pmatrix} 0 \\ 0 \end{pmatrix}.$$

Thus $k_1 = (1, -1, -1, 0, 0)$ belongs to $\operatorname{Ker}(A)$. On the other hand

$$\begin{pmatrix} 2 \\ 1 \end{pmatrix} + \begin{pmatrix} 1 \\ 2 \end{pmatrix} - 3\begin{pmatrix} 1 \\ 1 \end{pmatrix} = \begin{pmatrix} 0 \\ 0 \end{pmatrix}.$$

Thus $k_2 = (1, 1, 0, -3, 0)$ and $k_3 = (1, 1, 0, 0, -3)$ both belong to $\operatorname{Ker}(A)$.

Consequently the rank of $A$ has its maximal value of 2, while the kernel is three-dimensional. Hence for any $y \in \Re^2$ there is a set of solutions of the form $x_0 + \operatorname{Span}\{k_1, k_2, k_3\}$ to the equation $A(x) = y$.

Change the bases of $\Re^5$ and $\Re^2$ to

$$\begin{pmatrix} 1 \\ 0 \\ 0 \\ 0 \\ 0 \end{pmatrix}, \begin{pmatrix} 0 \\ 1 \\ 0 \\ 0 \\ 0 \end{pmatrix}, \begin{pmatrix} 1 \\ -1 \\ -1 \\ 0 \\ 0 \end{pmatrix}, \begin{pmatrix} 1 \\ 1 \\ 0 \\ -3 \\ 0 \end{pmatrix}, \begin{pmatrix} 1 \\ 1 \\ 0 \\ 0 \\ -3 \end{pmatrix}$$

## 2.2 Linear Transformations

and $\binom{2}{1}, \binom{1}{2}$ respectively, then

$$B = \begin{pmatrix} 2 & 1 \\ 1 & 2 \end{pmatrix}^{-1} \begin{pmatrix} 2 & 1 & 1 & 1 & 1 \\ 1 & 2 & -1 & 1 & 1 \end{pmatrix} \begin{pmatrix} 1 & 0 & 1 & 1 & 1 \\ 0 & 1 & -1 & 1 & 1 \\ 0 & 0 & -1 & 0 & 0 \\ 0 & 0 & 0 & -3 & 0 \\ 0 & 0 & 0 & 0 & -3 \end{pmatrix}$$

$$= \frac{1}{3} \begin{pmatrix} 2 & -1 \\ -1 & 2 \end{pmatrix} \begin{pmatrix} 2 & 1 & 0 & 0 & 0 \\ 1 & 2 & 0 & 0 & 0 \end{pmatrix} = \begin{pmatrix} 1 & 0 & 0 & 0 & 0 \\ 0 & 1 & 0 & 0 & 0 \end{pmatrix}.$$

*Example 2.3* Consider the matrix

$$Q = \begin{pmatrix} 1 & -1 & 0 & 0 \\ 5 & 0 & 0 & 0 \\ -1 & -4 & 1 & 0 \\ 3 & 2 & 0 & 1 \end{pmatrix}.$$

Since $|Q| = 5$ we can compute its inverse. The cofactor matrix $(Q_{ij})$ of $Q$ is

$$\begin{pmatrix} 0 & -5 & -20 & 10 \\ 1 & 1 & 5 & -5 \\ 0 & 0 & 5 & 0 \\ 0 & 0 & 0 & 5 \end{pmatrix}$$

and thus

$$Q^{-1} = \frac{1}{|Q|}(Q_{ij})^t = \begin{pmatrix} 0 & \frac{1}{5} & 0 & 0 \\ -1 & \frac{1}{5} & 0 & 0 \\ -4 & 1 & 1 & 0 \\ 2 & -1 & 0 & 1 \end{pmatrix}.$$

*Example 2.4* Let $T : \mathfrak{R}^3 \to \mathfrak{R}^4$ be the linear transformation represented by the matrix $A$ of Example 2.1, with respect to the standard bases for $\mathfrak{R}^3, \mathfrak{R}^4$. We seek to change the bases so as to represent $T$ by a diagonal matrix

$$B = \begin{pmatrix} I_r & \\ & 0 \end{pmatrix}.$$

By Example 2.1, the kernel is spanned by $(-3, 7, 5)$, and so we choose a new basis

$$e_1 = \begin{pmatrix} 1 \\ 0 \\ 0 \end{pmatrix}, \quad e_2 = \begin{pmatrix} 0 \\ 1 \\ 0 \end{pmatrix}, \quad k = \begin{pmatrix} -3 \\ 7 \\ 5 \end{pmatrix}$$

with basis change matrix $P = (e_1, e_2, k)$. Note that $|P| = 5$ and $P$ is non-singular. Thus $\{e_1, e_2, k\}$ form a basis for $\Re^3$. Now Im($A$) is spanned by the first two columns $a_1, a_2$, of $A$. Moreover $A(e_1) = a_1$ and $A(e_2) = a_2$. Thus choose

$$a_1 = \begin{pmatrix} 1 \\ 5 \\ -1 \\ 3 \end{pmatrix}, \quad a_2 = \begin{pmatrix} -1 \\ 0 \\ -4 \\ 2 \end{pmatrix}, \quad e_3' = \begin{pmatrix} 0 \\ 0 \\ 1 \\ 0 \end{pmatrix}, \quad e_4' = \begin{pmatrix} 0 \\ 0 \\ 0 \\ 1 \end{pmatrix}$$

as the new basis for $\Re^4$. Let $Q = (a_1, a_2, e_3', e_4')$ be the basis change matrix. The inverse $Q^{-1}$ is computed in Example 2.3. Thus we have $(B) = Q^{-1} \circ A \circ P$.

To check that this is indeed the case we compute:

$$Q^{-1} \circ A \circ P = \begin{pmatrix} 0 & \frac{1}{5} & 0 & 0 \\ -1 & \frac{1}{5} & 0 & 0 \\ -4 & 1 & 1 & 0 \\ 2 & -1 & 0 & 1 \end{pmatrix} \begin{pmatrix} 1 & 1 & 2 \\ 5 & 0 & 3 \\ -1 & -4 & 5 \\ 3 & 2 & -1 \end{pmatrix} \begin{pmatrix} 1 & 0 & -3 \\ 0 & 1 & 7 \\ 0 & 0 & 5 \end{pmatrix}$$

$$= \begin{pmatrix} 1 & 0 & 0 \\ 0 & 1 & 0 \\ 0 & 0 & 0 \\ 0 & 0 & 0 \end{pmatrix}$$

as required.

## 2.3 Canonical Representation

When considering a linear transformation $T : \Re^n \longrightarrow \Re^n$ it is frequently convenient to change the basis of $\Re^n$ to a new basis $V = \{v_1, \ldots, v_n\}$ such that $T$ is now represented by a matrix

$$M_2(T) = P^{-1} \circ M_1(T) \circ P.$$

In this case it is generally not possible to obtain $M_2(T)$ in the form $\begin{pmatrix} I_r & 0 \\ 0 & 0 \end{pmatrix}$ as before.

Under certain conditions however $M_2(T)$ can be written in a diagonal form

$$\begin{pmatrix} \lambda_1 & & 0 \\ & \ddots & \\ 0 & & \lambda_n \end{pmatrix},$$

where $\lambda_1, \ldots, \lambda_n$ are known as the *eigenvalues*.

More explicitly, a vector $x$ is called an *eigenvector* of the matrix $A$ iff there is a solution to the equation $A(x) = \lambda x$ where $\lambda$ is a real number. In this case, $\lambda$ is called the *eigenvalue* associated with the eigenvector $x$. (Note that we assume $x \neq \underline{0}$.)

## 2.3 Canonical Representation

### 2.3.1 Eigenvectors and Eigenvalues

Suppose that there are $n$ linearly independent eigenvectors $\{x_1, \ldots, x_n\}$ for $A$, where (for each $i = 1, \ldots, n$) $\lambda_i$ is the eigenvalue associated with $x_i$. Clearly the eigenvector $x_i$ belongs to Ker($A$) iff $\lambda_i = 0$. If rank($A$) = $r$ then there would a subset $\{x_1, \ldots, x_r\}$ of eigenvectors which form a basis for Im($A$), while $\{x_1, \ldots, x_n\}$ form a basis for $\Re^n$. Now let $Q$ be the ($n \times n$) matrix representing a basis change from the new basis to the original basis. That is to say the $i$th column, $v_i$, of $Q$ is the coordinate of $x_i$ with respect to the original basis.

After transforming, the original becomes

$$Q^{-1} \circ A \circ Q = \begin{pmatrix} \lambda_1 & \cdots & 0 \\ \vdots & \lambda_r & \vdots \\ 0 & & 0 \end{pmatrix} = \wedge,$$

where rank $\wedge$ = rank $A = r$.

In general we can perform this diagonalisation only if there are enough eigenvectors, as the following lemma indicates.

**Lemma 2.13** *If $A$ is an $n \times n$ matrix, then there exists a non-singular matrix $Q$, and a diagonal matrix $\wedge$ such that $\wedge = Q^{-1} A Q$ iff the eigenvectors of $A$ form a basis for $\Re^n$.*

*Proof*
1. Suppose the eigenvectors form a basis, and let $Q$ be the eigenvector matrix. By definition, if $v_i$ is the $i$th column of $Q$, then $A(v_i) = \lambda_i v_i$, where $\lambda_i$ is real. Thus $AQ = Q\wedge$. But since $\{v_1, \ldots, v_n\}$ is a basis, $Q^{-1}$ exists and so $\wedge = Q^{-1} A Q$.
2. On the other hand if $\wedge = Q^{-1} A Q$, where $Q$ is non-singular then $AQ = Q\wedge$. But this is equivalent to $A(v_1) = \lambda_i v_i$ for $i = 1, \ldots, n$ where $\lambda_i$ is the $i$th diagonal entry in $\wedge$, and $v_i$ is the $i$th column of $Q$.
   Since $Q$ is non-singular, the columns $\{v_1, \ldots, v_n\}$ are linearly independent, and thus the eigenvectors form a basis for $\Re^n$. □

If there are $n$ distinct (real) eigenvalues then this gives a basis, and thus a diagonalisation.

**Lemma 2.14** *If $\{v_1, \ldots, v_m\}$ are eigenvectors corresponding to distinct eigenvalues $\{\lambda_1, \ldots, \lambda_m\}$, of a linear transformation $T : \Re^n \to \Re^n$, then $\{v_1, \ldots, v_m\}$ are linearly independent.*

*Proof* Since $v_1$ is assumed to be an eigenvector, it is non-zero, and thus $\{v_1\}$ is a linearly independent set. Proceed by induction.

Suppose $V_k = \{v_1, \ldots, v_k\}$, with $k < m$, are linearly independent. Let $v_{k+1}$ be another eigenvector and suppose

$$v = \sum_{r=1}^{k+1} a_r v_r = \underline{0}.$$

Then $\underline{0} = T(v) = \sum_{r=1}^{k+1} a_r T(v_r) = \sum_{r=1}^{k+1} a_r \lambda_r v_r$.

If $\lambda_{k+1} = 0$, then $\lambda_i \neq 0$ for $i = 1, \ldots, k$ and by the linear independence of $V_k$, $a_r \lambda_r = 0$, and thus $a_r = 0$ for $r = 1, \ldots, k$.

Suppose $\lambda_{k+1} \neq 0$. Then

$$\lambda_{k+1} v = \sum_{r=1}^{k+1} \lambda_{k+1} a_r v_r = \sum_{r=1}^{k+1} a_r \lambda_r v_r = \underline{0}.$$

Thus $\sum_{r=1}^{k} (\lambda_{k+1} - \lambda_r) a_r v_r = \underline{0}$.

By the linear independence of $V_k$, $(\lambda_{k+1} - \lambda_r) a_r = 0$ for $r = 1, \ldots, k$.

But the eigenvalues are distinct and so $a_r = 0$, for $r = 1, \ldots, k$.

Thus $a_{k+1} v_{k+1} = \underline{0}$ and so $a_r = 0, r = 1, \ldots, k+1$. Hence

$$V_{k+1} = \{v_1, \ldots, v_{k+1}\}, \quad k < m,$$

is linearly independent.

By induction $V_m$ is a linearly independent set. $\square$

Having shown how the determination of the eigenvectors gives a diagonalisation, we proceed to compute eigenvalues.

Consider again the equation $A(x) = \lambda x$. This is equivalent to the equation $A'(x) = 0$, where

$$A' = \begin{pmatrix} a_{11} - \lambda & a_{12} & \cdots & a_{1n} \\ a_{21} & a_{22} - \lambda & \cdot & \\ \vdots & \vdots & \vdots & \vdots \\ a_{n1} & & & a_{nn} - \lambda \end{pmatrix}.$$

For this equation to have a non zero solution it is necessary and sufficient that $|A'| = 0$. Thus we obtain a polynomial equation (called the characteristic equation) of degree $n$ in $\lambda$, with $n$ roots $\lambda_1, \ldots, \lambda_n$ not necessarily all real. In the $2 \times 2$ case for example this equation is $\lambda^2 - \lambda(a_{11} + a_{22}) + (a_{11}a_{22} - a_{21}a_{12}) = 0$. If the roots of this equation are $\lambda_1, \lambda_2$ then we obtain

$$(\lambda - \lambda_1)(\lambda - \lambda_2) = \lambda^2 - \lambda(\lambda_1 + \lambda_2) + \lambda_1 \lambda_2.$$

## 2.3 Canonical Representation

Hence

$$\lambda_1\lambda_2 = (a_{11}a_{22} - a_{21}a_{22}) = |A|$$
$$\lambda_1 + \lambda_2 = a_{11} + a_{22}.$$

The sum of the diagonal elements of a matrix is called the *trace* of $A$. In the $2 \times 2$ case therefore

$$\lambda_1\lambda_2 = |A|, \qquad \lambda_1 + \lambda_2 = a_{11} + a_{22} = \text{trace}(A).$$

In the $3 \times 3$ case we find

$$(\lambda - \lambda_1)(\lambda - \lambda_2)(\lambda - \lambda_3)$$
$$= \lambda^3 - \lambda^2(\lambda_1 + \lambda_2 + \lambda_3) + \lambda(\lambda_1\lambda_2 + \lambda_1\lambda_3 + \lambda_2\lambda_3) - \lambda_1\lambda_2\lambda_3$$
$$= \lambda^3 - \lambda^2(\text{trace } A) + \lambda(A_{11} + A_{22} + A_{33}) - |A| = 0,$$

where $A_{ii}$ is the $i$th diagonal cofactor of $A$. Suppose all the roots are non-zero (this is equivalent to the non-singularity of the matrix $A$). Let

$$\wedge = \begin{pmatrix} \lambda_1 & 0 & 0 \\ 0 & \lambda_2 & 0 \\ 0 & 0 & \lambda_3 \end{pmatrix}$$

be the diagonal eigenvalue matrix, with $|\wedge| = \lambda_1\lambda_2\lambda_3$.

The cofactor matrix of $\wedge$ is then

$$\begin{pmatrix} \lambda_2\lambda_3 & 0 & 0 \\ 0 & \lambda_1\lambda_3 & 0 \\ 0 & 0 & \lambda_1\lambda_2 \end{pmatrix}.$$

Thus we see that the sum of the diagonal cofactors of $A$ and $\wedge$ are identical. Moreover trace $(A) = $ trace $(\wedge)$ and $|\wedge| = |A|$.

Now let $\sim$ be the equivalence relation defined on $L(\Re^n, \Re^n)$ by $B \sim A$ iff there exist basis change matrices $P, Q$ and a diagonal matrix $\wedge$ such that

$$\wedge = P^{-1}AP = Q^{-1}BQ.$$

On the set of matrices which can be diagonalised, $\sim$ is an equivalence relation, and each class is characterised by $n$ invariants, namely the trace, the determinant, and $(n - 2)$ other numbers involving the cofactors.

## 2.3.2 Examples

*Example 2.5* Let

$$A = \begin{pmatrix} 2 & 1 & -1 \\ 0 & 1 & 1 \\ 2 & 0 & -2 \end{pmatrix}.$$

The characteristic equation is

$$(2-\lambda)\big[(1-\lambda)(-2-\lambda)\big] - 1(-2) - (-2(1-\lambda)) = -\lambda(\lambda^2 - \lambda - 2)$$
$$= -\lambda(\lambda - 2)(\lambda + 1)$$
$$= 0.$$

Hence $(\lambda_1, \lambda_2, \lambda_3) = (0, 2, -1)$. Note that $\lambda_1 + \lambda_2 + \lambda_3 = \text{trace}(A) = 1$ and

$$\lambda_2 \lambda_3 = -2 = A_{11} + A_{22} + A_{33}.$$

Eigenvectors corresponding to these eigenvalues are

$$x_1 = \begin{pmatrix} 1 \\ -1 \\ 1 \end{pmatrix}, \quad x_2 = \begin{pmatrix} 2 \\ 1 \\ 1 \end{pmatrix}, \quad x_3 = \begin{pmatrix} 1 \\ -1 \\ 2 \end{pmatrix}.$$

Let $P$ be the basis change matrix given by these three column vectors. The inverse can be readily computed, to give

$$P^{-1}AP = \begin{pmatrix} 1 & -1 & -1 \\ \frac{1}{3} & \frac{1}{3} & 0 \\ -\frac{2}{3} & \frac{1}{3} & 1 \end{pmatrix} \begin{pmatrix} 2 & 1 & -1 \\ 0 & 1 & 1 \\ 2 & 0 & 2 \end{pmatrix} \begin{pmatrix} 1 & 2 & 1 \\ -1 & 1 & -1 \\ 1 & 1 & 2 \end{pmatrix} = \begin{pmatrix} 0 & 0 & 0 \\ 0 & 2 & 0 \\ 0 & 0 & -1 \end{pmatrix}.$$

Suppose we now compute $A^2 = A \circ A : \Re^3 \to \Re^3$. This can easily be seen to be

$$\begin{pmatrix} 2 & 3 & 1 \\ 2 & 1 & -1 \\ 0 & 2 & 2 \end{pmatrix}.$$

The characteristic function of $A^2$ is $(\lambda^3 - 5\lambda^2 + 4\lambda)$ with roots $\mu_1 = 0$, $\mu_2 = 4$, $\mu_3 = 1$.

In fact the eigenvectors of $A^2$ are $x_1, x_2, x_3$, the same as $A$, but with eigenvalues $\lambda_1^2, \lambda_2^2, \lambda_3^2$. In this case $\text{Im}(a) = \text{Im}(A^2)$ is spanned by $\{x_2, x_3\}$ and $\text{Ker}(A) = \text{Ker}(A^2)$ has basis $\{x_1\}$.

More generally consider a linear transformation $A : \Re^n \to \Re^n$. Then if $x$ is an eigenvector with a non-zero eigenvalue $\lambda$, $A^2(x) = A \circ A(x) = A[\lambda x] = \lambda A(x) = \lambda^2 x$, and so $x \in \text{Im}(A) \cap \text{Im}(A^2)$.

### 2.3 Canonical Representation

If there exist $n$ *distinct* real roots to the characteristic equation of $A$, then a basis consisting of eigenvectors can be found. Then $A$ can be diagonalized, and $\text{Im}(A) = \text{Im}(A^2)$, $\text{Ker}(A) = \text{Ker}(A^2)$.

*Example 2.6* Let

$$A = \begin{pmatrix} 3 & -1 & -1 \\ 1 & 3 & -7 \\ 5 & -3 & 1 \end{pmatrix}$$

Then $\text{Ker}(A)$ has basis $\{(1, 2, 1)\}$, and $\text{Im}(A)$ has basis $\{(3, 1, 5), (-1, 3, -3)\}$. The eigenvalues of $A$ are $0, 0, 7$. Since we cannot find three linearly independent eigenvectors, $A$ cannot be diagonalised. Now

$$A^2 = \begin{pmatrix} 3 & -3 & 3 \\ -29 & 29 & -29 \\ 17 & -17 & 17 \end{pmatrix}$$

and thus $\text{Im}(A^2)$ has basis $\{(3, -29, 17)\}$. Note that

$$\begin{pmatrix} 3 \\ -29 \\ 17 \end{pmatrix} = -2 \begin{pmatrix} 3 \\ 1 \\ 5 \end{pmatrix} - 9 \begin{pmatrix} -1 \\ 3 \\ -3 \end{pmatrix} \in \text{Im}(A)$$

and so $\text{Im}(A^2)$ is a *subspace* of $\text{Im}(A)$.

Moreover $\text{Ker}(A^2)$ has basis $\{(1, 2, 1), (1, -1, 0)\}$ and so $\text{Ker}(A)$ is a subspace of $\text{Ker}(A^2)$.

This can be seen more generally. Suppose $f : \mathfrak{R}^n \to \mathfrak{R}^n$ is linear, and $x \in \text{Ker}(f)$. Then $f^2(x) = f(f(x)) = \underline{0}$, and so $x \in \text{Ker}(f^2)$. Thus $\text{Ker}(f) \subset \text{Ker}(f^2)$. On the other hand if $v \in \text{Im}(f^2)$ then there exists $w \in \mathfrak{R}^n$ such that $f^2(w) = v$. But $f(w) \in \mathfrak{R}^n$ and so $f(f(w)) = v \in \text{Im}(f)$. Thus $\text{Im}(f^2) \subset \text{Im}(f)$.

#### 2.3.3 Symmetric Matrices and Quadratic Forms

Given two vectors $x = (x_1, \ldots, x_n)$ and $y = (y_1, \ldots, y_n)$ in $\mathfrak{R}^n$, let $\langle x, y \rangle = \sum_{i=1}^n x_i y_i \in \mathfrak{R}$ be the scalar product of $x$ and $y$. Note that $\langle \lambda x, y \rangle = \lambda \langle x, y \rangle = \langle x, \lambda y \rangle$ for any real $\lambda$. (We use $\langle -, - \rangle$ to distinguish the scalar product from a vector in $\mathfrak{R}^2$. However the notations $(x, y)$ or $x \cdot y$ are often used for scalar product.)

An $n \times n$ matrix $A = (a_{ij})$ may be regarded as a map $A^* : \mathfrak{R}^n \times \mathfrak{R}^n \to \mathfrak{R}$, where $A^*(x, y) = \langle x, A(y) \rangle$.

$A^*$ is linear in both $x$ and $y$ and is called *bilinear*. By definition $\langle x, A(y) \rangle = \sum_{i=1}^n \sum_{j=1}^n x_i a_{ij} y_j$.

Call an $n \times n$ matrix $A$ *symmetric* iff $A = A^t$ where $A^t = (a_{ji})$ is obtained from $A$ by exchanging rows and columns.

In this case $\langle A(x), y \rangle = \sum_{i=1}^n (\sum_{j=1}^n a_{ji} x_i) y_j = \sum_{i=1}^n \sum_{j=1}^n x_i a_{ij} y_j$, since $a_{ij} = a_{ji}$ for all $i, j$.

Hence $\langle A(x), y \rangle = \langle x, A(y) \rangle$ for any $x, y \in \Re^n$ whenever $A$ is symmetric.

**Lemma 2.15** *If $A$ is a symmetric $n \times n$ matrix, and $x, y$ are eigenvectors of $A$ corresponding to distinct eigenvalues then $\langle x, y \rangle = 0$, i.e., $x$ and $y$ are orthogonal.*

*Proof* Let $\lambda_1 \neq \lambda_2$ be the eigenvalues corresponding to the distinct eigenvectors $x, y$. Now

$$\langle A(x), y \rangle = \langle x, A(y) \rangle$$
$$= \langle \lambda_1 x, y \rangle = \langle x, \lambda_2 y \rangle$$
$$= \lambda_1 \langle x, y \rangle = \lambda_2 \langle x, y \rangle.$$

Here $\langle A(x), y \rangle = \langle x, A(y) \rangle$ since $A$ is symmetric. Moreover $\langle x, \lambda y \rangle = \sum_{i=1}^n x_i (\lambda y_i) = \lambda \langle x, y \rangle$. Thus $(\lambda_1 - \lambda_2)\langle x, y \rangle = 0$. If $\lambda_1 \neq \lambda_2$ then $\langle x, y \rangle = 0$. □

**Lemma 2.16** *If there exist $n$ distinct eigenvalues to a symmetric $n \times n$ matrix $A$, then the eigenvectors $X = \{x_1, \ldots, x_n\}$ form an orthogonal basis for $\Re^n$.*

*Proof* Directly by Lemmas 2.14 and 2.15.

We may also give a brief direct proof of Lemma 2.16 by supposing that $\sum_{i=1}^n \alpha_i x_i = 0$. But then for each $j = i, \ldots, n$,

$$0 = \langle x_j, 0 \rangle = \sum_{i=1}^n \alpha_i \langle x_j, x_i \rangle = \alpha_j \langle x_j, x_j \rangle.$$

But since $x_j \neq 0$, $\langle x_j, x_j \rangle > 0$ and so $\alpha_j = 0$ for each $j$. Thus $X$ is a frame. Since the vectors in $X$ are mutually orthogonal, $X$ is an orthogonal basis for $\Re^n$.

For a symmetric matrix the roots of the characteristic equation will all be real. To see this in the $2 \times 2$ case, consider the characteristic equation

$$(\lambda - \lambda_1)(\lambda - \lambda_2) = \lambda^2 - \lambda(a_{11} + a_{22}) = (a_{11}a_{22} - a_{21}a_{12}).$$

The roots of this equation are $\frac{-b + \sqrt{b^2 - 4c}}{2}$ with real roots iff $b^2 - 4c \geq 0$. But this is equivalent to

$$(a_{11} + a_{22})^2 - 4(a_{11}a_{22} - a_{21}a_{12}) = (a_{11} - a_{22})^2 + 4(a_{12})^2 \geq 0,$$

since $a_{12} = a_{21}$.

Both terms in this expression are non-negative, and so $\lambda_1, \lambda_2$ are real.

In the case of a symmetric matrix, $A$, let $E_\lambda$ be the set of eigenvectors associated with a particular eigenvalue, $\lambda$, of $A$ together with the zero vector. Suppose $x_1, x_2$ belong to $E_\lambda$. Clearly $A(x_1 + x_2) = A(x_1) + A(x_2) = \lambda(x_1 + x_2)$ and so $x_1 +$

## 2.3 Canonical Representation

$x_2 \in E_\lambda$. If $x \in E_\lambda$, then $A(\alpha x) = \alpha A(x) = \alpha(\lambda x) = \lambda(\alpha x)$ and $\alpha x \in E_\lambda$ for each non-zero real number, $\alpha$.

Since we also now suppose that for each eigenvalue, $\lambda$, the *eigenspace* $E_\lambda$ contains $\underline{0}$, then $E_\lambda$ will be a vector subspace of $\Re^n$. If $\lambda = \lambda_1 = \cdots = \lambda_r$ are repeated roots of the characteristic equation, then, in fact, the eigenspace, $E_\lambda$, will be of dimension $r$, and we can find $r$ mutually orthogonal vectors in $E_\lambda$, forming a basis for $E_\lambda$.

Suppose now that $A$ is a symmetric $n \times n$ matrix. As we shall show we may write $\wedge = P^{-1}AP$ where $P$ is the $n \times n$ basis change matrix whose columns are the $n$ linearly independent eigenvectors of $A$.

Now *normalise* each eigenvector $x_j$ by defining $z_j = \frac{1}{\|x_j\|}(x_{1j}, \ldots, x_{nj})$ where $\|x_j\| = \sqrt{\sum (x_{kj})^2} = \sqrt{\langle x_j, x_j \rangle}$ is called the *norm* of $x_j$.

Let $Q = (z_1, \ldots, z_n)$ be the $n \times n$ matrix whose columns consist of $z_1, \ldots, z_n$. Now

$$Q^t Q = \begin{pmatrix} z_{11} & z_{21} & z_{n1} \\ z_{1j} & & z_{nj} \\ z_{1n} & & z_{nn} \end{pmatrix} \begin{pmatrix} z_{11} & z_{1j} & z_{1n} \\ z_{21} & z_{2j} & z_{2n} \\ \vdots & \vdots & \vdots \\ z_{n1} & z_{nj} & z_{nn} \end{pmatrix}$$

$$= \begin{pmatrix} \langle z_1, z_1 \rangle & \cdots & \langle z_1, z_n \rangle \\ \langle z_2, z_1 \rangle & & \\ \vdots & \cdots & \langle z_n, z_n \rangle \end{pmatrix}$$

since the $(i,k)$th entry in $Q^t Q$ is $\sum_{r=1}^n z_{ri} z_{rk} = \langle z_i, z_k \rangle$.

But $\langle z_i, z_k \rangle = \langle \frac{x_i}{\|x_i\|}, \frac{x_k}{\|x_k\|} \rangle = \frac{1}{\|x_i\|\|x_k\|} \langle x_i, x_k \rangle = 0$ if $i \neq k$. On the other hand $\langle z_i, z_i \rangle = \frac{1}{\|x_i\|^2} \langle x_i, x_i \rangle = 1$, and $Q^t Q = I_n$, the $n \times n$ identity matrix. Thus $Q^t = Q^{-1}$.

Since $\{z_1, \ldots, z_n\}$ are eigenvectors of $A$ with real eigenvalues $\{\lambda_1, \ldots, \lambda_n\}$ we obtain

$$\wedge = \begin{pmatrix} \lambda_1 & & 0 \\ & \lambda_r & \\ 0 & & 0 \end{pmatrix} = Q^t A Q$$

where the last $(n-r)$ columns of $Q$ correspond to the kernel vectors of $A$.

When $A$ is a symmetric $n \times n$ matrix the function $A^* : \Re^n \times \Re^n \to \Re$ given by $A^*(x, y) = \langle x, A(y) \rangle$ is called a *quadratic form*, and in matrix notation is given by

$$(x_1, \ldots, x_n) \begin{pmatrix} a_{ij} \end{pmatrix} \begin{pmatrix} y_1 \\ \vdots \\ y_n \end{pmatrix}$$

Consider

$$A^*(x,x) = \langle x, A(x) \rangle$$
$$= \langle x, Q \wedge A^t(x) \rangle$$
$$= \langle Q^t(x), \wedge Q^t(x) \rangle.$$

Now $Q^t(x) = (x'_1, \ldots, x'_n)$ is the coordinate representation of the vector $x$ with respect to the new basis $\{z_1, \ldots, z_n\}$ for $\Re^n$. Thus

$$A^*(x,x) = (x'_1, \ldots, x'_n) \begin{pmatrix} \lambda_1 & & \\ & \ddots & \\ & \lambda_r & \\ & & 0 \end{pmatrix} \begin{pmatrix} x'_1 \\ \vdots \\ x'_n \end{pmatrix} = \sum_{i=1}^{r} \lambda_i (x'_i)^2.$$

Suppose that rank $A = r$ and all eigenvalues of $A$ are non-negative. In this case, $A^*(x,x) = \sum_{i=1}^{n} |\lambda_i|(x_i)^2 \geq 0$. Moreover if $x$ is a non-zero vector then $Q^t(x) \neq 0$, since $Q^t$ must have rank $n$.

Define the nullity of $A^*$ to be $\{x : A^*(x,x) = 0\}$. Clearly if $x$ is a non-zero vector in Ker($A$) then it is an eigenvector with eigenvalue 0. Thus the nullity of $A^*$ is a vector subspace of $\Re^n$ of dimension at least $n - r$, where $r = \text{rank}(A)$. If the nullity of $A^*$ is $\{0\}$ then call $A^*$ *non-degenerate*. If all eigenvalues of $A$ are strictly positive (so that $A^*$ is non-degenerate) then $A^*(x,x) > 0$ for all non-zero $x \in \Re^n$. In this case $A^*$ is called *positive definite*. If all eigenvalues of $A$ are non-negative but some are zero, then $A^*$ is called *positive semi-definite*, and in this case $A^*(x,x) > 0$, for all $x$ in a subspace of dimension $r$ in $\Re^n$. Conversely if $A^*$ is non-degenerate and all eigenvalues are strictly negative, then $A^*$ is called *negative definite*. If the eigenvalues are non-positive, but some are zero, then $A^*$ is called *negative semi-definite*.

The *index* of the quadratic form $A^*$ is the maximal dimension of the subspace on which $A^*$ is negative definite. Therefore index ($A^*$) is the number of strictly negative eigenvalues of $A$.

When $A$ has some eigenvalues which are strictly positive and some which are strictly negative, then we call $A^*$ a *saddle*.

We have not as yet shown that a symmetric $n \times n$ matrix has $n$ real roots to its characteristic equation. We can show however that any (symmetric) quadratic form can be diagonalised.

Let $A = (a_{ij})$ and $\langle x, A(x) \rangle = \sum_{i=1}^{n} \sum_{j=1}^{n} a_{ij} x_i x_j$. If $a_{ii} = 0$ for all $i = 1, \ldots, n$ then it is possible to make a linear transformation of coordinates such that $a_{ij} \neq 0$ for some $j$. After relabelling coordinates we can take $a_{11} \neq 0$. In this case the quadratic form can be written

$$\langle x, A(x) \rangle = a_{11} x_1^2 + 2 a_{12} x_1 x_2, \ldots$$
$$= a_{11} \left( x_1 + \frac{a_{12}}{a_{11}} x_2 \cdots \right)^2 + \left( a_{22} - \frac{a_{12}^2}{a_{11}} \right) (x_2 + \cdots)^2 + \cdots$$
$$= \sum_{i=1}^{n} \alpha_i y_i^2.$$

## 2.3 Canonical Representation

Here each $y_i$ is a linear combination of $\{x_1, \ldots, x_n\}$. Thus the transformation $x \to P(x) = y$ is non-singular and has inverse $Q$ say.

Letting $x = Q(y)$ we see the quadratic form becomes

$$\begin{aligned}\langle x, A(x)\rangle &= \langle Q(y), A \circ Q(y)\rangle \\ &= \langle y, Q^t A Q(y)\rangle \\ &= \langle y, D(y)\rangle,\end{aligned}$$

where $D$ is a diagonal matrix with real diagonal entries $(\alpha_1, \ldots, \alpha_n)$. Note that $D = Q^t A Q$ and so rank$(D)$ = rank$(A) = r$, say. Thus only $r$ of the diagonal entries may be non zero. Since the symmetric matrix, $A$, can be diagonalised, not only are all its eigenvalues real, but its eigenvectors form a basis for $\Re^n$. Consequently $\wedge = P^{-1} A P$ where $P$ is the $n \times n$ basis change matrix whose columns are these eigenvectors. Moreover, if $\lambda$ is an eigenvalue with multiplicity $r$ (i.e., $\lambda$ occurs as a root of the characteristic equation $r$ times) then the eigenspace, $E_\lambda$, has dimension $r$. □

### 2.3.4 Examples

*Example 2.7* To give an illustration of this procedure consider a matrix

$$A = \begin{pmatrix} 0 & 0 & 1 \\ 0 & 1 & 0 \\ 1 & 0 & 0 \end{pmatrix}$$

representing the quadratic form $x_2^2 + 2x_1 x_3$. Let

$$P_1(x) = \begin{pmatrix} 0 & 1 & 0 \\ 1 & 0 & 0 \\ 1 & 0 & 1 \end{pmatrix} \begin{pmatrix} x_1 \\ x_2 \\ x_3 \end{pmatrix} = \begin{pmatrix} z_1 \\ z_2 \\ z_3 \end{pmatrix}$$

giving the quadratic form $z_1^2 - 2(z_2 - \frac{1}{2}z_3)^2 + \frac{1}{2}z_3^2$ and

$$P_2(z) = \begin{pmatrix} 1 & 0 & 0 \\ 0 & 1 & -\frac{1}{2} \\ 0 & 0 & 1 \end{pmatrix} \begin{pmatrix} z_1 \\ z_2 \\ z_3 \end{pmatrix} = \begin{pmatrix} y_1 \\ y_2 \\ y_3 \end{pmatrix}.$$

Then $\langle x, A(x)\rangle = \langle y, D(y)\rangle$, where

$$D = \begin{pmatrix} 1 & 0 & 0 \\ 0 & -2 & 0 \\ 0 & 0 & \frac{1}{2} \end{pmatrix} = \begin{pmatrix} \alpha_1 & 0 & 0 \\ 0 & \alpha_2 & 0 \\ 0 & 0 & \alpha_3 \end{pmatrix}, \quad \text{and} \quad A = P_1^t P_2^t D P_2 P_1.$$

Consequently the matrix $A$ can be diagonalised. $A$ has characteristic equation $(1 - \lambda)(\lambda^2 - 1)$ with eigenvalues $1, 1, -1$.

Then normalized eigenvectors of $A$ are

$$\frac{1}{\sqrt{2}}\begin{pmatrix} 1 \\ 0 \\ 1 \end{pmatrix}, \quad \begin{pmatrix} 0 \\ 1 \\ 0 \end{pmatrix}, \quad \frac{1}{\sqrt{2}}\begin{pmatrix} 1 \\ 0 \\ -1 \end{pmatrix},$$

corresponding to the eigenvalues $1, 1, -1$.

Thus $A^*$ is a non-degenerate saddle of index 1. Let $Q$ be the basis change matrix

$$\frac{1}{\sqrt{2}}\begin{pmatrix} 1 & 0 & 1 \\ 0 & 2 & 0 \\ 1 & 0 & -1 \end{pmatrix}.$$

Then

$$Q^t A Q = \frac{1}{2}\begin{pmatrix} 1 & 0 & 1 \\ 0 & 2 & 0 \\ 1 & 0 & -1 \end{pmatrix}\begin{pmatrix} 0 & 0 & 1 \\ 0 & 1 & 0 \\ 1 & 0 & 0 \end{pmatrix}\begin{pmatrix} 1 & 0 & 1 \\ 0 & 2 & 0 \\ 1 & 0 & -1 \end{pmatrix} = \begin{pmatrix} 1 & 0 & 0 \\ 0 & 1 & 0 \\ 0 & 0 & -1 \end{pmatrix}.$$

As a quadratic form

$$(x_1, x_2, x_3) A \begin{pmatrix} x_1 \\ x_2 \\ x_3 \end{pmatrix} = (x_1, x_2, x_3)\begin{pmatrix} x_3 \\ x_2 \\ x_1 \end{pmatrix} = x_1 x_3 + x_2^2.$$

We can also write this as

$$(x_1, x_2, x_3)\frac{1}{2}\begin{pmatrix} 1 & 0 & 1 \\ 0 & \sqrt{2} & 0 \\ 1 & 0 & -1 \end{pmatrix}\begin{pmatrix} 1 & 0 & 0 \\ 0 & 1 & 0 \\ 0 & 0 & -1 \end{pmatrix}\begin{pmatrix} 1 & 0 & 1 \\ 0 & \sqrt{2} & 0 \\ 1 & 0 & -1 \end{pmatrix}\begin{pmatrix} x_1 \\ x_2 \\ x_3 \end{pmatrix}$$

$$= \frac{1}{2}(x_1 + x_3, \sqrt{2x_2}, x_1 - x_3)\begin{pmatrix} 1 & 0 & 0 \\ 0 & 1 & 0 \\ 0 & 0 & -1 \end{pmatrix}\begin{pmatrix} x_1 + x_3 \\ \sqrt{2x_2} \\ x_1 - x_3 \end{pmatrix}$$

$$= \frac{1}{2}(x_1 + x_3)^2 + 2x_2^2 - (x_1 - x_3)^2.$$

Note that $A$ is positive definite on the subspace $\{(x_1, x_2, x_3) \in \Re^3 : (x_1 = x_3)\}$ spanned by the first two eigenvectors.

We can give a geometric interpretation of the behaviour of a matrix $A$ with both positive and negative eigenvalues. For example

$$A = \begin{pmatrix} 1 & 2 \\ 2 & 1 \end{pmatrix}$$

**Fig. 2.2** A translation followed by a reflection

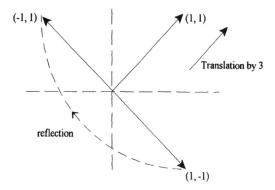

has eigenvectors

$$z_1 = \begin{pmatrix} 1 \\ 1 \end{pmatrix} \quad z_2 = \begin{pmatrix} 1 \\ -1 \end{pmatrix}$$

corresponding to the eigenvalues $3, -1$ respectively. Thus $A$ maps the vector $z_1$ to $3z_1$ and $z_2$ to $-z_2$. The second operation can be regarded as a reflection of the vector $z_2$ in the line $\{(x, y) : x - y = 0\}$, associated with the first eigenvalue. The first operation $z_1 \to 3z_1$ is a translation of $z_1$ to $3z_1$. Consider now any point $x \in \Re^2$. We can write $x = \alpha z_1 + \beta z_2$. Thus $A(x) = 3\alpha z_1 - \beta z_2$. In other words $A$ may be decomposed into two operations: a translation in the direction $z_1$, followed by a reflection about $z_1$ as in Fig. 2.2.

## 2.4 Geometric Interpretation of a Linear Transformation

More generally suppose $A$ has real roots to the characteristic equation and has eigenvectors $\{x_1, \ldots, x_s, z_1, \ldots, z_t, k_1, \ldots, k_p\}$.

The first $s$ vectors correspond to positive eigenvalues, the next $t$ vectors to negative eigenvalues, and the final $p$ vectors belong to the kernel, with zero eigenvalues.

Then $A$ may be described in the following way:
1. collapse the kernel vectors on to the image spanned by $\{x_1, \ldots, x_s, z_1, \ldots, z_t\}$.
2. translate each $x_i$ to $\lambda_i x_i$.
3. reflect each $z_j$ to $-z_j$, and then translate to $-|\mu_j| z_j$ (where $\mu_j$ is the negative eigenvalue associated with $z_j$).

These operations completely describe a symmetric matrix or a matrix, $A$, which is diagonalisable. When $A$ is non-symmetric then it is possible for $A$ to have complex roots.

For example consider the matrix

$$\begin{pmatrix} \cos\theta & -\sin\theta \\ \sin\theta & \cos\theta \end{pmatrix}.$$

**Fig. 2.3** A translation followed by a rotation

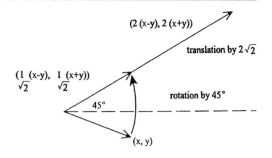

As we have seen this corresponds to a rotation by $\theta$ in an anticlockwise direction in the plane $\mathfrak{R}^2$. To determine the eigenvalues, the characteristic equation is $(\cos\theta - \lambda)^2 + \sin^2\theta = \lambda^2 - 2\lambda\cos\theta + (\cos^2\theta + \sin^2\theta) = 0$. But $\cos^2\theta + \sin^2\theta = 1$. Thus $\lambda = 2\cos\frac{\theta \pm 2\sqrt{\cos^2\theta - 1}}{2} = \cos\theta \pm i\sin\theta$.

More generally a $2 \times 2$ matrix with complex roots may be regarded as a transformation $\lambda e^{i\theta}$ where $\lambda$ corresponds to a translation by $\lambda$ and $e^{i\theta}$ corresponds to rotation by $\theta$.

*Example 2.8* Consider $A = \begin{pmatrix} 2 & -2 \\ 2 & 2 \end{pmatrix}$ with trace $(A) = \text{tr}(A) = 4$ and $|A| = 8$.

As we have seen the characteristic equation for $A$ is $(\lambda^2 - (\text{trace } A)) + |A| = 0$, with roots $\frac{\text{trace}(A) \pm \sqrt{(\text{trace } A)^2 - 4|A|}}{2}$. Thus the roots are $2 \pm \frac{1}{2}\sqrt{16 - 32} = 2 \pm 2i = 2\sqrt{2}[\frac{1}{\sqrt{2}} + \frac{i}{\sqrt{2}}]$ where $\cos\theta = \sin\theta = \frac{1}{\sqrt{2}}$ and so $\theta = 45°$. Thus

$$A : \begin{pmatrix} x \\ y \end{pmatrix} \to 2\sqrt{2}\begin{pmatrix} x\cos 45 & -y\sin 45 \\ x\sin 45 & +y\cos 45 \end{pmatrix}.$$

Consequently $A$ first sends $(x, y)$ by a translation to $(2\sqrt{2}x, 2\sqrt{2}y)$ and then rotates this vector through an angle $45°$.

More abstractly if $A$ is an $n \times n$ matrix with two complex conjugate eigenvalues $(\cos\theta + i\sin\theta), (\cos\theta - i\sin\theta)$, then there exists a two dimensional eigenspace $E^\theta$ such that $A(x) = \lambda e^{i\theta}(x)$ for all $x \in E_\theta$, where $\lambda e^{i\theta}(x)$ means rotate $x$ by $\theta$ within $E_\theta$ and then translate by $\lambda$.

In some cases a linear transformation, $A$, can be given a *canonical form* in terms of rotations, translations and reflections, together with a collapse onto the kernel. What this means is that there exists a number of distinct *eigenspaces*

$$\{E_1, \ldots, E_p, X_1, \ldots, X_s, K\}$$

where $A$ maps
1. $E_j$ to $E_j$ by rotating any vector in $E_j$ through an angle $\theta_j$;
2. $X_j$ to $X_j$ by translating a vector $x$ in $X_j$ to $\lambda_j x$, for some non-zero real number $\lambda_j$;

3. the kernel $K$ to $\{0\}$.

In the case that the dimensions of these eigenspaces sum to $n$, then the canonical form of the matrix $A$ is

$$\begin{pmatrix} e^{i\theta} & 0 & 0 \\ 0 & \wedge & 0 \\ 0 & 0 & 0 \end{pmatrix}$$

where $e^{i\theta}$ consists of $p$ different $2 \times 2$ matrices, and $\wedge$ is a diagonal $s \times s$ matrix, while 0 is an $(n-r) \times (n-r)$ zero matrix, where $r = \text{rank}(A) = 2p + s$.

However, even when all the roots of the characteristic equation are real, it need not be possible to obtain a diagonal, canonical form of the matrix.

To illustrate, in Example 2.6 it is easy to show that the eigenvalue $\lambda = 0$ occurs twice as a root of the characteristic equation for the non-symmetric matrix $A$, even though the kernel is of dimension 1. The eigenvalue $\lambda = 7$ occurs once. Moreover the vector $(3, -29, 17)$ clearly must be an eigenvector for $\lambda = 7$, and thus span the image of $A^2$. However it is also clear that the vector $(3, -29, 17)$ does not span the image of $A$. Thus the eigenspace $E_7$ does not provide a basis for the image of $A$, and so the matrix $A$ cannot be diagonalised.

However, as we have shown, for any *symmetric* matrix the dimensions of the eigenspaces sum to $n$, and the matrix can be expressed in canonical, diagonal, form.

In Chap. 4 below we consider smooth functions and show that "locally" such a function can be analysed in terms of the canonical form of a particular symmetric matrix, known as the Hessian.

# Topology and Convex Optimisation   3

## 3.1 A Topological Space

In the previous chapter we introduced the notion of the scalar product of two vectors in $\Re^n$. More generally if a scalar product is defined on some space, then this permits the definition of a *norm*, or length, associated with a vector, and this in turn allows us to define the *distance* between two vectors. A distance function or *metric* may be defined on a space, $X$, even when $X$ admits no norm. For example let $X$ be the surface of the earth. Clearly it is possible to say what is the shortest distance, $d(x, y)$, between two points, $x, y$, on the earth's surface, although it is not meaningful to talk of the "length" of a point on the surface. More general than the notion of a metric is that of a *topology*. Essentially a topology on a space is a mathematical notion for making more precise the idea of "nearness". The notion of topology can then be used to precisely define the property of continuity of a function between two topological spaces. Finally continuity of a preference gives proof of existence of a social choice and of an economic equilibrium in a world that is bounded.

### 3.1.1 Scalar Product and Norms

In Sect. 2.3 we defined the Euclidean *scalar product* of two vectors

$$x = \sum_{i=1}^{n} x_i e_i, \quad \text{and} \quad y = \sum_{i=1}^{n} y_i e_i \quad \text{in } \Re^n,$$

where $\{e_1, \ldots, e_n\}$ is the standard basis, to be

$$\langle x, y \rangle = \sum_{i=1}^{n} x_i y_i.$$

N. Schofield, *Mathematical Methods in Economics and Social Choice*,
Springer Texts in Business and Economics, DOI 10.1007/978-3-642-39818-6_3,
© Springer-Verlag Berlin Heidelberg 2014

More generally suppose that $\{v_1, \ldots, v_n\}$ is a basis for $\Re^n$, and $(x_1, \ldots, x_n)$, $(y_1, \ldots, y_n)$ are the coordinates of $x, y$ with respect to this basis. Then

$$\langle x, y \rangle = \sum_{i=1}^{n} \sum_{j=1}^{n} x_i y_j \langle v_i, v_j \rangle,$$

where $\langle v_i, v_j \rangle$ is the scalar product of $v_i$ and $v_j$. Thus

$$\langle x, y \rangle = (x_1, \ldots, x_n) \begin{pmatrix} \vdots \\ a_{ij} \\ \vdots \end{pmatrix} \begin{pmatrix} y_1 \\ \vdots \\ y_n \end{pmatrix}$$

where $A = (a_{ij}) = \langle v_i, v_j \rangle_{i=1,\ldots,n; j=1,\ldots,n}$. If we let $v_i = \sum_{k=1}^{n} v_{ik} e_k$, then clearly $\langle v_i, v_i \rangle = \sum_{k=1}^{n} (v_{ik})^2 > 0$. Moreover $\langle v_i, v_j \rangle = \langle v_j, v_i \rangle$. Thus the matrix $A$ is symmetric. Since $A$ must be of rank $n$, it can be diagonalized to give a matrix $\wedge = Q^t A Q$, all of whose diagonal entries are positive. Here $Q$ represents the orthogonal basis change matrix and $Q^t(x_1, \ldots, x_n) = (x'_1, \ldots, x'_n)$ gives the coordinates of $x$ with respect to the basis of eigenvectors of $A$. Hence

$$\langle x, y \rangle = \langle x, A(y) \rangle = \langle x, Q \wedge Q^t(y) \rangle$$
$$= \langle Q^t(x), \wedge Q^t(y) \rangle$$
$$= \sum_{i=1}^{n} \lambda_i x'_i y'_i.$$

Thus a scalar product is a non-degenerate positive definite quadratic form. Note that the scalar product is *bilinear* since

$$\langle x_1 + x_2, y \rangle = \langle x_1, y \rangle + \langle x_2, y \rangle \quad \text{and} \quad \langle x, y_1 + y_2 \rangle = \langle x, y_1 \rangle + \langle x, y_2 \rangle,$$

and *symmetric* since

$$\langle x, y \rangle = \langle x, A(y) \rangle = \langle y, A(x) \rangle = \langle y, x \rangle.$$

We shall call the scalar product given by $\langle x, y \rangle = \sum_{i=1}^{n} x_i y_i$ the *Euclidean* scalar product.

We define the *Euclidean norm*, $\| \ \|_E$, by $\|x\|_E = \sqrt{\langle x, x \rangle} = \sqrt{\sum_{i=1}^{n} x_i^2}$. Note that $\|x\|_E \geq 0$ if and only if $x = (0, \ldots, 0)$. Moreover, if $a \in \Re$, then

$$\|ax\|_E = \sqrt{\sum_{i=1}^{n} a^2 x_i^2} = |a| \|x\|_E.$$

## 3.1 A Topological Space

**Fig. 3.1** Projection

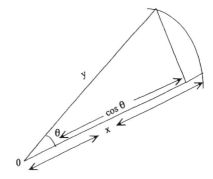

**Lemma 3.1** *If $x, y \in \Re^n$, then $\|x + y\|_E \leq \|x\|_E + \|y\|_E$.*

*Proof* For convenience write $\|x\|_E$ as $\|x\|$. Now

$$\|x + y\|^2 = \langle x + y, x + y \rangle = \langle x, x \rangle + \langle x, y \rangle + \langle y, x \rangle + \langle y, y \rangle.$$

But the scalar product is symmetric. Therefore

$$\|x + y\|^2 = \|x\|^2 + \|y\|^2 + 2\langle x, y \rangle.$$

Furthermore $(\|x\| + \|y\|)^2 = \|x\|^2 + \|y\|^2 + 2\|x\|\|y\|$. Thus $\|x + y\| \leq \|x\| + \|y\|$ iff $\langle x, y \rangle \leq \|x\|\|y\|$. To show this note that $\sum_{i<j}(x_i y_j - x_j y_i)^2 \geq 0$. Thus $\sum_{i<j}(x_i^2 y_j^2 + x_j^2 y_i^2) \geq 2\sum_{i<j} x_i y_i x_j y_j$. Add $\sum_{i=1}^n x_i^2 y_i^2$ to both sides. This gives $(\sum_{i=1}^n x_i^2)(\sum_{i=1}^n y_i^2) \geq (\sum_{i=1}^n x_i y_i)^2$. Therefore $\|x\|^2 \|y\|^2 \geq \langle x, y \rangle^2$ and so $\frac{\langle x, y \rangle}{\|x\|\|y\|} \leq 1$, or $\|x + y\| \leq \|x\| + \|y\|$. □

In this lemma we have shown that

$$-1 \leq \frac{\langle x, y \rangle}{\|x\| \, \|y\|} \leq 1.$$

This ratio can be identified with $\cos \theta$, where $\theta$ is the angle between the two vectors $x, y$. In the case of unit vectors, $\langle x, y \rangle$ can be identified with the perpendicular projection of $y$ onto $x$ as in Fig. 3.1.

The property $\|x + y\| \leq \|x\| + \|y\|$ is known as the *triangle inequality* (see Fig. 3.2).

**Definition 3.1** Let $X$ be a vector space over the field $\Re$. A *norm*, $\| \ \|$, on $X$ is a mapping $\| \ \| : X \to \Re$ which satisfies the following three properties:
**N1.** $\|x\| \geq 0$ for all $x \in X$, and $\|x\| = 0$ iff $x = \underline{0}$.
**N2.** $\|ax\| = |a|\|x\|$ for all $x \in X$, and $a \in \Re$.
**N3.** $\|x + y\| \leq \|x\| + \|y\|$ for all $x, y \in X$.

**Fig. 3.2** Triangle inequality

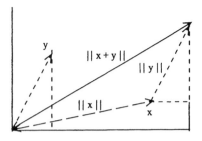

There are many different norms on a vector space. For example if $A$ is a non-degenerate positive definite symmetric matrix, we could define $\| \ \|_A$ by $\|x\|_A = \sqrt{\langle x, Ax \rangle}$.

The *Cartesian* norm is $\|x\|_c = \|(x_1, \ldots, x_n)\|_c = \max\{|x_1|, \ldots, |x_n|\}$.

Clearly $\|x\|_c \geq 0$ and $\|x\|_c = 0$ iff $x_i = 0$ for all $i = 1, \ldots, n$. Moreover $\|ax\|_c = \max\{|ax_1|, \ldots, |ax_n|\} = \max\{|a|\,|x_1|, \ldots, |a|\,|x_n|\} = |a|\,|x_i|$, for some $i$.

Thus $\|ax\|_c = |a|\max\{|x_1|, \ldots, |x_n|\} = |a|\,\|x\|_c$.

Finally, $\|x + y\|_c = |x_i + y_i|$ for some $i \leq |x_i| + |y_i| \leq \max\{|x_1|, \ldots, |x_n|\} + \max\{|y_1|, \ldots, |y_n|\} = \|x\|_c + \|y\|_c$. Define the *city block* norm $\|x\|_B$ to be $\|x\|_B = \sum_{i=1}^n |x_i|$. Clearly $\|x + y\|_B = \sum_{i=1}^n |x_i + y_i| \leq \sum_{i=1}^n (|x_i| + |y_i|) = \|x\|_B + \|y\|_B$.

If $\| \ \|$ is a norm on the vector space $X$, the *distance function* or *metric* $d$ on $X$ induced by $\| \ \|$ is the function $d : X \times X \to \Re : d(x, y) = \|x - y\|$.

Note that $d(x, y) \geq 0$ for all $x, y \in X$ and that $d(x, y) = 0$ iff $x - y = 0$, i.e., $x = y$. Moreover,

$$d(x, y) + d(y, z) = \|x - y\| + \|y - z\| \geq \|(x - y) + (y - z)\|$$
$$= \|x - z\| = d(x, z).$$

Hence $d(x, z) \leq d(x, y) + d(y, z)$.

**Definition 3.2** A *metric* on a *set* $X$ is a function $d : X \times X \to \Re$ such that
**D1.** $d(x, y) \geq 0$ for all $x, y \in X$ and $d(x, y) = 0$ iff $x = y$
**D2.** $d(x, z) \leq d(x, y) + d(y, z)$ for all $x, y, z \in X$.

Note that a metric $d$ may be defined on a set $X$ even when $X$ is not a vector space. In particular $d$ may be defined without reference to a particular norm. At the beginning of the chapter for example we mentioned that the surface of the earth, $S^2$, admits a metric $d$, where the distance between two points $x, y$ on the surface is measured along a great circle through $x, y$. A second metric on $S^2$ is obtained by defining $d(x, y)$ to be the angle, $\theta$, subtended at the centre of the earth by the two radii to $x, y$ (see Fig. 3.3).

Any set $X$ which admits a metric, $d$, we shall call *metrisable*, or a *metric space*. To draw attention to the metric, $d$, we shall sometimes write $(X, d)$ for a metric space.

In a metric space $(X, d)$ the *open ball* at $x$ of radius $r$ in $X$ is

$$B_d(x, r) = \{y \in X : d(x, y) < r\},$$

## 3.1 A Topological Space

**Fig. 3.3** Spherical metric

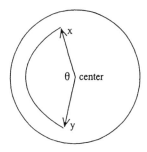

and the closed ball centre $x$ of radius $r$ is

$$\text{Clos } B_d(x,r) = \{y \in X : d(x,y) \leq r\}.$$

The *sphere* of radius $r$ at $x$ is

$$S_d(x,r) = \{y \in X : d(x,y) = r\}.$$

In $\Re^n$, the Euclidean sphere of radius $r$ is therefore

$$S(x,r) = \left\{y : \sum_{i=1}^{n}(x_i - y_i)^2 = r_i^2\right\}.$$

For convenience a sphere in $\Re^n$ is often written as $S^{n-1}$. Here the superfix is $n-1$ because as we shall see the sphere in $\Re^n$ is $(n-1)$-dimensional, even though it is not a vector space. If $(X,d)$ is a metric space, say a set $V$ in $X$ is $d$-open iff for any $x \in V$ there is some radius $r_x$ (dependent on $x$) such that

$$B_d(x, r_x) \subset V.$$

**Lemma 3.2** *Let $\Gamma_d$ be the family of all sets in $X$ which are $d$-open. Then $\Gamma_d$ satisfies the following properties*:
**T1.** *If $U, V \in \Gamma_d$, then $U \cap V \in \Gamma_d$.*
**T2.** *If $U_j \in \Gamma_d$ for all $j$ belonging to an index set $J$ (which is possibly infinite), then $\bigcup_{j \in J} U_j \in \Gamma_d$.*
**T3.** *Both $X$ and the empty set, $\Phi$, belong to $\Gamma_d$.*

*Proof* Clearly $X$ and $\Phi$ are $d$-open. If $U$ and $V \in \Gamma_d$, but $U \cap V = \Phi$ then $U \cap V$ is $d$-open. Suppose on the other hand that $x \in U \cap V$. Since both $U$ and $V$ are open, there exist $r_1, r_2$ such that $B_d(x, r_1) \subset U$ and $B_d(x_1, r_2) \subset V$. Let $r = \min(r_1, r_2)$. By definition

$$B_d(x,r) = B_d(x,r_1) \cap B_d(x,r_2) \subset U \cap V.$$

Thus there is an open ball, centre $x$, of radius $r$ contained in $U \cap V$.

Finally suppose $x \in \bigcup_{j \in J} U_j = U$. Since $x$ belongs to at least one $U_j$, say $U_1$, there is an open ball $B = B(x, r_1)$ contained in $U_1$. Since $U_1$ is open so is $U$. □

Note that by **T1** the *finite* intersection of open sets is an open set, but *infinite* intersection of open sets need not be open. To see this consider a set of the form

$$I = (a, b) = \{x \in \mathfrak{R} : a < x < b\}.$$

For any $x \in I$ it is possible to find an $\varepsilon$ such that $a + \varepsilon < x < b - \varepsilon$. Hence the open ball $B(x, \varepsilon) = \{y : x - \varepsilon < y < x + \varepsilon\}$ belongs to $I$, and so $I$ is open.

Now consider the family $\{U_r : r = 1, \ldots, \infty\}$ of sets of the form $U_r = (-\frac{1}{r}, \frac{1}{r})$.

Clearly the origin, 0, belongs to each $U_r$, and so $0 \in \cap U_r = U$. Suppose that $U$ is open. Since $0 \in U$, there must be some open ball $B(0, \varepsilon)$ belonging to $U$. Let $r_0$ be an integer such that $r_0 > \frac{1}{\varepsilon}$, so $\frac{1}{r_0} < \varepsilon$. But then $U_{r_0} = (-\frac{1}{r_0}, \frac{1}{r_0})$ is strictly contained in $(-\varepsilon, \varepsilon)$.

Therefore the ball $B(0, \varepsilon) = \{y \in \mathfrak{R} : |y| < \varepsilon\} = (-\varepsilon, \varepsilon)$ is not contained in $U_{r_0}$, and so cannot be contained in $U$. Hence $U$ is not open.

### 3.1.2 A Topology on a Set

We may define a *topology* on a set $X$ to be any collection of sets in $X$ which satisfies the three properties **T1'**, **T2'**, **T3'**.

**Definition 3.3** A *topology* $\Gamma$ on a set $X$ is a collection of sets in $X$ which satisfies the following properties:
**T1'**   If $U, V \in \Gamma$ then $U \cap V \in \Gamma$.
**T2'**   If $J$ is any index set and $U_j \in \Gamma$ for each $j \in J$, then $\bigcup_J U_j \in \Gamma$.
**T3'**   Both $X$ and the empty set belong to $\Gamma$.

A set $X$ which has a topology $\Gamma$ is called a *topological space*, and written $(X, \Gamma)$. The sets in $\Gamma$ are called $\Gamma$-open, or simply *open*. An open set in $\Gamma$ which contains a point $x$ is called a *neighbourhood* (or nbd.) of $x$.

A *base* $\mathcal{B}$ for a topology $\Gamma$ is a collection of $\Gamma$-open sets such that any member $U$ of $\Gamma$ can be written as a union of members of $\mathcal{B}$. Equivalently if $x$ belongs to a $\Gamma$-open set $U$, then there is a member $V$ of the base such that $x \in V \subset U$.

By Lemma 3.2, the collection $\Gamma_d$ of sets which are open with respect to the metric $d$ comprise a topology called the *metric topology* $\Gamma_d$ on $X$. We also said that a set $U$ belonged to $\Gamma_d$ iff each point $x \in U$ had a neighbourhood $B_d(x, r)$ which belonged to $U$. Thus the family of sets

$$\mathcal{B} = \{B_d(x, r) : x \in X, r > 0\}$$

forms a base for the metric topology $\Gamma_d$ on $X$.

**Fig. 3.4** Accumulation point

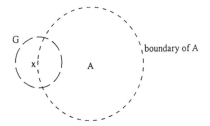

Consider again the metric topology on $\Re$. As we have shown any set of the form $(a, b)$ is open. Indeed a set of the form $(-\infty, a)$ or $(b, \infty)$ is also open.

In general if $U$ is an open set (in the topology $\Gamma$ for the topological space $X$), then the complement $\overline{U_X} = X \backslash U$ of $U$ in $X$ is called *closed*.

Thus in $\Re$, the set $[a, b] = \{x \in \Re : a \leq x \leq b\}$ is closed since it is the complement of the open set $(-\infty, a) \cup (b, \infty)$. Note that the sets $(-\infty, a]$ and $[b, \infty)$ are also closed since they are complements of the open sets $(a, \infty)$ and $(-\infty, b)$ respectively.

If $A$ is any set in a topological space, $(X, \Gamma)$, then define the open set, Int($A$), called the *interior* of $A$, by $x \in$ Int($A$) iff $x$ is in $A$ and there exists an open set $G$ containing $x$ such that $G \subset A$.

Conversely, define the closed set, Clos($A$), or *closure* of $A$, by $x \in$ Clos($A$) iff $x$ is in $X$ and for any open set $G$ containing $x$, $G \cap A$ is non empty.

Clearly Int($A$) $\subset A \subset$ Clos($A$). (See Exercise 3.1 at the end of the book.) The *boundary* of $A$, written $\delta A$, is Clos($A$) $\cap$ Clos($\overline{A}_X$), where $A_X = X \backslash A$.

For example, if $A$ is an open set, the complement $(\overline{A}_X)$ of $A$ in $X$ is a closed set containing $\delta A$ (*i.e.*, Clos($\overline{A}_X$) $= \overline{A}_X$). The closure, Clos($A$), on the other hand is the closed set which intersects $(\overline{A}_X)$ precisely in $\delta A$. Clearly if $x$ belongs to the boundary of $A$, then any neighbourhood of $x$ intersects both $A$ and its complement.

A point $x$ is an *accumulation* or *limit point* of a set $A$ if any open set $U$ containing $x$ also contains points of $A$ different from $x$. If $A$ is closed then it contains its limit points. If $A$ is a subset of $X$ and Clos($A$) $= X$ then call $A$ *dense* in $X$. Note that this means that any point $x \in X$ either belongs to $A$ or, if it belongs to $X \backslash A$, has the property that for any neighborhood $U$ of $x$, $U \cap A \neq \Phi$. Thus if $x \in X \backslash A$ it is an accumulation point of $A$. If $A$ is dense in $X$ a point outside $A$ may be 'approximated arbitrarily closely' by points inside $A$. See Fig. 3.4.

For example the set of non-integer real numbers $\mathcal{Z}$ is dense (as well as open) in $\Re$. The set of rational numbers $\mathcal{Q} = \{\frac{p}{q} : p, q \in \mathcal{Z}\}$ is also dense in $\Re$, but not open, since any neighbourhood of a rational number must contain irrational numbers. For each rational $q \in \mathcal{Q}$, note that $\Re \backslash \{q\}$ is open, and dense. Thus

$$\Re \backslash \mathcal{Q} = \bigcap_{q \in \mathcal{Q}} \{\Re \backslash \{q\}\} :$$

the set of irrationals is the "countable" intersection of open dense sets. Moreover $\Re \backslash \mathcal{Q}$ is itself dense. Such a set is called a *residual* set, and is in a certain sense "more dense" than a dense set.

**Fig. 3.5** Open sets in the product topology

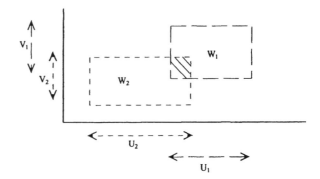

We now consider different topologies on a set $X$.

**Definition 3.4**
1. If $(X, \Gamma)$ is a topological space, and $Y$ is a subset of $X$, the *relative topology* $\Gamma_Y$ on $Y$ is the topology whose open sets are $\{U \cap Y : U \in \Gamma\}$.
2. If $(X, \Gamma)$ and $(Y, \mathcal{S})$ are topological spaces the *product topology* $\Gamma \times \mathcal{S}$ on the set $X \times Y$ is the topology whose base consists of all sets of the form $\{U \times V : U \in \Gamma, V \in \mathcal{S}\}$.

We already introduced the relative topology in Sect. 1.1.2, and showed that this formed a topology. To show that $\Gamma \times \mathcal{S}$ is a topology for $X \times Y$ we need to show that the union and intersection properties are satisfied. Suppose that

$$W_i = U_i \times V_i$$

for $i = 1, 2$, where $U_i \in \Gamma$, $V_i \in \mathcal{S}$. Now $W_1 \cap W_2 = (U_1 \times V_1) \cap (U_2 \times V_2) = (U_1 \cap U_2) \times (V_1 \cap V_2)$. But $U_1 \cap U_2 \in \Gamma$, $V_1 \cap V_2 \in \mathcal{S}$, since $\mathcal{S}$ and $\Gamma$ are topologies. Thus $W_1 \cap W_2 \in \Gamma \times \mathcal{S}$. Suppose now that $x \in W_1 \cup W_2$. Then $x$ belongs either to $U_1 \times V_1$ or $U_2 \times V_2$ (or both). In either case $x$ belongs to a member of the base for $\Gamma \times \mathcal{S}$.

Another way of expressing the product topology is that $W$ is open in the product topology iff for any $x \in W$ there exist open sets, $U \in \Gamma$ and $V \in \mathcal{S}$, such that $x \in U \times V$ and $U \times V \subset W$.

For example consider the metric topology $\Gamma$ induced by the norm $\| \; \|$ on $\Re$. This gives the product topology $\Gamma^n$ on $\Re^n$, where $U$ is open in $\Gamma^n$ iff for each $x = (x_1, \ldots, x_n) \in U$ there exists an open interval $B(x_i, r_i)$ of radius $r_i$ about the $i$th coordinate $x_i$ such that

$$B(x_1, r_1) \times \cdots \times B(x_n, r_n) \subset U.$$

Consider now the *Cartesian norm* on $\Re^n$, where

$$\|x\|_C = \max\{|x_1|, \ldots, |x_n|\}.$$

## 3.1 A Topological Space

**Fig. 3.6** The product topology in $\Re^2$

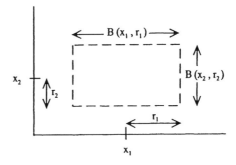

**Fig. 3.7** A Cartesian open ball of radius $r$ in $\Re^2$

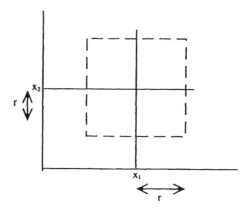

This induces a *Cartesian metric*

$$d_C(x, y) = \max\{|x_1 - y_1|, \ldots, |x_n - y_n|\}.$$

A *Cartesian* open ball of radius $r$ about $x$ is then the set

$$B_C(x, r) = \{y \in \Re^n : |y_i - x_i| < r \; \forall i = 1, \ldots, n\}.$$

A set $U$ is open in the *Cartesian topology* $\Gamma_C$ for $\Re^n$ iff for every $x \in U$ there exists some $r > 0$ such that the ball $B_C(x, r) \subset U$. See Fig. 3.7.

Suppose now that $U$ is an open set in the product topology $\Gamma^n$ for $\Re^n$. At any point $x \in U$, there exist $r_1 \ldots r_n$ all $> 0$ such that

$$B = B(x_1, r_1) \times, \ldots, B(x_n, r_n) \subset U.$$

Now let $r = \min(r_1, \ldots, r_n)$. Then clearly the Cartesian ball $B_C(x, r)$ belongs to the product ball $B$, and hence to $U$. Thus $U$ is open in the Cartesian topology.

On the other hand if $U$ is open in the Cartesian topology, for each point $x$ in $U$, the Cartesian ball $B_C(x, r)$ belongs to $U$. But this means the product ball $B(x_1, r) \times, \ldots, B(x_n, r)$ also belongs to $U$. Hence $U$ is open in the product topology.

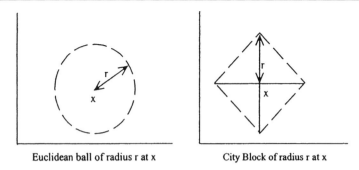

Euclidean ball of radius r at x    City Block of radius r at x

**Fig. 3.8** Balls under different metrics

We have therefore shown that a set $U$ is open in the product topology $\Gamma^n$ on $\Re^n$ iff it is open in the Cartesian topology $\Gamma_C$ on $\Re^n$. Thus the two topologies are identical.

We have also introduced the *Euclidean* and *city block norms* on $\Re^n$. These norms induce metrics and thus the *Euclidean topology* $\Gamma_E$ and *city block topology* $\Gamma_B$ on $\Re^n$. As before a set $U$ is open in $\Gamma_E$ (resp. $\Gamma_B$) iff for any point $x \in U$, there is an open neighborhood

$$B_E(x,r) = \left\{ y \in \Re^n : \sum_{i=1}^n (y_i - x_i)^2 < r^2 \right\},$$

resp. $B_B(x,r) = \{ y \in \Re^n : \sum |y_i - x_i| < r \}$ of $x$ which belongs to $U$.

(The reason we use the term "city block" should be obvious from the nature of a ball under this metric, so displayed in Fig. 3.8.)

In fact these three topologies $\Gamma_C$, $\Gamma_E$ and $\Gamma_B$ on $\Re^n$ are all identical. We shall show this in the case $n = 2$.

**Lemma 3.3** *The Cartesian, Euclidean and city block topologies on $\Re^2$ are identical.*

*Proof* Suppose that $U$ is an open set in the Euclidean topology $\Gamma_E$ for $\Re^2$. Thus at $x \in U$, there is an $r > 0$, such that the set

$$B_E(x,r) = \left\{ y \in \Re^2 : \sum_{i=1}^2 (y_i - x_i)^2 < r^2 \right\} \subset U.$$

From Fig. 3.9, it is obvious that the city block ball $B_B(x,r)$ also belongs to $B_E(x,r)$ and thus $U$. Thus $U$ is open in $\Gamma_B$.

On the other hand the Cartesian ball $B_C(x, \frac{r}{2})$ belongs to $B_B(x,r)$ and thus to $U$. Hence $U$ is open in $\Gamma_C$.

Finally the Euclidean ball $B_E(x, \frac{r}{2})$ belongs to $B_C(x, \frac{r}{2})$. Hence if $U$ is open in $\Gamma_C$ it is open in $\Gamma_E$.

**Fig. 3.9** Balls under different metrics

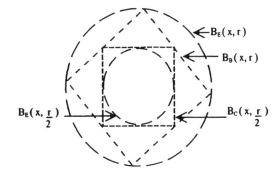

Thus $U$ open in $\Gamma_E \Rightarrow U$ open in $\Gamma_B \Rightarrow U$ open in $\Gamma_C \Rightarrow U$ open in $\Gamma_E$. Consequently all three topologies are identical in $\Re^2$. □

Suppose that $\Gamma_1$ and $\Gamma_2$ are two topologies on a space $X$. If any open set $U$ in $\Gamma_1$ is also an open set in $\Gamma_2$ then say that $\Gamma_2$ is *as fine as* $\Gamma_1$ and write $\Gamma_1 \subset \Gamma_2$. If $\Gamma_1 \subset \Gamma_2$ and $\Gamma_2 \subset \Gamma_1$ then $\Gamma_1$ and $\Gamma_2$ are identical, and we write $\Gamma_1 = \Gamma_2$. If $\Gamma_1 \subset \Gamma_2$ but $\Gamma_2$ contains an open set that is not open in $\Gamma_1$ then say $\Gamma_2$ is *finer* than $\Gamma_1$. We also say $\Gamma_1$ is *coarser* than $\Gamma_2$.

If $d_1$ and $d_2$ are two metrics on a space $X$, then under some conditions the topologies $\Gamma_1$ and $\Gamma_2$ induced by $d_1$ and $d_2$ are identical. Say the metrics $d_1$ and $d_2$ are *equivalent* iff for each $\varepsilon > 0$ there exist $\eta_1 > 0$ and $\eta_2 > 0$ such that $d_1(x, y) < \eta_1 \Rightarrow d_2(x, y) < \varepsilon$, and $d_2(x, y) < \eta_2 \Rightarrow d_1(x, y) < \varepsilon$. Another way of expressing this is that $B_1(x, \eta_1) \subset B_2(x, \varepsilon)$, and $B_2(x, \eta_2) \subset B_2(x, \varepsilon)$, where $B_i(x, r) = \{y : d_i(x, y) < r\}$ for $i = 1$ or $2$.

Just as in Lemma 3.3, the Cartesian, Euclidean and city block metrics on $\Re^n$ are equivalent. We can use this to show that the induced topologies are identical.

We now show that equivalent metrics induce identical topologies.

If $f : X \to \Re$ is a function, and $V$ is a set in $X$, define $\sup(f, V)$, the *supremum* (from the Latin, supremus) of $f$ on $V$ to be the smallest number $M \in \Re$ such that $f(x) \leq M$ for all $x \in V$.

Similarly define $\inf(f, V)$, the *infimum* (again from the Latin, infimus) of $f$ on $V$ to be the largest number $m \in \Re$ such that $f(x) \geq m$ for all $x \in V$.

Let $d : X \times X \to \Re$ be a metric on $X$. Consider a point, $x$, in $X$, and a subset of $X$. Then define the distance from $x$ to $V$ to be $d(x, V) = \inf(d(x, -), V)$, where $d(x, -) : V \to \Re$ is the function $d(x, -)(y) = d(x, y)$.

Suppose now that $U$ is an open set in the topology $\Gamma_1$ induced by the metric $d_1$. For any point $x \in U$ there exists $r > 0$ such that $B_1(x, r) \subset U$, where $B_1(x, r) = \{y \in X : d_1(x, y) < r\}$. Since we assume the metrics $d_1$ and $d_2$ are equivalent, there must exist $s > 0$, say, such that

$$B_2(x, s) \subset B_1(x, r)$$

**Fig. 3.10** Balls centered at $x$

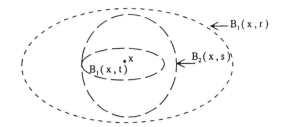

where $B_2(x, s) = \{y \in X : d_2(x, y) < s\}$. Indeed one may choose $s = d_2(x, \overline{B_1(x, r)})$ where $\overline{B_1(x, r)}$ is the complement of $B_1(x, r)$ in $X$ (see Fig. 3.10). Clearly the set $U$ must be open in $\Gamma_2$ and so $\Gamma_2$ is as fine as $\Gamma_1$. In the same way, however, there exists $t > 0$ such that $B_1(x, t) \subset B_2(x, s)$, where again one may choose $t = d_1(x, \overline{B_2(x, s)})$. Hence if $U$ is open in $\Gamma_2$ it is open in $\Gamma_1$. Thus $\Gamma_1$ is as fine as $\Gamma_2$. As a consequence $\Gamma_1$ and $\Gamma_2$ are identical.

Thus we obtain the following lemma.

**Lemma 3.4** *The product topology, $\Gamma^n$, Euclidean topology $\Gamma_E$, Cartesian topology $\Gamma_C$ and city block topology $\Gamma_B$ are all identical on $\Re^n$.*

As a consequence we may use, as convenient, any one of these three metrics, or any other equivalent metric, on $\Re^n$ knowing that topological results are unchanged.

## 3.2 Continuity

Suppose that $(X, \Gamma)$ and $(Y, \mathcal{S})$ are two topological spaces, and $f : X \to Y$ is a function between $X$ and $Y$. Say that $f$ is *continuous* (with respect to the topologies $\Gamma$ and $\mathcal{S}$) iff for any set $U$ in $\mathcal{S}$ (i.e., $U$ is $\mathcal{S}$-open) the set $f^{-1}(U) = \{x \in X : f(x) \in U\}$ is $\Gamma$-open.

This definition can be given an alternative form in the case when $X$ and $Y$ are metric spaces, with metrics $d_1, d_2$, say.

Consider a point $x_0$ in the domain of $f$. For any $\varepsilon > 0$ the ball

$$B_2(f(x_0), \varepsilon) = \{y \in Y : d_2(f(x_0), y) < \varepsilon\}$$

is open. For continuity, we require that the inverse image of this ball is open. That is to say there exists some $\delta$, such that the ball

$$B_1(x_0, \delta) = \{x \in X : d_1(x_0, x) < \delta\}$$

belongs to $f^{-1}(B_2(f(x_0), \varepsilon))$. Thus $x \in B_1(x_0, \delta) \Rightarrow f(x) \in B_2(f(x_0), \varepsilon)$.

Therefore say $f$ is continuous at $x_0 \in X$ iff for any $\varepsilon > 0, \exists \delta > 0$ such that

$$f(B_1(x_0, \delta)) \subset B_2(f(x_0), \varepsilon).$$

**Fig. 3.11** A discontinuity

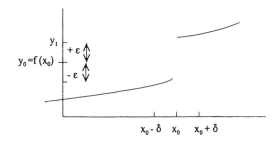

In the case that $X$ and $Y$ have norms $\|\ \|_X$ and $\|\ \|_Y$, then we may say that $f$ is *continuous at* $x_0$ iff for any $\varepsilon > 0$, $\exists\, \delta > 0$ such that

$$\|x - x_0\|_x < \delta \Rightarrow \|f(x) - f(x_0)\|_y < \varepsilon.$$

Then $f$ is continuous on $X$ iff $f$ is continuous at every point $x$ in its domain.

If $X, Y$ are vector spaces then we may check the continuity of a function $f : X \to Y$ by using the metric or norm form of the definition.

For example suppose $f : \Re \to \Re$ has the graph given in Fig. 3.11. Clearly $f$ is not continuous. To see this formally let $f(x_0) = y_0$ and choose $\varepsilon$ such that $|y_1 - y_0| > \varepsilon$. If $x \in (x_0 - \delta, x_0)$ then $f(x) \in (y_0 - \varepsilon, y_0)$.

However for any $\delta > 0$ it is the case that $x \in (x_0, x_0 + \delta)$ implies $f(x) > y_0 + \varepsilon$. Thus there exists no $\delta > 0$ such that

$$x \in (x_0 - \delta, x_0 + \delta) \quad \Rightarrow \quad f(x) \in (y_0 - \varepsilon, y_0 + \varepsilon).$$

Hence $f$ is not continuous at $x_0$.

We can give an alternative definition of continuity. A *sequence* of points in a space $X$ is a collection of points $\{x_k : k \in \mathcal{Z}\}$, indexed by the positive integers $\mathcal{Z}$. The sequence is written $(x_k)$. The sequence $(x_k)$ in the metric space, $X$, has a *limit*, $x$, iff $\forall \varepsilon > 0\ \exists k_0 \in \mathcal{Z}$ such that $k > k_0$ implies $\|x_k - x\|_X < \varepsilon$.

In this case write $x_k \to x$ as $k \to \infty$, or $\mathrm{Lim}_{k \to \infty} x_k = x$.

Note that $x$ is then an accumulation point of the sequence $\{x_1, \ldots\}$.

More generally $(x_n) \to x$ iff for any open set $G$ containing $x$, all but a finite number of points in the sequence $(x_k)$ belong to $G$.

Thus say $f$ is continuous at $x_0$ iff

$$x_k \to x_0 \quad \text{implies}\ f(x_k) \to f(x_0).$$

*Example 3.1* Consider the function $f : \Re \to \Re$ given by

$$f : x \to \begin{cases} x \sin \frac{1}{x} & \text{if } x \neq 0 \\ 0 & \text{if } x = 0. \end{cases}$$

Now $x \sin \frac{1}{x} = \frac{\sin y}{y}$ where $y = \frac{1}{x}$. Consider a sequence $(x_k)$ where $\text{Lim}_{k \to \infty} x_k = 0$. We shall write this limit as $x \to 0$. $\text{Lim}_{x \to 0} x \sin \frac{1}{x} = \frac{\sin y}{y} = 0$ since $|\sin y|$ is bounded above by 1, and $\text{Lim}_{y \to \infty} \frac{1}{y} = 0$. Thus $\text{Lim}_{x \to 0} f(x) = 0$. But $f(0) = 0$, and so $x_k \to 0$ implies $f(x_k) \to 0$. Hence $f$ is continuous at 0. On the other hand suppose $g(x) = \sin \frac{1}{x}$. Clearly $g(x)$ has no limit as $x \to 0$. To see this observe that for any sequence $(x_k) \to 0$ it is impossible to find a neighborhood $G$ of some point $y \in [-1, 1]$ such that $g(x_k) \in G$ whenever $k > k_o$.

Any linear function between finite-dimensional vector spaces is continuous. Thus the set of continuous functions contains the set of linear functions, when the domain is finite-dimensional. To see this suppose that $f : V \to W$ is a linear transformation between normed vector spaces. (Note that $V$ and $W$ may be infinite-dimensional.) Let $\| \|_v$ and $\| \|_w$ be the norms on $V$, $W$ respectively.

Say that $f$ is *bounded* iff $\exists B > 0$ such that $\|f(x)\|_w \le B \|x\|_v$ for all $x \in V$. Suppose now that

$$\|f(x) - f(x_0)\|_w < \varepsilon.$$

Now

$$\|f(x) - f(x_0)\|_w = \|f(x - x_0)\|_w \le B \|x - x_0\|_v,$$

since $f$ is linear and bounded. Choose $\delta = \frac{\varepsilon}{B}$. Then

$$\|x - x_0\|_v < \delta \Rightarrow \|f(x) - f(x_0)\|_w \le B \|x - x_0\|_v.$$

Thus if $f$ is linear and bounded it is continuous.

**Lemma 3.5** *Any linear transformation $f : V \to W$ is bounded and thus continuous if $V$ is finite-dimensional (of dimension $n$).*

*Proof* Use the Cartesian norm $\| \|_c$ on $V$, and let $\| \|_w$ be the norm on $W$. Let $e_1 \ldots e_n$ be a basis for $V$,

$$\|x\|_c = \sup\{|x_i| : i = 1, \ldots, n\}, \quad \text{and}$$
$$e = \sup_n \{\|f(e_i)\|_w : i = 1, \ldots, n\}.$$

Now $f(x) = \sum_{i=1}^n x_i f(e_i)$. Thus $\|f(x)\|_w \le \sum_i^n \|f(x_i e_i)\|_w$, by the triangle inequality, and $n \le \sum_{i=1}^n |x_i| \|f(e_i)\|_w$, since $\|ay\|_w = |a| \|y\|_w \le n e \|x\|_c$. Thus $f$ is bounded, and hence continuous with respect to the norms $\| \|_c$, $\| \|_w$. But for any other norm $\| \|_v$ it can be shown that there exists $B' > 0$ such that $\|x\|_c \le B' \|x\|_v$. Thus $f$ is bounded, and hence continuous, with respect to the norms $\| \|_v$, $\| \|_w$. □

Consider now the set $L(\mathfrak{R}^n, \mathfrak{R}^m)$ of linear functions from $\mathfrak{R}^n$ to $\mathfrak{R}^m$. Clearly if $f, g \in L(\mathfrak{R}^n, \mathfrak{R}^m)$ then the sum $f + g$ defined by $(f + g)(x) = f(x) + g(x)$ is also linear, and for any $\alpha \in \mathfrak{R}$, $\alpha f$, defined by $(\alpha f)(x) = \alpha(f(x))$ is linear.

## 3.2 Continuity

Hence $L(\Re^n, \Re^m)$ is a vector space over $\Re$. Since $\Re^n$ is finite dimensional, by Lemma 3.5 any member of $L(\Re^n, \Re^m)$ is bounded. Therefore for any $f \in L(\Re^n, \Re^m)$ we may define

$$\|f\| = \sup_{x \in \Re^n} \left\{ \frac{\|f(x)\|}{\|x\|} : \|x\| \neq 0 \right\}.$$

Since $f$ is bounded this is defined. Moreover $\| \ \| = 0$ only if $f$ is the zero function. By definition $\|f\|$ is the real number such that $\|f\| \leq B$ for all $B$ such that $\|f(x)\| \leq B\|x\|$. In particular $\|f(x)\| \leq \|f\| \ \|x\|$. If $f, g \in L(\Re^n, \Re^m)$, then $\|(f+g)(x)\| = \|f(x) + g(x)\| \leq \|f(x)\| + \|g(x)\| \leq \|f\| \ \|x\| + \|g\| \ \|x\| = (\|f\| + \|g\|)\|x\|$. Thus $\|f + g\| \leq \|f\| + \|g\|$.

Hence $\| \ \|$ on $L(\Re^n, \Re^m)$ satisfies the triangle inequality, and so $L(\Re^n, \Re^m)$ is a normed vector space. This in turn implies that $L(\Re^n, \Re^m)$ has a metric and thus a topology. It is often useful to use the metrics

$$d_1(f, g) = \sup\{\|f(x) - g(x)\| : x \in \Re^n\}, \quad \text{or}$$
$$d_2(f, g) = \sup\{|f_i(x) - g_i(x)| : i = 1, \ldots, m, x \in \Re^n\}$$

where $f = (f_1, \ldots, f_m)$, $g = (g_1, \ldots, g_m)$. We write $\mathcal{L}^1(\Re^n, \Re^m)$ and $\mathcal{L}^2(\Re^n, \Re^m)$ for the set $L(\Re^n, \Re^m)$ with the topologies induced by $d_1$ and $d_2$ respectively. Clearly these two topologies are identical.

Alternatively, choose bases for $\Re^n$ and $\Re^m$ and consider the matrix representation function

$$M : \left(L(\Re^n, \Re^m), +\right) \to \left(M(n, m), +\right).$$

On the right hand side we add matrices element by element under the rule $(a_{ij}) + (b_{ij}) = (a_{ij} + b_{ij})$. With this operation $M(n, m)$ is also a vector space. Clearly we may choose a basis for $M(n, m)$ to be $\{E_{ij} : i = 1, \ldots, n; j = 1, \ldots, m\}$ where $E_{ij}$ is the elementary matrix with 1 in the $i$th column and $j$th row.

Thus $M(n, m)$ is a vector space of dimension $nm$. Since $M$ is a bijection, $L(\Re^n, \Re^m)$ is also of dimension $nm$.

A norm on $M(n, m)$ is given by

$$\|A\| = \sup\{|a_{ij}| : i = 1, \ldots, n; j = 1, \ldots, m\},$$

where $A = (a_{ij})$.

This in turn defines a metric and thus a topology on $M(n, m)$. Finally this defines a *topology* on $L(\Re^n, \Re^m)$ as follows. For any open set $U$ in $M(n, m)$, let $V = M^{-1}(U)$ and call $V$ open. The base for the topology on $L(\Re^n, \Re^m)$ then consists of all sets of this form. One can show that the topology induced on $L(\Re^n, \Re^m)$ in this way is independent of the choice of bases. We call this the induced topology on $L(\Re^n, \Re^m)$ and write $\mathcal{L}(\Re^n, \Re^m)$ for this topological space. If we consider the norm topology on $M(n, m)$ and the induced topology $\mathcal{L}(\Re^n, \Re^m)$ then the representation map is continuous. Moreover the two topologies induced by the metrics $d_1$ and $d_2$ on $L(\Re^n, \Re^m)$ are identical to the induced topology $\mathcal{L}(\Re^n, \Re^m)$. Thus $M$ is also continuous when these metric topologies are used for its domain. (Exercise 3.3 at the end of the book is devoted to this observation.)

If $V$ is an infinite dimensional vector space and $f \in L(V, W)$, then $f$ need not be continuous or bounded. However the subset of $L(V, W)$ consisting of bounded, and thus continuous, maps in $L(V, W)$ admits a norm and thus a topology. So we may write this subset as $\mathcal{L}(V, W)$.

Now let $C_o(\Re^n, \Re^m)$ be the set of continuous functions from $\Re^n$ to $\Re^m$. We now show that $C_o(\Re^n, \Re^m)$ is a vector space.

**Lemma 3.6** $C_o(\Re^n, \Re^m)$ *is a vector space over* $\Re$.

*Proof* Suppose that $f, g$ are both continuous maps. At any $x_0 \in \Re^n$, and any $\varepsilon > 0$, $\exists \delta_1, \delta_2 > 0$ such that

$$\|x - x_0\| < \delta_1 \Rightarrow \|f(x) - f(x_0)\| < \left(\frac{1}{2}\right)\varepsilon$$

$$\|x - x_0\| < \delta_2 \Rightarrow \|g(x) - g(x_0)\| < \left(\frac{1}{2}\right)\varepsilon.$$

Let $\delta = \min(\delta_1, \delta_2)$. Then $\|x - x_0\| < \delta \Rightarrow \|(f+g)(x) - (f+g)(x_0)\| = \|f(x) - f(x_0) + g(x) - g(x_0)\| \leq \|f(x) - f(x_0)\| + \|g(x) - g(x_0)\| < (\frac{1}{2})\varepsilon + (\frac{1}{2})\varepsilon = \varepsilon$. Thus $f + g \in C_o(\Re^n, \Re^m)$.

Also for any $\alpha \in \Re$, any $\varepsilon > 0$ $\exists \delta > 0$ such that

$$\|x - x_0\| < \delta \Rightarrow \|f(x) - f(x_0)\| < \frac{\varepsilon}{|\alpha|}.$$

Therefore $\|(\alpha f)(x) - (\alpha f)(x_0)\| = |\alpha| \|f(x) - f(x_0)\| < \varepsilon$. Thus $\alpha f \in C_o(\Re^n, \Re^m)$. Hence $C_o(\Re^n, \Re^m)$ is a vector space. $\square$

Since $[L(\Re^n, \Re^m)]$ is closed under addition and scalar multiplication, it is a *vector subspace* of dimension $nm$ of the vector space $C_o(\Re^n, \Re^m)$. Note however that $C_o(\Re^n, \Re^m)$ is an *infinite*-dimensional vector space (a *function* space).

**Lemma 3.7** *If* $(X, \Gamma), (Y, \mathcal{S})$ *and* $(Z, \mathcal{R})$ *are topological spaces and* $C_o((X, \Gamma), (Y, \mathcal{S})), C_o((Y, \mathcal{S}), (Z, \mathcal{R}))$ *and* $C_o((X, \Gamma), (Z, \mathcal{R}))$ *are the sets of functions which are continuous with respect to these topologies, then the composition operator,* $\circ$, *maps* $C_o((X, \Gamma), (Y, \mathcal{S})) \times C_o((Y, \mathcal{S}), (Z, \mathcal{R}))$ *to* $C_o((X, \Gamma), (Z, \mathcal{R}))$.

*Proof* Suppose $f : (X, \Gamma) \to (Y, \mathcal{S})$ and $g : (Y, \mathcal{S}) \to (Z, \mathcal{R})$ are continuous. We seek to show that $g \circ f : (X, \Gamma) \to (Z, \mathcal{R})$ is continuous. Choose any open set $U$ in $Z$. By continuity of $g$, $g^{-1}(U)$ is an $\mathcal{S}$-open set in $Y$. But by the continuity of

$f$, $f^{-1}(g^{-1}(U))$ is a $\Gamma$-open set in $X$. However $f^{-1}g^{-1}(U) = (g \circ f)^{-1}(U)$. Thus $g \circ f$ is continuous. □

Note therefore that if $f \in L(\mathfrak{R}^n, \mathfrak{R}^m)$ and $g \in L(\mathfrak{R}^m, \mathfrak{R}^k)$ then $g \circ f \in L(\mathfrak{R}^n, \mathfrak{R}^k)$ will also be continuous.

## 3.3 Compactness

Let $(X, \Gamma)$ be a topological space. An *open cover* for $X$ is a family $\{U_j : j \in J\}$ of $\Gamma$-open sets such that $X = \bigcup_{j \in J} U_j$.

If $U = \{U_j : j \in J\}$ is an open cover for $X$, *a subcover* of $U$ is an open cover $U'$ of $X$ where $U' = \{U_j : j \in J'\}$ and the index set $J'$ is a subset of $J$. The subcover is called *finite* if $J'$ is a finite set (*i.e.*, $|J'|$, the number of elements of $J'$, is finite).

**Definition 3.5** A topological space $(X, \Gamma)$ is called *compact* iff any open cover of $X$ has a finite subcover. If $Y$ is a subset of the topological space $(X, \Gamma)$ then $Y$ is compact iff the topological space $(Y, \Gamma_Y)$ is compact. Here $\Gamma_Y$ is the topology induced on $Y$ by $\Gamma$. (See Sect. 1.1.3.)

Say that a family $C_J = \{C_j : j \subset J\}$ of closed sets in a topological space $(X, \Gamma)$ has the *finite intersection property* (FIP) iff whenever $J'$ is a finite subset of $J$ then $\bigcap_{j \in J'} C_j$ is non-empty.

**Lemma 3.8** *A topological space $(X, \Gamma)$ is compact iff whenever $C_J$ is a family of closed sets with the finite intersection property then $\bigcap_{j \in J} C_j$ is non-empty.*

*Proof* We establish first of all that $U_J = \{U_j : j \in J\}$ is an open cover of $X$ iff the family $C_J = \{C_j = X \setminus U_j : j \in J\}$ of closed sets has empty intersection. Now

$$\bigcup_J U_j = \bigcup_J (X \setminus C_j) = \bigcup_J (X \cap \overline{C_j})$$

$$= X \cap \left(\bigcup_j \overline{C_j}\right) = X \cap \left(\overline{\bigcap_j C_j}\right).$$

Thus

$$\bigcup_J U_j = X \quad \text{iff} \quad \bigcap_J C_j = \Phi.$$

To establish necessity, suppose that $X$ is compact and that the family $C_J$ has the *FIP*. If $C_J$ in fact has empty intersection then $U_J = \{X \setminus C_j : j \in J\}$ must be an open cover. Since $X$ is compact there exists a finite set $J' \subset J$ such that $U_{J'}$ is a cover. But then $C_{J'}$ has empty intersection contradicting *FIP*. Thus $C_J$ has non-empty intersection. To establish sufficiency, suppose that any family $C_J$, satisfying *FIP*, has non-empty intersection. Let $U_J = \{U_j : j \in J\}$ be an open cover for $X$. If $U_J$ has no finite subcover, then for any finite $J'$, the family $C_{J'} = \{X \backslash U_j : j \in J'\}$ must have non-empty intersection. By the *FIP*, the family $C_J$ must have non-empty intersection, contradicting the assumption that $U_J$ is a cover. Thus $(X, \Gamma)$ is compact. □

This lemma allows us to establish conditions under which a preference relation $P$ on $X$ has a non-empty choice $C_P(X)$. (See Lemma 1.8 for the case with $X$ finite.) Say the preference relation $P$ on the topological space $X$ is lower demi-continuous (LDC) iff the inverse correspondence $\phi_P^{-1} : X \to X : x \to \{y \in X : xPy\}$ is open for all $x$ in $X$.

**Lemma 3.9** *If $X$ is a compact topological space and $P$ is an acyclic and lower demi-continuous preference on $X$, then there exists a choice $\bar{x}$ of $P$ in $X$.*

*Proof* Suppose on the contrary that there is no choice. Thus for all $x \in X$ there exists some $y \in X$ such that $yPx$. Thus $x \in \phi_P^{-1}(y)$. Hence $U = \{\phi_P^{-1}(y) : y \in X\}$ is a cover for $X$. Moreover for each $y \in X$, $\phi_P^{-1}(y)$ is open. Since $X$ is compact, there exists a finite subcover of $U$. That is to say there exists a finite set $A$ in $X$ such that $U' = \{\phi_P^{-1}(y) : y \in A\}$ is a cover for $X$. In particular this implies that for each $x \in A$, there is some $y \in A$ such that $x \in P_P^{-1}(y)$, or that $yPx$. But then $C_P(A) = \{x \in A : \phi_P(x) = \Phi\} = \Phi$. Now $P$ is acyclic on $X$, and thus acyclic on $A$. Hence, by the acyclicity of $P$ and Lemma 1.8, $C_P(A) \neq \Phi$. By the contradiction $U = \{\phi_P^{-1}(y) : y \in X\}$ cannot be a cover. That is to say there is some $\bar{x} \in X$ such that $\bar{x} \in \phi_P^{-1}(y)$ for no $y \in X$. But then $yP\bar{x}$ for no $y \in X$, and $\bar{x} \in C_P(X)$, or $\bar{x}$ is the choice on $X$. □

This lemma can be used to show that a continuous function $f : X \to \Re$ from a compact topological space $X$ into the reals attains its bounds. Remember that we defined the supremum and infimum of $f$ on a set $Y$ to be
1. $\sup(f, Y) = M$ such that $f(x) \leq M$ for all $x \in Y$ and if $f(x) \leq M'$ for all $x \in Y$ then $M' \geq M$
2. $\inf(f, Y) = m$ such that $f(x) \geq m$ for all $x \in Y$ and if $f(x) \geq m'$ for all $x \in Y$ then $m \geq m'$.

Say $f$ attains its upper bound on $Y$ iff there is some $x_s$ in $Y$ such that $f(x_s) = \sup(f, Y)$. Similarly say $f$ attains its lower bound on $Y$ iff there is some $x_i$ in $Y$ such that $f(x_i) = \inf(f, Y)$.

Given the function $f : X \to \Re$, define a preference $P$ on $X \times X$ by $xPy$ iff $f(x) > f(y)$. Clearly $P$ is acyclic, since $>$ on $\Re$ is acyclic. Moreover for any $x \in X$,

$$\phi_P^{-1}(x) = \{y : f(y) < f(x)\}$$

is open, when $f$ is continuous.

## 3.3 Compactness

To see this let $U = (-\infty, f(x))$. Clearly $U$ is an open set in $\mathfrak{R}$. Moreover $f(y)$ belongs to the open interval $(-\infty, f(x))$ iff $y \in \phi_P^{-1}(x)$. But $f$ is continuous, and so $f^{-1}(U)$ is open in $X$. Since $y \in f^{-1}(U)$ iff $y \in \phi_P^{-1}(x)$, $\phi_P^{-1}(x)$ is open for any $x \in X$.

**Weierstrass Theorem** *Let $(X, \Gamma)$ be a topological space and $f : X \to \mathfrak{R}$ a continuous real-valued function. If $Y$ is a compact subset of $X$, then $f$ attains its bounds on $Y$.*

*Proof* As above, for each $x \in Y$, define $U_x = (-\infty, f(x))$. Then $\phi_P^{-1}(x) = \{y \in Y : f(y) < f(x)\} = f^{-1}(U_x) \cap Y$ is an open set in the induced topology on $Y$. By Lemma 3.9 there exists a choice $\overline{x}$ in $Y$ such that $\phi_P^{-1}(\overline{x}) = \Phi$. But then $f(y) > f(\overline{x})$ for no $y \in Y$. Hence $f(y) \le f(\overline{x})$ for all $y \in Y$. Thus $f(\overline{x}) = \sup(f, Y)$.

In the same way let $Q$ be the relation on $X$ given by $xQy$ iff $f(x) < f(y)$. Then there is a choice $\overline{\overline{x}} \in Y$ such that $f(y) < f(\overline{\overline{x}})$ for no $y \in Y$. Hence $f(y) \ge f(\overline{\overline{x}})$ for all $y \in Y$, and so $f(\overline{\overline{x}}) = \inf(f, Y)$. Thus $f$ attains its bounds on $Y$. □

We can develop this result further.

**Lemma 3.10** *If $f : (X, \Gamma) \to (Z, \mathcal{S})$ is a continuous function between topological spaces, and $Y$ is a compact subset of $X$, then*

$$f(Y) = \{f(y) \in Z : y \in Y\}$$

*is compact.*

*Proof* Let $\{W_\alpha\}$ be an open cover for $f(Y)$. Then each member $W_\alpha$ of this cover may be expressed in the form $W_\alpha = U_\alpha \cap f(Y)$ where $U_\alpha$ is an open set in $Z$. For each $\alpha$, let $V_\alpha = f^{-1}(U_\alpha) \cap Y$. Now each $V_\alpha$ is open in the induced topology on $Y$. Moreover, for each $y \in Y$, there exists some $W_\alpha$ such that $f(y) \in W_\alpha$. Thus $\{V_\alpha\}$ is an open cover for $Y$. Since $Y$ is compact, $\{V_\alpha\}$ has a finite subcover $\{V_\alpha : \alpha \in J\}$, and so $\{f(V_\alpha) : \alpha \in J\}$ is a finite subcover of $\{W_\alpha\}$. Thus $f(Y)$ is compact. □

Now a real-valued continuous function $f$ is bounded on a compact set, $Y$ (by the Weierstrass Theorem). So $f(Y)$ will be contained in $[f(\overline{\overline{x}}), f(\overline{x})]$ say, for some $\overline{x} \in Y$. Since $f(Y)$ must also be compact, this suggests that a closed set of the form $[a, b]$ must be compact.

For a set $Y \subset \mathfrak{R}$ define $\sup(Y) = \sup(\text{id}, Y)$, the *supremum* of $Y$, and $\inf(Y) = \inf(\text{id}, Y)$, the *infimum* of $Y$. Here id : $\mathfrak{R} \to \mathfrak{R}$ is the identity on $\mathfrak{R}$. The set $Y \subset \mathfrak{R}$ is *bounded above* (or below) iff its supremum (or infimum) is finite. The set is *bounded* iff it is both bounded above and below. Thus a set of the form $[a, b]$, say with $-\infty < a < b < +\infty$ is bounded.

**Heine Borel Theorem** *A closed bounded interval, $[a, b]$, of the real line is compact.*

*Sketch of Proof* Consider a family $C = \{[a, c_i], [d_j, b] : i \in I, j \in J\}$ of subsets of $[a, b]$ with the *finite intersection property*. Suppose that neither $I$ nor $J$ is empty. Let $d = \sup(\{d_j : j \in J\})$ and suppose that for some $k \in I$, $c_k < d$. Then there exists $i \in I$ and $j \in J$ such that $[a, c_i] \cap [d_j, b] = \Phi$, contradicting the finite intersection property. Thus $c_i \geq d$, and so $[a, c_i] \cap [d, b] \neq \Phi$, for all $i \in I$. Hence the family $C$ has non empty intersection. By Lemma 3.8, $[a, b]$ is compact $\square$

**Definition 3.6** A topological space $(X, \Gamma)$ is called *Hausdorff* iff any two distinct points $x, y \in X$ have $\Gamma$-open neighbourhoods $U_x, U_y$ such that $U_x \cap U_y = \Phi$.

**Lemma 3.11** *If $(X, d)$ is a metric space then $(X, \Gamma_d)$ is Hausdorff, where $\Gamma_d$ is the topology on $X$ induced by the metric $d$.*

*Proof* For two points $x \neq y$, let $\varepsilon = d(x, y) \neq 0$. Define $U_x = B(x, \frac{\varepsilon}{3})$ and $U_x = B(y, \frac{\varepsilon}{3})$.

Clearly, by the triangle inequality, $B(x, \frac{\varepsilon}{3}) \cap B(y, \frac{\varepsilon}{3}) = \Phi$. Otherwise there would exist a point $z$ such that $d(x, z) < \frac{\varepsilon}{3}, d(z, y) < \frac{\varepsilon}{3}$, which would imply that $d(x, y) < d(x, z) + d(z, y) = \frac{2\varepsilon}{3}$. By contradiction the open balls of radius $\frac{\varepsilon}{3}$ do not intersect. Thus $(x, \Gamma_d)$ is Hausdorff. $\square$

A Hausdorff topological space is therefore a natural generalisation of a metric space.

**Lemma 3.12** *If $(X, \Gamma)$ is a Hausdorff topological space, then any compact subset $Y$ of $X$ is closed.*

*Proof* We seek to show that $X \setminus Y$ is open, by showing that for any $x \in X \setminus Y$, there exists a neighbourhood $G$ of $x$ and an open set $H$ containing $Y$ such that $G \cap H = \Phi$. Let $x \in X \setminus Y$, and consider any $y \in Y$. Since $X$ is Hausdorff, there exists a neighbourhood $V(y)$ of $y$ and a neighbourhood $U(y)$, say, of $x$ such that $V(y) \cap U(y) = \Phi$. Since the family $\{V(y) : y \in Y\}$ is an open cover of $Y$, and $Y$ is compact, there exists a finite subcover $\{V(y) : y \in A\}$, where $A$ is a finite subset of $Y$.

Let $H = \bigcup_{y \in A} V(y)$ and $G = \bigcap_{y \in A} U(y)$.

Suppose that $G \cap H \neq \Phi$. Then this implies there exists $y \in A$ such that $V(y) \cap U(y)$ is non-empty. Thus $G \cap H = \Phi$. Since $A$ is finite, $G$ is open. Moreover $Y$ is contained in $H$. Thus $X \setminus Y$ is open and $Y$ is closed. $\square$

**Lemma 3.13** *If $(X, \Gamma)$ is a compact topological space and $Y$ is a closed subset of $X$, then $Y$ is compact.*

*Proof* Let $\{U_\alpha\}$ be an open cover for $Y$, where each $U_\alpha \subset X$. Then $\{V_\alpha = U_\alpha \cap Y\}$ is also an open cover for $Y$.

Since $X \setminus Y$ is open, $\{X \setminus Y\} \cup \{V_\alpha\}$ is an open cover for $X$.

Since $X$ is compact there is a finite subcover, and since each $V_\alpha \subset Y$, $X\backslash Y$ must be a member of this subcover. Hence the subcover is of the form $\{X\backslash Y\} \cup \{V_j : j \in J\}$. But then $\{V_j : j \in J\}$ is a finite subcover of $\{V_\alpha\}$ for $Y$. Hence $Y$ is compact. □

**Tychonoff's Theorem** *If $(X, \Gamma)$ and $(Y, \mathcal{S})$ are two compact topological spaces then $(X \times Y, \Gamma \times \mathcal{S})$, with the product topology, is compact.*

*Proof* To see this we need only consider a cover for $X \times Y$ of the form $\{U_\alpha \times V_\beta\}$ for $\{U_\alpha\}$ an open cover for $X$ and $\{V_\beta\}$ an open cover for $Y$. Since both $X$ and $Y$ are compact, both $\{U_\alpha\}$ and $\{V_\alpha\}$ have finite subcovers $\{U_j\}_{j \in J}$ and $\{V_k\}_{k \in K}$, and so $\{U_j \times V_k : (j, k) \in J \times K\}$ is a finite subcover for $X \times Y$. □

As a corollary of this theorem, let $I_k = [a_k, b_k]$ for $k = 1, \ldots, n$ be a family of closed bounded intervals in $\Re$. Each interval is compact by the Heine Borel Theorem. Thus the closed *cube* $I^n = I_1 \times I_2, \ldots, I_n$ in $\Re^n$ is compact, by Tychonoff's Theorem. Say that a set $Y \subset \Re^n$ is bounded iff for each $y \in Y$ there exists some finite number $K(y)$ such that $\|x - y\| < K(y)$ for all $x \in Y$. Here $\| \ \|$ is any convenient norm on $\Re^n$. If $Y$ is bounded then clearly there exists some closed cube $I^n \subset \Re^n$ such that $Y \subset I^n$. Thus we obtain:

**Lemma 3.14** *If $Y$ is a closed bounded subset of $\Re^n$ then $Y$ is compact.*

*Proof* By the above, there is a closed cube $I^n$ such that $Y \subset I^n$. But $I^n$ is compact by Tychonoff's Theorem. Since $Y$ is a closed subset of $I^n$, $Y$ is compact, by Lemma 3.13. □

In $\Re^n$ a compact set $Y$ is one that is both closed and bounded. To see this note that $\Re^n$ is certainly a metric space, and therefore Hausdorff. By Lemma 3.12, if $Y$ is compact, then it must be closed. To see that it must be bounded, consider the unbounded closed interval $A = [0, \infty)$ in $\Re$, and an open cover $\{U_k = (k-2, k); k = 1, \ldots, \infty\}$. Clearly $\{(-1, 1), (0, 2), (1, 3), \ldots\}$ cover $[0, \infty)$. A finite subcover must be bounded above by $K$, say, and so the point $K$ does not belong to the subcover. Hence $[0, \infty)$ is non-compact.

**Lemma 3.15** *A compact subset $Y$ of $\Re$ contains its bounds.*

*Proof* Let $s = \sup(\text{id}, Y)$ and $i = \inf(\text{id}, Y)$ be the supremum and infimum of $Y$. Here $\text{id} : \Re \to \Re$ is the identity function. By the discussion above, $Y$ must be bounded, and so $i$ and $s$ must be finite. We seek to show that $Y$ contains these bounds, *i.e.*, that $i \in Y$ and $s \in Y$. Suppose for example that $s \notin Y$. By Lemma 3.12, $Y$ must be closed and hence $\Re \backslash Y$ must be open. But then there exists a neighbourhood $(s - \varepsilon, s + \varepsilon)$ of $s$ in $\Re \backslash Y$, and so $s - \frac{\varepsilon}{2} \notin Y$. But this implies that $y \leq s - \frac{\varepsilon}{2}$ for all $y \in Y$, which contradicts the assumption that $s = \sup(\text{id}, Y)$. Hence $s \in Y$. A similar argument shows that $i \in Y$. Thus $Y = [i, y_1] \cup, \ldots, \cup [y_r, s]$ say, and so $Y$ contains its bounds. □

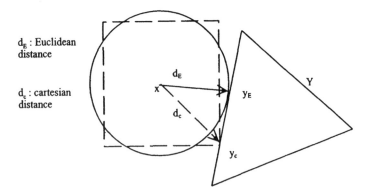

**Fig. 3.12** Nearest points under different metrics

**Lemma 3.16** *Let $(X, \Gamma)$ be a topological space and $f : X \to \Re$ a continuous real-valued function. If $Y \subset X$ is compact then there exist points $x_0$ and $x_1$ in $Y$ such that $f(x_0) \leq f(y) \leq f(x_1)$ for all $y \in Y$.*

*Proof* By Lemma 3.10, $f(Y)$ is compact. By Lemma 3.15, $f(Y)$ contains its infimum and supremum. Thus there exists $x_0, x_1 \in Y$ such that

$$f(x_0) \leq f(y) \leq f(x_1) \quad \text{for all } y \in Y.$$

Note that $f(Y)$ must be bounded, and so $f(x_0)$ and $f(x_1)$ must be finite. □

We have here obtained a second proof of the Weierstrass Theorem that a continuous real-valued function on a compact set attains its bounds, and shown moreover that these bounds are finite. A useful application of this theorem is that if $Y$ is a compact set in $\Re^n$ and $x \notin Y$ then there is some point $y$ in $Y$ which is nearest to $x$. Remember that we defined the *distance* from a point $x$ in a metric space $(X, d)$ to a subset $Y$ of $X$ to be $d(x, Y) = \inf(f_x, Y)$ where $f_x : Y \to \Re$ is defined by $f_x(y) = d(x, -)(y) = d(x, y)$. See Fig. 3.12.

**Lemma 3.17** *Suppose $Y$ is a subset of a metric space $(X, d)$ and $x \in X$. Then the function $f_x : Y \to \Re$ given by $f_x(y) = d(x, y)$ is continuous.*

*Proof* Consider $y_1, y_2 \in Y$ and suppose that $d(x, y_1) \geq d(x, y_2)$. Then $|d(x, y_1) - d(x, y_2)| = d(x, y_1) - d(x, y_2)$.

By the triangle inequality $d(x, y_1) \leq d(x, y_2) + d(y_2, y_1)$. Hence $|d(x, y_1) - d(x, y_2)| \leq d(y_1, y_2)$ and so $d(y_1, y_2) < \varepsilon \Rightarrow |d(x, y_1) - d(x, y_2)| < \varepsilon$, for any $\varepsilon > 0$.

Thus for any $\varepsilon > 0$, $d(y_1, y_2) < \varepsilon \Rightarrow |f_x(y_1) - f_x(y_2)| < \varepsilon$, and so $f_x$ is continuous. □

## 3.4 Convexity

**Lemma 3.18** *If $Y$ is a compact subset of a compact metric space $X$ and $x \in X$, then there exists a point $y_0 \in Y$ such that $d(x, Y) = d(x, y_0) < \infty$.*

*Proof* By Lemma 3.17, the function $d(x, -) : Y \to \Re$ is continuous. By Lemma 3.16, this function attains its lower and upper bounds on $Y$. Thus there exists $y_0 \in Y$ such that $d(x, y_0) = \inf(d(x, -), Y) = d(x, Y)$, where $d(x, y_0)$ is finite. □

The point $y_0$ in $Y$ such that $d(x, y_0) = d(x, Y)$ is the nearest point in $Y$ to $x$. Note of course that if $x \in Y$ then $d(x, Y) = 0$.

More importantly, when $Y$ is compact $d(x, Y) = 0$ *if and only if* $x \in Y$. To see this necessity, suppose that $d(x, Y) = 0$. Then by Lemma 3.18, there exists $y_0 \in Y$ such that $d(x, y_0) = 0$. By the definition of a metric $d(x, y_0) = 0$ iff $x = y_0$ and so $x \in Y$. The point $y \in Y$ that is nearest to $x$ is dependent on the metric of course, and may also not be unique.

## 3.4 Convexity

### 3.4.1 A Convex Set

If $x, y$ are two points in a vector space, $X$, then the *arc*, $[x, y]$, is the set $\{z \in X : \exists \lambda \in [0, 1] \text{ s.t. } z = \lambda x + (1 - \lambda)y\}$. A point in the arc $[x, y]$ is called a *convex combination* of $x$ and $y$. If $Y$ is a subset of $X$, then the *convex hull*, con($Y$), of $Y$ is the smallest set in $X$ that contains, for every pair of points $x, y$ in $Y$, the arc $[x, y]$. The set $Y$ is called *convex* iff con($Y$) $= Y$. The set $Y$ is *strictly convex* iff for any $x, y \in Y$ the combination $\lambda x + (1 - \lambda)y$, for $\lambda \in (0, 1)$, belongs to the interior of $Y$.

Note that if $Y$ is a vector subspace of the real vector space $X$ then $Y$ must be convex. For then if $x, y \in Y$ both $\lambda, (1 - \lambda) \in \Re$ and so $\lambda x + (1 - \lambda) y \in Y$.

**Definition 3.7** Let $Y$ be a real vector space, or a convex subset of a real vector space, and let $f : Y \to \Re$ be a function. Then $f$ is said to be
1. *convex* iff $f(\lambda x + (1 - \lambda)y) \leq \lambda f(x) + (1 - \lambda) f(y)$
2. *concave* iff $f(\lambda x + (1 - \lambda)y) \geq \lambda f(x) + (1 - \lambda) f(y)$
3. *quasi-concave* iff $f(\lambda x + (1 - \lambda)y) \geq \min[f(x), f(y)]$ for any $x, y \in Y$ and any $\lambda \in [0, 1]$.

Suppose now that $f : Y \to \Re$ and consider the preference $P \subset Y \times Y$ induced by $f$. For notational convenience from now on we regard $P$ as a correspondence $P : Y \to Y$. That is define $P$ by

$$P(x) = \{y \in Y : f(y) > f(x)\}.$$

If $f$ is quasi-concave then when $y_1, y_2, \in P(x)$,

$$f(\lambda y_1 + (1 - \lambda) y_2) \geq \min[f(y_1), f(y_2)] > f(x).$$

Hence $\lambda y_1 + (1-\lambda)y_2 \in P(x)$. Thus for all $x \in Y$, $P(x)$ is convex.

We shall call a preference correspondence $P : Y \to Y$ *convex* when $Y$ is convex and $P$ is such that $P(x)$ is convex for all $x \in Y$.

When a function $f : Y \to \Re$ is quasi-concave then the strict preference correspondence $P$ defined by $f$ is convex. Note also that the weak preference $R : Y \to Y$ given by

$$R(x) = \{y \in Y : f(y) \geq f(x)\}$$

will also be convex.

If $f : Y \to \Re$ is a concave function then it is quasi-concave. To see this consider $x, y \in Y$, and suppose that $f(x) \leq f(y)$. By concavity,

$$\begin{aligned} f(\lambda x + (1-\lambda)y) &\geq \lambda f(x) + (1-\lambda) f(y) \\ &\geq \lambda f(x) + (1-\lambda) f(x) \\ &\geq \min[f(x), f(y)]. \end{aligned}$$

Thus $f$ is quasi-concave. Note however that a quasi-concave function need be neither convex nor concave. However if $f$ is a linear function then it is convex, concave and quasi-concave. There is a partial order $>$ on $\Re^n$ given by $x > y$ iff $x_i > y_i$ where $x = (x_1, \ldots, x_n)$, $y = (y_1, \ldots, y_n)$. A function $f : \Re^n \to \Re$ is *weakly monotonically increasing* iff $f(x) \geq f(y)$ for any $x, y \in \Re^n$ such that $x > y$. A function $f : \Re^n \to \Re$ has *decreasing returns to scale* iff $f$ is weakly monotonically increasing and concave. A very standard assumption in economic theory is that feasible production of an output has decreasing returns to scale of inputs, and that consumers' utility or preference has decreasing returns to scale in consumption. We shall return to this point below.

### 3.4.2 Examples

*Example 3.2* (i) Consider the set $X_1 = \{(x_1, x_2) \in \Re^2 : x_2 \geq x_1\}$. Clearly if $x_2 \geq x_1$ and $x_2' \geq x_1'$ then $\lambda x_2 + (1-\lambda)x_2' \geq \lambda x_1 + (1-\lambda)x_1'$, for $\lambda \in [0, 1]$. Thus $\lambda(x_1, x_2) + (1-\lambda)(x_1', x_2') = \lambda x_1 + (1-\lambda)x_1', \lambda x_2 + (1-\lambda)x_2' \in X_1$. Hence $X_1$ is convex.

On the other hand consider the set $X_2 = \{(x_1, x_2) \in \Re^2 : x_2 \geq x_1^2\}$.

As Fig. 3.13a indicates, this is a strictly convex set. However the set $X_3 = \{(x_1, x_2) \in \Re^2 : |x_2| \geq x_1^2\}$ is not convex. To see this suppose $x_2 < 0$. Then $(x_1, x_2) \in X_3$ implies that $x_2 \leq -x_1^2$. But then $-x_2 \geq x_1^2$. Clearly $(x_1, 0)$ belongs to the convex combination of $(x_1, x_2)$ and $(x_1, -x_2)$ yet $(x_1, 0) \notin X_3$.

(ii) Consider now the set $X_4 = \{(x_1, x_2) : x_2 > x_1^3\}$. As Fig. 3.13b shows it is possible to choose $(x_1, x_2)$ and $(x_1', x_2')$ with $x_1 < 0$, so that the convex combination of $(x_1, x_2)$ and $(x_1', x_2')$ does not belong to $X_4$. However $X_5 = \{(x_1, x_2) \in \Re^2 : x_2 \geq x_1^3$ and $x_1 \geq 0\}$ and $X_6 = \{(x_1, x_2) \in \Re^2 : x_2 \leq x_1^3$ and $x_1 \leq 0\}$ are both convex sets.

## 3.4 Convexity

**Fig. 3.13a** Example 3.2(i)

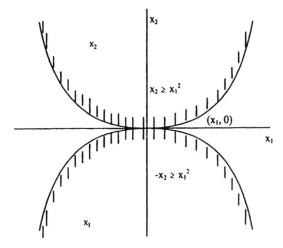

**Fig. 3.13b** Example 3.2(ii)

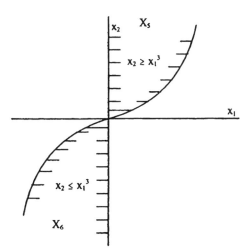

(iii) Now consider the set $X_7 = \{(x_1, x_2) \in \Re^2 : x_1 x_2 \geq 1\}$. From Fig. 3.13c it is clear that the restriction of the set $X_7$ to the positive quadrant $\Re_+^2 = \{(x_1, x_2) \in \Re^2 : x_1 \geq 0 \text{ and } x_2 \geq 0\}$ is strictly convex, as is the restriction of $x_7$ to the negative quadrant $\Re_-^2 = \{(x_1, x_2) \in \Re^2 : x_1 \leq 0 \text{ and } x_2 \leq 0\}$. However if $(x_1, x_2) \in X_7 \cap \Re_+^2$ then $(-x_1, -x_2) \in X_7 \cap \Re_-^2$. Clearly the origin $(0, 0)$ belongs to the convex hull of $(x_1, x_2)$ and $(-x_1, -x_2)$, yet does not belong to $X_7$. Thus $X_7$ is not convex.

(iv) Finally a set of the form

$$X_8 = \{(x_1, x_2) \in \Re_+^2 : x_2 \leq x_1^\alpha \text{ for } \alpha \in (0, 1)\}$$

is also convex. See Fig. 3.13d.

**Fig. 3.13c** Example 3.2(iii)

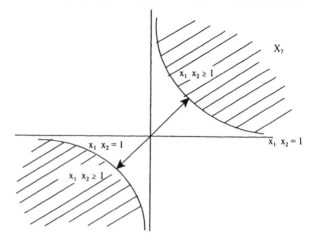

**Fig. 3.13d** Example 3.2(iv)

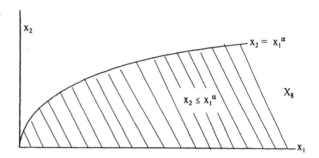

**Fig. 3.13e** The Euclidean ball

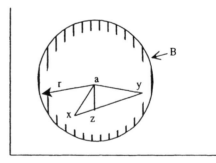

*Example 3.3* (i) Consider the set $B = \{(x_1, x_2) \in \Re^2 : (x_1 - a_1)^2 + (x_2 - a_2)^2 \leq r^2\}$. See Fig. 3.13e. This is the closed ball centered on $(a_1, a_2) = a$, of radius $r$. Suppose that $x, y \in B$ and $z = \lambda x + (1 - \lambda) y$ for $\lambda \in [0, 1]$.

Let $\| \ \|$ stand for the Euclidean norm. Then $x, y$ both satisfy $\|x - a\| \leq r, \|y - a\| \leq r$. But $\|z - a\| \leq \lambda \|x - a\| + (1 - \lambda) \|y - a\|$. Thus $\|z - a\| \leq r$ and so $z \in B$. Hence $B$ is convex. Moreover $B$ is a closed and bounded subset of $\Re^2$ and is thus compact. For a general norm on $\Re^n$, the closed ball $B = \{x \in \Re^n : \|x - a\| \leq r\}$

## 3.4 Convexity

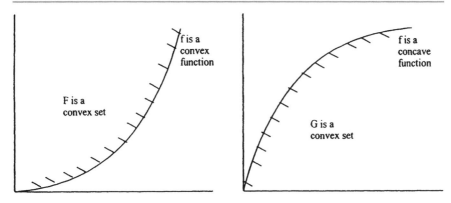

**Fig. 3.14** Convex and concave functions

will be compact and convex. In particular, if the Euclidean norm is used, then $B$ is strictly convex.

(ii) In the next section we define the hyperplane $H(\rho, \alpha)$ *normal* to a vector $\rho$ in $\Re^n$ to be $\{x \in \Re^n : \langle \rho, x \rangle = \alpha\}$ where $\alpha$ is some real number. Suppose that $x, y \in H(\rho, x)$.

Now

$$\langle \rho, \lambda x + (1 - \lambda)y \rangle = \langle \lambda(\rho, x) + (1 - \lambda)(\rho, y) \rangle$$
$$= \alpha, \quad \text{whenever } \lambda \in [0, 1].$$

Thus $H(\rho, \alpha)$ is a convex set. We also define the closed half-space $H_+(\rho, \alpha)$ by $H_+(\rho, \alpha) = \{x \in \Re^n : \langle \rho, x \rangle \geq \alpha\}$. Clearly if $x, y \in H_+(\rho, \alpha)$ then $\langle \rho, \lambda x + (1 - \lambda)y \rangle = (\lambda \langle \rho, x \rangle + (1 - \lambda)\langle \rho, y \rangle) \geq \alpha$ and so $H_+(\rho, \alpha)$ is also convex.

Notice that if $B$ is the compact convex ball in $\Re^n$ then there exists some $\rho \in \Re^n$ and some $\alpha \in \Re$ such that $B \subset H_+(\rho, \alpha)$.

If $A$ and $B$ are two convex sets in $\Re^n$ then $A \cap B$ must also be a convex set, while $A \cup B$ need not be. For example the union of two disjoint convex sets will not be convex.

We have called a function $f : Y \to \Re$ convex on $Y$ iff $f(\lambda x + (1 - \lambda)y) \leq \lambda f(x) + (1 - \lambda)f(y)$ for $x, y \in Y$.

Clearly this is equivalent to the requirement that the set $F = \{(z, x) \in \Re \times Y : z \geq f(x)\}$ is convex. (See Fig. 3.14.)

To see this suppose $(z_1, x_1)$ and $(z_2, x_2) \in F$.

Then $\lambda(z_1, x_1) + (1 - \lambda)(z_2, x_2) \in F$ iff $\lambda z_1 + (1 - \lambda)z_2 \geq f(\lambda x_1 + (1 - \lambda)x_2)$. But $(f(x_1), x_1)$ and $(f(x_2), x_2) \in F$, and so $\lambda z_1 + (1 - \lambda)z_2 \geq \lambda f(x_1) + (1 - \lambda)f(x_2) \geq f(\lambda x_1 + (1 - \lambda)x_2)$ for $\lambda \in [0, 1]$.

In the same way $f$ is concave on $Y$ iff $G = \{(z, x) \in \Re \times Y : z \leq f(x)\}$ is convex. If $f : Y \to \Re$ is concave then the function $(-f) : Y \to \Re$, given by $(-f)(x) = -f(x)$, is convex and vice versa.

To see this note that if $z \leq f(x)$ then $-z \geq -f(x)$, and so $G = \{(z, x) \in \Re \times Y : z \leq f(x)\}$ is convex implies that $F = \{(z, x) \in \Re \times Y : z \geq (-f)(x)\}$ is convex.

Finally $f$ is quasi-concave on $Y$ iff, for all $z \in \Re$, the set $G(z) = \{x \in Y : z \leq f(x)\}$ is convex.

Notice that $G(x)$ is the image of $G$ under the projection mapping $p_z : \Re \times Y \to Y : (z, x) \to x$. Since the image of a convex set under a projection is convex, clearly $G(z)$ is convex for any $z$ whenever $G$ is convex. As we know already this means that a concave function is quasi-concave. We now apply these observations.

*Example 3.4*
(i) Let $f : \Re \to \Re$ by $x \to x^2$.
   As Example 3.2(i) showed, the set $F = \{(x, z) \in \Re \times \Re : z \geq f(x) = x^2\}$ is convex. Hence $f$ is convex.
(ii) Now let $f : \Re \to \Re$ by $x \to x^3$. Example 3.2(ii) showed that the set $F = \{(x, z) \in \Re_+ \times \Re : z \geq f(x) = x^3\}$ is convex and so $f$ is convex on the convex set $\Re_+ = \{x \in \Re : x \geq 0\}$.
   On the other hand $F = \{(x, z) \in \Re_- \times \Re : z \leq f(x) = x^3\}$ is convex and so $f$ is *concave* on the convex set $\Re_- = \{x \in \Re : x \leq 0\}$.
(iii) Let $f : \Re \to \Re$ by $x \to \frac{1}{x}$. By Example 3.2(iii) the set $F = \{(x, z) \in \Re_+ \times \Re : z \geq f(x) = \frac{1}{x}\}$ is convex, and so $f$ is *convex* on $\Re_+$ and *concave* on $\Re_-$.
(iv) Let $f(x) = x^\alpha$ where $0 < \alpha < 1$. Then $F = \{(x, z) \in \Re_+ \times \Re : z \leq f(x) = x^\alpha\}$ is convex, and so $f$ is concave.
(v) Consider the exponential function $\exp : \Re \to \Re : x \to e^x$. Figure 3.15a demonstrates that the exponential function is convex. Another way of showing this is to note that $e^x > f(x)$ for any geometric function $f : x \to x^r$ for $r > 1$, for any $x \in \Re_+$.
   Since the geometric functions are convex, so is $e^x$. On the other hand as Fig. 3.15b shows the function $\log_e : \Re_+ \to \Re$, inverse to $\exp$, is concave.
(vi) Consider now $f : \Re^2 \to \Re : (x, y) \to xy$. Just as in Example 3.2(iii) the set $\{(x, y) \in \Re_+^2 : xy \geq t\} = \Re_+^2 \cap f^{-1}[t, \infty)\}$ is convex and so $f$ is a quasi-concave function on $\Re_+^2$. Similarly $f$ is quasi-concave on $\Re_-^2$. However $f$ is not quasi-concave on $\Re^2$.
(vii) Let $f : \Re^2 \to \Re : (x_1, x_2) \to r^2 - (x_1 - a_1)^2 - (x_2 - a_2)^2$. Since the function $g(x) = x^2$ is convex, $(-g)(x) = -x^2$ is concave, and so clearly $f$ is a concave function. Moreover it is obvious that $f$ has a supremum in $\Re^2$ at the point $(x_1, x_2) = (a_1, a_2)$. On the other hand the functions in Examples 3.4(iv) to (vi) are monotonically increasing.

### 3.4.3 Separation Properties of Convex Sets

Let $X$ be a vector space of dimension $n$ with a scalar product $\langle, \rangle$. Define $H(\rho, \alpha) = \{x \in X : \langle \rho, x \rangle = \alpha\}$ to be the hyperplane in $X$ *normal* to the vector $\rho \in \Re^n$. It should be clear that $H(\rho, \alpha)$ is an $(n-1)$ dimensional plane displaced some distance from the origin. To see this suppose that $x = \lambda \rho$ belongs to $H(\rho, \alpha)$. Then $\langle \rho, \lambda \rho \rangle =$

## 3.4 Convexity

**Fig. 3.15** The exponential and logarithmic functions

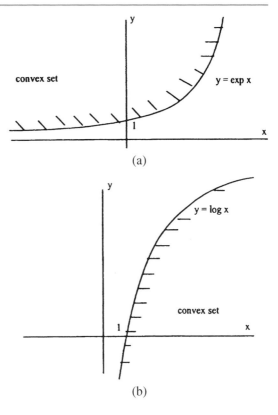

$\lambda \|\rho\|^2 = \alpha$. Thus $\lambda = \frac{\alpha}{\|\rho\|^2}$. Hence the length of $x$ is $\|x\| = |\lambda| \|\rho\| = \frac{|\alpha|}{\|\rho\|}$ and so $x = \pm \frac{|\alpha|}{\|\rho\|^2} \rho$.

Clearly if $y = \lambda \rho + y_0$ belongs to $H(\rho, \alpha)$ then $\langle \rho, y \rangle = \langle \rho, \lambda \rho + y_0 \rangle = \alpha + \langle \rho, y_0 \rangle$ and so $\langle \rho, y_0 \rangle = 0$.

Thus any vector $y$ in $H(\rho, \alpha)$ can be written in the form $y = \lambda \rho + y_0$ where $y_0$ is orthogonal to $\rho$. Since there exist $(n-1)$ linearly independent vectors $y_1, \ldots, y_{n-1}$, all orthogonal to $\rho$, any vector $y \in H(\rho, \alpha)$ can be written in the form

$$y = \lambda \rho + \sum_{i=1}^{n-1} a_i y_i,$$

where $\lambda \rho$ is a vector of length $\frac{|\alpha|}{\|\rho\|}$. Now let $\{\rho^\perp\} = \{x \in \mathfrak{R}^n : \langle \rho, x \rangle = 0\}$. Clearly $\{\rho^\perp\}$ is a vector subspace of $\mathfrak{R}^n$, of dimension $(n-1)$ through the origin.

Thus $H(\rho, \alpha) = \lambda \rho + \{\rho^\perp\}$ has the form of an $(n-1)$-dimensional vector subspace displaced a distance $\frac{|\alpha|}{\|\rho\|}$ along the vector $\rho$. Clearly if $\rho_1$ and $\rho_2$ are colinear vectors (*i.e.*, $\rho_2 = a\rho_1$ for some $a \in \mathfrak{R}$) then $\{\rho^\perp\} = \{(\rho_2)^\perp\}$.

Suppose that $\frac{\alpha_1 \rho_1}{\|\rho\|^2} = \frac{\alpha_2 \rho_2}{\|\rho\|^2}$, then both $H(\rho_1, \alpha_1)$ and $H(\rho_2, \alpha_2)$ contain the same point and are thus identical. Thus $H(\rho_1, \alpha_1) = H(a\rho_1, a\alpha_1) = H(\frac{\rho_1}{\|\rho_2\|}, \frac{\alpha_1}{\|\rho_1\|})$.

The hyperplane $H(\rho, \alpha)$ separates $X$ into two *closed* half-spaces:

$$H_+(\rho, \alpha) = \{x \in X : \langle \rho, x \rangle \geq \alpha\}, \quad \text{and} \quad H_-(\rho, \alpha) = \{x \in X : \langle \rho, x \rangle \leq \alpha\}.$$

We shall also write

$$H_+^0(\rho, \alpha) = \{x \in X : \langle \rho, x \rangle > \alpha\}, \quad \text{and} \quad H_-^0(\rho, \alpha) = \{x \in X : \langle \rho, x \rangle < \alpha\}$$

for the open half-spaces formed by taking the interiors of $H_+(\rho, \alpha)$ and $H_-(\rho, \alpha)$, in the case $\rho \neq 0$.

**Lemma 3.19** *Let $Y$ be a non-empty compact convex subset of a finite dimensional real vector space $X$, and let $x$ be a point in $X \backslash Y$. Then there is a hyperplane $H(\rho, \alpha)$ through a point $y_0 \in Y$ such that*

$$\langle \rho, x \rangle < \alpha = \langle \rho, y_0 \rangle \leq \langle \rho, y \rangle \quad \text{for all } y \in Y.$$

*Proof* As in Lemma 3.17 let $f_x : Y \to \Re$ be the function $f_x(y) = \|x - y\|$, where $\| \|$ is the norm induced from the scalar product $\langle, \rangle$ in $X$.

By Lemma 3.18 there exists a point $y_0 \in Y$ such that $\|x - y_0\| = \inf(f_x, Y) = d(x, Y)$. Thus $\|x - y_0\| \leq \|x - y\|$ for all $y \in Y$. Now define $\rho = y_0 - x$ and $\alpha = \langle \rho, y_0 \rangle$. Then

$$\langle \rho, x \rangle = \langle \rho, y_0 \rangle - \langle \rho, (y_0 - x) \rangle = \langle \rho, y_0 \rangle - \|\rho\|^2 < \langle \rho, y_0 \rangle.$$

Suppose now that there is a point $y \in Y$ such that $\langle \rho, y_0 \rangle > \langle \rho, y \rangle$. By convexity, $w = \lambda y + (1 - \lambda) y_0 \in Y$, where $\lambda$ belongs to the interval $(0, 1)$. But

$$\|x - y_0\|^2 - \|x - w\|^2 = \|x - y_0\|^2 - \|x - \lambda y - y_0 + \lambda y_0\|^2$$
$$= 2\lambda \langle \rho, (y_0 - y) \rangle - \lambda^2 \|y - y_0\|^2.$$

Now $\langle \rho, y_0 \rangle > \langle \rho, y \rangle$ and so, for sufficiently small $\lambda$, the right hand side is positive. Thus there exists a point $w$ in $Y$, close to $y_0$, such that $\|x - y_0\| > \|x - w\|$. But this contradicts the assumption that $y_0$ is the nearest point in $Y$ to $x$. Thus $\langle \rho, y \rangle \geq \langle \rho, y_0 \rangle$ for all $y \in Y$. Hence $\langle \rho, x \rangle < \alpha = \langle \rho, y_0 \rangle \leq \langle \rho, y \rangle$ for all $y \in Y$. □

Note that the point $y_0$ belongs to the hyperplane $H(\rho, \alpha)$, the set $Y$ belongs to the closed half-space $H_+(\rho, \alpha)$, while the point $x$ belongs to the *open* half-space

$$H_-^0(\rho, \alpha) = \{z \in X : \langle \rho, z \rangle < \alpha\}.$$

Thus the hyperplane *separates* the point $x$ from the compact convex set $Y$ (see Fig. 3.16).

### 3.4 Convexity

**Fig. 3.16** A separating hyperplane

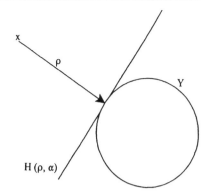

**Fig. 3.17** Non-existance of a separating hyperplane

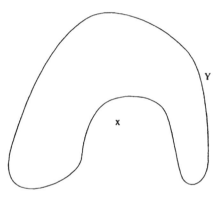

While convexity is necessary for the proof of this theorem, the compactness requirement may be weakened to $Y$ being closed. Suppose however that $Y$ is an open set. Then it is possible to choose a point $x$ outside $Y$, which is, nonetheless, an accumulation point of $Y$ such that $d(x, Y) = 0$.

On the other hand if $Y$ is compact but not convex, then a situation such as Fig. 3.17 is possible. Clearly no hyperplane separates $x$ from $Y$.

If $A$ and $B$ are two sets, and $H(\rho, \alpha) = H$ is a hyperplane such that $A \subseteq H_-(\rho, \alpha)$ and $B \subseteq H_+(\rho, \alpha)$ then say that $H$ *weakly separates* $A$ and $B$. If $H$ is such that $A \subset H_-(\rho, \alpha)$ and $B \subset H_+(\rho, \alpha)$ then say $H$ *strongly separates* $A$ and $B$. Note in the latter case that it is necessary that $A \cap B = \Phi$.

In Lemma 3.19 we found a hyperplane $H(\rho, \alpha)$ such that $\langle \rho, x \rangle < \alpha$. Clearly it is possible to find $\alpha_- < \alpha$ such that $\langle \rho, x \rangle < \alpha_-$.

Thus the hyperplane $H(\rho, \alpha_-)$ *strongly* separates $x$ from the compact convex set $Y$.

If $Y$ is convex but not compact, then it is possible to find a hyperplane $H$ that weakly separates $X$ from $Y$.

We now extend this result to the separation of convex sets.

**Separating Hyperplane Theorem** *Suppose that A and B are two disjoint non-empty convex sets of a finite dimensional vector space X. Then there exists a hyperplane H that weakly separates A from B. If both A and B are compact then H strongly separates A from B.*

*Proof* Since $A$ and $B$ are convex the set $A - B = \{w \in X : w = a - b \text{ where } a \in A, b \in B\}$ is also convex.

To see this suppose $a_1 - b_1$ and $a_2 - b_2 \in A - B$. Then $\lambda(a_1 - b_1) + (1 - \lambda)(a_2 - b_2) = [\lambda a_1 + (1 - \lambda)a_2] + [\lambda b_1 + (1 - \lambda)b_2] \in A - B$.

Now $A \cap B = \Phi$. Thus there exists no point in both $A$ and $B$, and so $\underline{0} \notin A - B$. By Lemma 3.19, there exists a hyperplane $H(-\rho, \underline{0})$ weakly separating $\underline{0}$ from $A - B$, i.e., $\langle \rho, \underline{0} \rangle \leq \langle \rho, w \rangle$ for all $w \in B - A$. But then $\langle \rho, a \rangle \leq \langle \rho, b \rangle$ for all $a \in A, b \in B$.

Choose $\alpha \in [\sup_{a \in A}\langle \rho, a \rangle, \inf_{b \in B}\langle \rho, b \rangle]$. In the case that $A, B$ are non-compact, it is possible that

$$\sup_{a \in A}\langle \rho, a \rangle = \inf_{b \in B}\langle \rho, b \rangle.$$

Thus $\langle \rho, a \rangle \leq \alpha \leq \langle \rho, b \rangle$ and so $H(\rho, \alpha)$ weakly separates $A$ and $B$.

Consider now the case when $A$ and $B$ are compact.

The function $\rho^* : X \to \Re$ given by $\rho^*(x) = \langle \rho, x \rangle$ is clearly continuous. By Lemma 3.16, since both $A$ and $B$ are compact, there exist points $\overline{a} \in A$ and $\overline{b} \in B$ such that

$$\langle \rho, \overline{a} \rangle = \sup_{a \in A}\langle \rho, a \rangle \quad \text{and} \quad \langle \rho, \overline{b} \rangle = \inf_{b \in B}\langle \rho, b \rangle.$$

If $\sup_{a \in A}\langle \rho, a \rangle = \inf_{b \in B}\langle \rho, b \rangle$, then $\langle \rho, \overline{a} \rangle = \langle \rho, \overline{b} \rangle$, and so $\overline{a} = \overline{b}$, contradicting $A \cap B = \Phi$.

Thus $\langle \rho, \overline{a} \rangle < \langle \rho, \overline{b} \rangle$ and we can choose $\alpha$ such that $\langle \rho, a \rangle \leq \langle \rho, \overline{a} \rangle < \alpha < \langle \rho, \overline{b} \rangle \leq \langle \rho, b \rangle$ for all $a, b$ in $A, B$.

Thus $H(\rho, \alpha)$ strongly separates $A$ and $B$. (See Fig. 3.18b.) □

*Example 3.5* Hildenbrand and Kirman (1976) have applied this theorem to find a price vector which supports a final allocation of economic resources. Consider a society $M = \{1, \ldots, m\}$ in which each individual $i$ has an initial *endowment* $e_i = (e_{i1}, \ldots, e_{in}) \in \Re^n$ of $n$ commodities. At price vector $p = (p_1, \ldots, p_n)$, the budget set of individual $i$ is

$$B_i(p) = \{x \in \Re^n : \langle p, x \rangle \leq \langle p, e_i \rangle\}.$$

Each individual has a preference relation $P_i$ on $\Re^n \times \Re^n$, and at the price vector $p$ the *demand* $D_i(p)$ by $i$ is the set

$$\{x \in B_i(p) : y P_i x \text{ for no } y \in B_i(p)\}.$$

### 3.4 Convexity

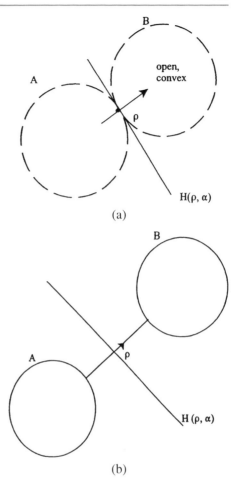

**Fig. 3.18** a Weak separation.
b Strong separation

Let $f_i = (f_{i1}, \ldots, f_{in}) \in \Re^n$ be the final allocation to individual $i$, for $i = 1, \ldots, m$. Suppose there exists a price vector $p = (p_1, \ldots, p_n)$ with the property (∗) $x P_i f_i \Rightarrow \langle p, x \rangle > \langle p, e_i \rangle$. Then this would imply that $f_i \in B_i(p) \Rightarrow f_i \in D_i(p)$. If property (∗) holds at some price vector $p$, for each $i$, then $f_i \in D_i(p)$ for each $i$.

To show existence of such a price vector, let

$$\pi_i = P_i(f_i) - e_i \in \Re^n.$$

Here as before $P_i(f_i) = \{x \in \Re^n : x P_i f_i\}$. Suppose that there exists a hyperplane $H(p, 0)$ strongly separating 0 from $\pi_i$. In this case $0 < \langle p, x - e_i \rangle$ for all $x \in P_i(f_i)$. But this is equivalent to $\langle p, x \rangle > \langle p, e_i \rangle$ for all $x \in P_i(f_i)$.

Let $\pi = \text{Con}[\bigcup_{i \in N} \pi_i]$ be the convex hull of the sets $\pi_i$, $i \in M$. Clearly if $\underline{0} \notin \pi$ and there is a hyperplane $H(p, 0)$ strongly separating $\underline{0}$ from $\pi$, then $p$ is a price vector which supports the final allocation $f_1, \ldots, f_n$.

## 3.5 Optimisation on Convex Sets

A key notion underlying economic theory is that of the maximisation of an objective function subject to one or a number of constraints. The most elementary case of such a problem is the one addressed by the Weierstrass Theorem: if $f : X \to \Re$ is a continuous function, and $Y$ is a compact constraint set then there exists some point $\bar{y}$ such that $f(\bar{y}) = \sup(f, Y)$. Here $\bar{y}$ is a *maximum* point of $f$ on $Y$.

Using the Separating Hyperplane Theorem we can extend this analysis to the optimisation of a convex preference correspondence on a compact convex constraint set.

### 3.5.1 Optimisation of a Convex Preference Correspondence

Suppose that $Y$ is a compact, convex constraint set in $\Re^n$ and $P : \Re^n \to \Re^n$ is a preference correspondence which is convex (*i.e.*, $P(x)$ is convex for all $x \in \Re^n$). A choice for $P$ on $Y$ is a point $\bar{y} \in Y$ such that $P(\bar{y}) \cap Y = \Phi$.

We shall say that $P$ is *non-satiated* in $\Re^n$ iff for no $y \in \Re^n$ is it the case that $P(y) = \Phi$. A sufficient condition to ensure non-satiation for example is the assumption of monotonicity, *i.e.*, $x > y$ (where as before this means $x_i > y_i$, for each of the coordinates $x_i, y_i, i = 1, \ldots, n$) implies that $x \in P(y)$.

Say that $P$ is *locally non-satiated* in $\Re^n$ iff for each $y \in \Re^n$ and any neighbourhood $U_y$ of $y$ in $\Re^n$, then $P(y) \cap U_y \neq \Phi$.

Clearly monotonicity implies local non-satiation implies non-satiation.

Suppose that $y$ belongs to the interior of the compact constraint set $Y$. Then there is a neighbourhood $U_y$ of $y$ within $Y$. Consequently $P(y) \cap U_y \neq \Phi$ and so $y$ cannot be a choice from $Y$. On the other hand, since $Y$ is compact it is closed, and so if $y$ belongs to the boundary $\delta Y$ of $Y$, it belongs to $Y$ itself. By definition if $y \in \delta Y$ then any neighbourhood $U_y$ of $y$ intersects $\Re^n \backslash Y$. Thus when $P(y) \subset \Re^n \backslash Y$, $y$ will be a choice from $Y$. Alternatively if $y$ is a choice of $P$ from $Y$, then $y$ must belong to the boundary of $P$.

**Lemma 3.20** *Let $Y$ be a compact, convex constraint set in $\Re^n$ and let $P : \Re^n \to \Re^n$ be a preference correspondence which is locally non-satiated, and is such that, for all $x \in \Re^n$, $P(x)$ is open and convex. Then $\bar{y}$ is a choice of $P$ from $Y$ iff there is a hyperplane $H(p, \alpha)$ through $\bar{y}$ in $Y$ which separates $Y$ from $P(\bar{y})$ in the sense that*

$$\langle p, y \rangle \leq \alpha = \langle p, \bar{y} \rangle < \langle p, x \rangle \quad \text{for all } y \in Y \text{ and all } x \in P(\bar{y}).$$

*Proof* Suppose that the hyperplane $H(p, \alpha)$ contains $\bar{y}$ and separates $Y$ from $P(\bar{y})$ in the above sense. Clearly $\bar{y}$ must belong to the boundary of $Y$. Moreover $\langle p, y \rangle < \langle p, x \rangle$ for all $y \in Y, x \in P(\bar{y})$. Thus $Y \cap P(\bar{y}) = \Phi$ and so $\bar{y}$ is the choice of $P$ from $Y$.

On the other hand suppose that $\bar{y}$ is a choice. Then $P(\bar{y}) \cap Y = \Phi$.

## 3.5 Optimisation on Convex Sets

Moreover the local non-satiation property, $P(\bar{y}) \cap U_{\bar{y}} \neq \Phi$ for $U_{\bar{y}}$ a neighborhood of $\bar{y}$ in $\Re^n$, guarantees that $\bar{y}$ must belong to the boundary of $Y$. Since $Y$ and $P(\bar{y})$ are both convex, there exists a hyperplane $H(p, \alpha)$ through $\bar{y}$ such that

$$\langle p, y \rangle \leq \alpha = \langle p, \bar{y} \rangle \leq \langle p, x \rangle$$

for all $y \in Y$, all $x \in P(\bar{y})$. But $P(\bar{y})$ is open, and so the last inequality can be written $\langle p, \bar{y} \rangle < \langle p, x \rangle$. □

Note that if either the constraint set, $Y$, or the correspondence $P$ is such that $P(y)$ is strictly convex, for all $y \in \Re^n$, then the choice $\bar{y}$ is unique.

If $f : \Re^n \to \Re$ is a concave or quasi-concave function then application of this lemma to the preference correspondence $P : \Re^n \to \Re$, where $P(x) = \{y \in \Re^n : f(y) > f(x)\}$, characterises the maximum point $\bar{y}$ of $f$ on $Y$. Here $\bar{y}$ is a maximum point of $f$ on $Y$ if $f(\bar{y}) = \sup(f, Y)$. Note that local non-satiation of $P$ requires that for any point $x$ in $\Re^n$, and any neighbourhood $U_x$ of $x$ in $\Re^n$, there exists $y \in U_x$ such that $f(y) > f(x)$.

The vector $p = (p_1, \ldots, p_n)$ which characterises the hyperplane $H(p, \alpha)$ is called in economic applications the vector of *shadow prices*. The reason for this will become clear in the following example.

*Example 3.6* As an example suppose that optimal use of $(n-1)$ different inputs $(x_1, \ldots, x_{n-1})$ gives rise to an output $y$, say, where $y = y(x_1, \ldots, x_{n-1})$. Any $n$ vector $(x_1, \ldots, x_{n-1}, x_n)$ is *feasible* as long as $x_n \leq y(x_1, \ldots, x_{n-1})$. Here $x_n$ is the output. Write $g(x_1, \ldots, x_{n-1}, x_n) = y(x_1, \ldots, x_{n-1}) - x_n$. Then a vector $x = (x_1, \ldots, x_{n-1}, x_n)$ is feasible iff $g(x) \geq 0$.

Suppose now that $y = y(x_1, \ldots, x_{n-1})$ is a concave function in $x_1, \ldots, x_{n-1}$. Then clearly the set $G = \{x \in \Re^n : g(x) \geq 0\}$ is a convex set. Now let $\pi(x_1, \ldots, x_{n-1}, x_n) = -\sum p_i x_i + p_n x_n$ be the *profit function* of the producer, when prices for inputs and outputs are given *exogenously* by $(-p_1, \ldots, -p_{n-1}, p_n)$. Again let $P : \Re^n \to \Re^n$ be the preference correspondence $P(x) = \{z \in \Re^n : \pi(z) > \pi(x)\}$. Since for each $x$, $P(x)$ is convex, and locally non-satiated, there is a choice $\bar{x}$ and a hyperplane $H(\rho, \alpha)$ separating $P(\bar{x})$ from $G$.

Indeed it is clear from the construction that $P(\bar{x}) \subset H^0_+(\rho, \alpha)$ and $G \subset H_-(\rho, \alpha)$.

Moreover the hyperplane $H(\rho, \alpha)$ must coincide with the set of points $\{x \in \Re^n : \pi(x) = \pi(\bar{x})\}$. Thus the hyperplane $H(\rho, \alpha)$ has the form

$$\{x \in \Re^n : \langle p, x \rangle = \pi(\bar{x})\}$$

and so may be written $H(p, \pi(x))$. Note that the intercept on the $x_n$ axis is $\frac{\pi(\bar{x})}{p_n}$ while the distance of the hyperplane from the origin is $\frac{\pi(\bar{x})}{\|p\|} = \frac{\pi(\bar{x})}{\sqrt{\sum p_i^2}}$. Thus the intercept gives the profit measured in units of output, while the distance from the

**Fig. 3.19** Separation from a feasible set

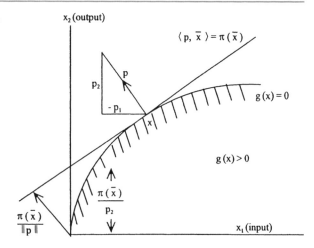

origin of the profit hyperplane gives the profit in terms of a normalized price vector ($\|p\|$).

Figure 3.19 illustrates the situation with one input ($x_1$) and one output ($x_2$).

Precisely the same analysis can be performed when optimal production is characterised by a general production function $F : \mathfrak{R}^n \to \mathfrak{R}$.

Here $x_1, \ldots, x_m$ are inputs, with prices $-p_1, \ldots, -p_m$ and $x_{m+1}, \ldots, x_n$ are outputs with prices $p_{m+1}, \ldots, p_n$. Let $p = (-p_1, \ldots, -p_m, -p_{m+1}, \ldots, p_n) \in \mathfrak{R}^n$.

Define $F$ so that a vector $x \in \mathfrak{R}^n$ is *feasible* iff $F(x) \geq 0$. Note that we also need to restrict all inputs and outputs to be non-negative. Therefore define

$$\mathfrak{R}^n_+ = \{x : x_i \geq 0 \text{ for } i = 1, \ldots, n\}.$$

Assume that the feasible set (or *production set*)

$$G = \{x \in \mathfrak{R}^n_+ : F(x) \geq 0\} \quad \text{is convex.}$$

As before let $P : \mathfrak{R}^n \to \mathfrak{R}^n$ where $P(x) = \{z \in \mathfrak{R}^n : \pi(z) > \pi(x)\}$. Then the point $\bar{x}$ is a choice of $P$ from $G$ iff $\bar{x}$ maximises the profit function

$$\pi(x) = \sum_{j=1}^{n-m} p_{m+j}\, x_{m+j} - \sum_{j=1}^{m} p_j x_j.$$

By the previous example $\bar{x}$ is a choice iff the hyperplane $H(p, \pi(\bar{x}))$ separates $P(\bar{x})$ and $G$: i.e., $P(\bar{x}) \subset H^0_+(p, \pi(\bar{x}))$ and $G \subset H_-(p, \pi(\bar{x}))$.

In the next chapter we shall use this optimality condition to characterise the choice $\overline{m}$ more fully in the case when $F$ is "smooth".

*Example 3.7* Consider now the case of a *consumer* maximising a preference correspondence $P : \mathfrak{R}^n \to \mathfrak{R}^n$ subject to a budget constraint $B(p)$ which is dependent on a set of exogenous prices $p_1, \ldots, p_n$.

## 3.5 Optimisation on Convex Sets

**Fig. 3.20** Example 3.7

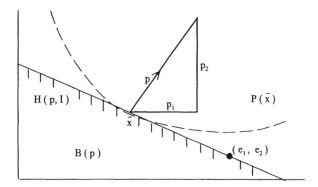

For example the consumer may hold an initial set of endowments $(e_1, \ldots, e_n)$, so let

$$I = \sum_{i=1}^{n} p_i e_i = \langle p, e \rangle.$$

The budget set is then

$$B(p) = \{x \in \Re_+^n : \langle p, x \rangle \leq I\},$$

where for convenience we assume the consumer only buys a non-negative amount of each commodity. Suppose that $P$ is monotonic, and $P(x)$ is open, convex for all $x \in \Re^n$. As before $\bar{x}$ is a choice from $B(p)$ iff there is a hyperplane $H(\rho, \alpha)$ separating $P(\bar{x})$ from $B(p)$.

Under these conditions the choice must belong to the upper boundary of $B(p)$ and so satisfy $(p, \bar{x}) = (p, e) = I$. Thus the hyperplane has the form $H(p, I)$, and so the optimality condition is $P(\bar{x}) \subset H_+^0(p, I)$ and $B(p) \subset H_-(p, I)$; i.e., $(p, x) \leq I = (p, \bar{x}) < (p, y)$ for all $x \in B(p)$ and all $y \in P(\bar{x})$.

Figure 3.20 illustrates the situation with two commodities $x_1$ and $x_2$.

In the next chapter we use this optimality condition to characterise a choice when preference is given by a smooth utility function $f : \Re^n \to \Re$.

In the previous two examples we considered
1. optimisation of a profit function, which is determined by exogenously given prices, subject to a fixed production constraint, and
2. optimisation of a fixed preference correspondence, subject to a budget constraint, which is again determined by exogenous prices.

Clearly at a given price vector each producer will "demand" a certain input vector and supply a particular output vector, so that the combination is his choice in the environment determined by $p$. In the same way a consumer will respond to a price vector $p$ by demanding optimal amounts of each commodity, and possibly supplying other commodities such as labor, or various endowments. In contrast to Example 3.7, regard all prices as positive, and consider a commodity $x_j$ demanded by an agent $i$ as an input to be negative, and positive when supplied as an output.

Let $\bar{x}_{ij}(p)$ be the optimal demand or supply of commodity $j$ by agent $i$ at the price vector $p$, with $m$ agents and $n$ commodities, then *market equilibrium* of supply and demand in the economy occurs when $\sum_{i=1}^{m}\sum_{j=1}^{n}\bar{x}_{ij}(p)=0$. A price vector which leads to market equilibrium in demand and supply is called an *equilibrium price vector*.

*Example 3.8* To give a simple example, consider two agents. The first agent controls a machine which makes use of labor, $x$, to produce a single output $y$. Regard $x \in (-\infty, 0]$ and consider a price vector $p \in \mathfrak{R}_+^2$, where $p=(w,r)$ and $w$ is the price of labor, and $r$ the price of the output. An output is feasible from Agent One iff $F(x,y) \geq 0$.

Agent Two is the only supplier of labor, but is averse to working. His preference is described by a quasi-concave utility function $f: \mathfrak{R}^2 \to \mathfrak{R}$ and we restrict attention to a domain

$$D = \{(x,y) \in \mathfrak{R}^2 : x \leq 0, y \geq 0\}.$$

Assume that $f$ is monotonic, i.e., if $x_1 < x_2$ and $y_1 < y_2$ then $f(x_1, y_1) < f(x_2, y_2)$. The budget constraint of Agent Two at $(w, r)$ is therefore

$$B(w, r) = \{(x_1, y_2) \in D : ry_2 \leq w|x|\},$$

where $|x|$ is the quantity of labor supplied, and $y_2$ is the amount of commodity $y$ consumed. Profit for Agent One is $\pi(x,y) = ry - wx$, and we shall assume that this agent then consumes an amount $y_1 = \frac{\pi(x,y)}{r}$ of commodity $y$.

For equilibrium of supply and demand at prices $(\bar{w}, \bar{r})$
1. $\bar{y} = \bar{y}_1 + \bar{y}_2$;
2. $(\bar{x}, \bar{y})$ maximises $\pi(x, y) = \bar{r}y - \bar{w}x$ subject to $F(x, y) \geq 0$;
3. $(\bar{x}, \bar{y}_2)$ maximises $f(x, y_2)$ subject to $\bar{r}y_2 = \bar{w}x$.

At any point $(x, y) \in D$, and vector $(w, r)$ define

$$P(x, y) = \{(x', y') \in D : f(x', y' - y_1) > f(x, y - y_1)\}$$

where as above $y_1 = \frac{ry - wx}{r}$ is the amount of commodity $y$ consumed by the producer. Thus $P(x, y)$ is the preference correspondence of Agent One displaced by the distance $y_1$ up the $y$-axis.

Figure 3.21 illustrates that it is possible to find a vector $(w, r)$ such that $H = H((w, r), \pi(\bar{x}, \bar{y}))$ separates $P(\bar{x}, \bar{y})$ and the production set

$$G = \{(x, y) \in D : F(x, y) \geq 0\}.$$

As in Example 3.6, the intersect of $H$ with the $y$-axis is $\bar{y}_1 = \frac{\pi(\bar{x}, \bar{y})}{r}$, the consumption of $y$ by Agent One.

The hyperplane $H$ is the set

$$\{(x, y) : ry + wx = r\bar{y}_1\}.$$

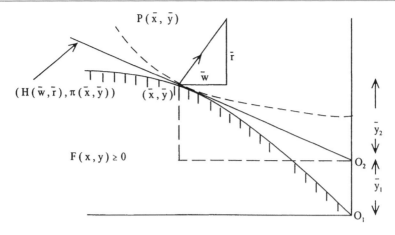

**Fig. 3.21** Example 3.8

Hence for all $(x, y) \in H$, $(x, y - \overline{y}_1)$ satisfies $r(y - \overline{y}_1) + wx = 0$. Thus $H$ is the boundary of the second agent's budget set $\{(x, y_2) : ry_2 + wx = 0\}$ displaced up the $y$-axis by $\overline{y}_1$. Consequently $\overline{y} = \overline{y}_1 + \overline{y}_2$ and so $(\overline{x}, \overline{y})$ is a market equilibrium.

Note that the hyperplane separation satisfies:

$$py - rx \leq \pi(\overline{x}, \overline{y}) < ry' - wx'$$

for all $(x, y) \in G$, and all $(x', y') \in P(\overline{x}, \overline{y})$.

As above $(x', y') \in P(\overline{x}, \overline{y})$ iff $f(x', y' - \overline{y}_1) > f(\overline{x}, \overline{y}_2)$. So the right hand side implies that if $(x', y') \in P(\overline{x}, \overline{y})$, then $ry' - wx' > \pi(\overline{x}, \overline{y})$.

Since $y_2' = y' - \overline{y}_1, ry_2' - wx' > 0$ or $(x', y' - \overline{y}_1) \in H_+^0((\overline{w}, \overline{r}), 0)$ and so $(x', y' - \overline{y}_1)$ is infeasible for Agent Two.

Finally $(\overline{x}, \overline{y})$ maximises $\pi(x, y)$ subject to $(x, y) \in G$, and so $(\overline{x}, \overline{y}) \in G$ and so $(\overline{x}, \overline{y})$ results from optimising behaviour by both agents at the price vector $(w, r)$.

As this example illustrates it is possible to show existence of a market equilibrium in economies characterised by compact convex constraint sets, and convex preference correspondences. To do this in a general context however requires more powerful mathematical tools, which we shall introduce in Sect. 3.8 below. Before this however we consider one further application of the hyperplane separation theorem to a situation where we wish to optimise a concave objective function subject to a number of concave constraints.

## 3.6 Kuhn-Tucker Theorem

Here we consider a family of constraints in $\Re^n$. Let $g = (g_1, \ldots, g_m) : \Re^n \to \Re^m$. As before let $\Re_+^m = \{(y_1, \ldots, y_m) : y_i \geq 0 \text{ for } i = 1, \ldots, n\}$. A point $x$ is *feasible* iff $x \in \Re_+^n$ and $g(x) \in \Re_+^m$ (i.e., $g_i(x) \geq 0$ for $i = 1, \ldots, m$). Let $f : \Re^n \to \Re$ be the objective function. Say that $x^* \in \Re^n$ is an *optimum* of the constrained optimisation problem $(f, g)$ iff $x^*$ solves the problem: maximise $f(x)$ subject to the feasibility constraint $g(x) \in \Re^m_+$, $x \in \Re^n_+$.

Call the problem $(f, g)$ *solvable* iff there is some $x \in \Re^n_+$ such that $g_i(x) > 0$ for $i = 1, \ldots, m$. The *Lagrangian* to the problem $(f, g)$ is:

$$L(x, \lambda) = f(x) + \sum_{i=1}^{m} \lambda_i g_i(x) = f(x) + (\lambda, g(x))$$

where $x \in \Re^n_+$ and $\lambda = (\lambda_1, \ldots, \lambda_m) \in \Re^m_+$. The pair $(x^*, \lambda^*) \in \Re^{n+m}_+$ is called a *global saddle point* for the problem $(f, g)$ iff

$$L(x, \lambda^*) \leq L(x^*, \lambda^*) \leq L(x^*, \lambda)$$

for all $x \in \Re^n_+$, $\lambda \in \Re^m_+$.

**Kuhn-Tucker Theorem 1** *Suppose $f, g_1, \ldots, g_m : \Re^n \to \Re^m$ are concave functions for all $x \in \Re^n_+$. Then if $x^*$ is an optimum to the solvable problem $(f, g) : \Re^n \to \Re^{m+1}$ there exists a $\lambda^* \in \Re^m_+$ such that $(x^*, \lambda^*)$ is a saddle point for $(f, g)$.*

*Proof* Let $A = \{y \in \Re^{m+1} : \exists\, x \in \Re^n_+ : y \leq (f, g)(x)\}$. Here $(f, g)(x) = (f(x), g(x), \ldots, g_m(x))$. Thus $y = (y_1, \ldots, y_{m+1}) \in A$ iff $\exists x \in \Re^n_+$ such that

$$y_1 \leq f(x)$$
$$y_{j+1} \leq g_j(x) \quad \text{for } j = 1, \ldots, m.$$

Let $x^*$ be an optimum and

$$B = \{z = (z_1, \ldots, z_{m+1}) \in \Re^m : z_1 > f(x^*) \text{ and } (z_2, \ldots, z_{m+1}) > 0\}.$$

Since $f, g$ are concave, $A$ is convex. To see this suppose $y_1, y_2 \in A$. But since both $f$ and $g$ are concave $af(x_1) + (1-a)f(x_2) \leq f(ax_1 + (1-a)x_2)$ and similarly for $g$, for any $a \in [0, 1]$. Thus

$$ay_1 + (1-a)y_2 \leq a(f, g)(x_1) + (1-a)(f, g)(x_2) \leq (f, g)(ax_1 + (1-a)x_2).$$

Since $x_1, x_2 \in \Re^n_+$, $ax_1 + (1-a)x_2 \in \Re^n_+$, and so $ay_1 + (1-a)y_2 \in \Re^n_+$.

Clearly $B$ is convex, since $az_{11} + (1-a)z_{12} > f(x^*)$ if $a \in [0, 1]$ and $z_{11}, z_{12} > f(x^*)$.

To see $A \cap B = \Phi$, consider $x \in \Re^n$ such that $g(x) < 0$. Then $(y_2, \ldots, y_{m+1}) \leq g(x) < \underline{0} \leq (z_2, \ldots, z_{m+1})$.

If $g(x) \in \Re^m_+$ then $x$ is feasible. In this case $y_1 \leq f(x) \leq f(x^*) < z_1$.

## 3.6 Kuhn-Tucker Theorem

By the separating hyperplane theorem, there exists $(p_1, \ldots, p_{m+1}) \in \Re^{m+1}$ and $\alpha \in \Re$ such that $H(p, \alpha) = \{w \in \Re^{m+1} : \sum_{j=1}^{m+1} w_j p_j = \alpha\}$ separates $A$ and $B$, i.e., $\sum_{j=1}^{m+1} p_j y_j \leq \alpha \leq \sum_{j=1}^{m+1} p_j z_j$ for any $y \in A$ and $z \in B$.

Moreover $p \in \Re_+^{m+1}$. By the definition of $A$, for any $y \in A$, $\exists\, x \in \Re_+^n$ such that $y \leq (f, g)(x)$.

Thus for any $x \in \Re_+^n$,

$$p_1 f(x) + \sum_{j=2}^{m} p_j g_j(x) \leq \sum_{j=1}^{m+1} p_j z_j.$$

Since $(f(x^*), 0, \ldots, 0)$ belongs to the boundary of $B$,

$$p_1 f(x) + \sum_{j=2}^{m} p_j g_j(x) \leq p_1 f(x^*).$$

Suppose $p_1 = 0$. Since $p \in \Re_+^{m+1}$, there exists $p_j > 0$.

Since the problem is solvable, $\exists\, x \in \Re_+^n$ such that $g_j(x) > 0$. But this gives $\sum_{j=2}^{m} p_j g_j(x) > 0$, contradicting $p_1 = 0$. Hence $p_1 > 0$.

Let $\lambda_j^* = \frac{p_{j+1}}{p_1}$ for $j = 1, \ldots, m$.

Then $L(x, \lambda^*) = f(x) + \sum_{j=1}^{m} \lambda_j^* g_j(x) \leq f(x^*)$ for all $x \in \Re_+^n$, where $\lambda^* = (\lambda_1^*, \ldots, \lambda_m^*) \in \Re_+^m$. Since $x^*$ is feasible, $g(x^*) \in \Re_+^m$, and $\langle \lambda^*, g(x^*) \rangle \geq 0$. But $f(x^*) + \langle \lambda^*, g(x^*) \rangle \leq f(x^*)$ implying $\langle \lambda^*, g(x^*) \rangle \leq 0$. Thus $\langle \lambda^*, g(x^*) \rangle = 0$. Clearly $\langle \lambda, g(x^*) \rangle \geq 0$ if $\lambda \in \Re_+^m$. Thus $L(x, \lambda^*) \leq L(x^*, \lambda^*) \leq L(x^*, \lambda)$ for any $x \in \Re_+^n, \lambda \in \Re_+^m$. $\square$

**Kuhn-Tucker Theorem 2** *If the pair $(x^*, \lambda^*)$ is a global saddle point for the problem $(f, g)$, then $x^*$ is an optimum.*

*Proof* By the assumption

$$L(x, \lambda^*) \leq L(x^*, \lambda^*) \leq L(x^*, \lambda)$$

for all $x \in \Re_+^n, \lambda \in \Re_+^m$.

Choose $\lambda = (\lambda_1^*, \ldots, 2\lambda_i^*, \ldots, \lambda_m^*)$.

Then $L(x^*, \lambda) \geq L(x^*, \lambda^*)$ implies $g_i(x^*)\lambda_i^* \geq 0$. If $\lambda_i^* \neq 0$ then $g_i(x^*) \geq 0$, and so $\langle \lambda^*, g(x^*) \rangle \geq 0$.

On the other hand, $L(x^*, \lambda^*) \leq L(x^*, 0)$ implies $\langle \lambda^*, g(x^*) \rangle \leq 0$. Thus $\langle \lambda^*, g(x^*) \rangle = 0$. Hence $f(x) + \langle \lambda^*, g(x) \rangle \leq f(x^*) \leq f(x^*) + \langle \lambda, g(x^*) \rangle$. If $x$ is feasible, $g(x) \geq 0$ and so $\langle \lambda^*, g(x) \rangle \geq 0$. Thus $f(x) \leq f(x) + \langle \lambda^*, g(x) \rangle \leq f(x^*)$ for all $x \in \Re_+^n$, whenever $g(x) \in \Re_+^m$. Hence $x^*$ is an optimum for the problem $(f, g)$. Note that for a concave optimisation problem $(f, g)$, $x^*$ is an optimum for $(f, g)$ iff $(x^*, \lambda^*)$ is a global saddle point for the Lagrangian $L(x, \lambda), \lambda \in \Re_+^m$.

Moreover $(x^*, \lambda^*)$ are such that $\langle \lambda^*, g(x^*) \rangle$ *minimises* $\langle \lambda, g(x) \rangle$ for all $\lambda \in \Re_+^m, x \in \Re_+^n, g(x) \in \Re_+^m$. Since $\langle \lambda^*, g(x^*) \rangle = 0$ this implies that if $g_i(x^*) > 0$ then $\lambda_i^* = 0$ and if $\lambda_i^* > 0$ then $g_i(x^*) = 0$.

The coefficients $(\lambda_1^*, \ldots, \lambda_m^*)$ are called *shadow prices*. If the optimum is such that $g_i(x^*) > 0$ then the shadow price $\lambda_1^* = 0$. In other words if the optimum does not lie in the boundary of the $i$th constraint set $\partial B_i = \{x : g_i(x) = 0\}$, then this constraint is *slack*, with zero shadow price. If the shadow price is non zero then the constraint cannot be slack, and the optimum lies on the boundary of the constraint set.

In the case of a single constraint, the assumption of non-satiation was sufficient to guarantee that the constraint was not slack.

In this case

$$f(x) + \frac{p_2}{p_1} g(x) \leq f(x^*) \leq f(x^*) + \lambda g(x^*)$$

for any $x \in \Re_+$, and $\lambda \in \Re_+$, where $\frac{p_2}{p_1} > 0$.

The Kuhn-Tucker theorem is of particular use when objective and constraint functions are smooth. In this case the Lagrangean permits computation of the optimal points of the problem. We deal with these procedures in Chap. 4. □

## 3.7 Choice on Compact Sets

In Lemma 3.9 we showed that when a preference relation is acyclic and lower demi-continuous (LDC) on a compact space, $X$, then $P$ admits a choice. Lemma 3.20 gives a different kind of result making use of compactness and convexity of the constraint set, and convexity and openness of the image $P(x)$ at each point $x$. We now present a class of related results using compactness, convexity and lower demi-continuity. These results are essentially all based on the Brouwer Fixed Point Theorem (Brouwer 1912) and provide the technology for proving existence of a market equilibrium. We first introduce the notion of retraction and contractibility, and then show that any continuous function on the closed ball in $\Re^n$ admits a fixed point.

**Definition 3.8** Let $X$ be a topological space.
(i) Suppose $Y$ is a subset of $X$. Say $Y$ has the *fixed point property* iff whenever $f : Y \to X$ is a continuous function (with respect to the induced topology on $Y$) such that $f(Y) = \{f(y) \in X : y \in Y\} \subset Y$, then there exists a point $\overline{x} \in Y$ such that $f(\overline{x}) = \overline{x}$.
(ii) If $Y$ is a *topological space*, and there exists a bijective continuous function $h : X \to Y$ such that $h^{-1}$ is also continuous then $h$ is called a *homeomorphism* and $X, Y$ are said to be homeomorphic (see Sect. 1.2.3 for the definition of bijective).
(iii) If $Y$ is a topological space and there exist continuous functions $g : Y \to X$ and $h : X \to Y$ such that $h \circ g : Y \to X \to Y$ is the identity on $Y$, then $h$ is called an $r$-map (for $g$).

## 3.7 Choice on Compact Sets

(iv) If $Y \subset Z \subset X$ and $g = \text{id} : Y \to Y$ is the (continuous) identity map and $h : Z \to Y$ is an $r$-map for $g$ then $h$ is called a *retraction* (of $Z$ on $Y$) and $Y$ is called a *retract* of $Z$.

(v) If $Y \subset X$ and there exists a continuous function $f : Y \times [0, 1] \to Y$ such that $f(y, 0) = y$ (so $f(\,,0)$ is the identity on $Y$) and $f(y, 1) = y_0 \in Y$, for all $y \in Y$, then $Y$ is said to be *contractible*.

(vi) Suppose that $Y \subset Z \subset X$ and there exists a continuous function $f : Z \times [0, 1] \to Z$ such that $f(z, 0) = z \; \forall z \in Z$, $f(y, t) = y \; \forall y \in Y$ and $f(z, 1) = h(z)$ where $h : Z \to Y$ is a retraction, then $f$ is called a strong retraction of $Z$ on $Y$, and $Y$ is called a *deformation retract* of $Z$.

To illustrate the idea of contractibility, observe that the closed *Euclidean* ball in $\Re^n$ of radius $\xi$, centered at $x$, namely

$$B^n = \text{clos}\big(B_d(x, \xi)\big) = \big\{y \in \Re^n : d(x, y) \leq \xi\big\},$$

is obviously strictly convex and compact. Moreover the center, $\{x\}$, is clearly a deformation retract of $B^n$. To see this let $f(y, t) = (1 - t)y + tx$ for $y \in B^n$. Clearly $f(y, 1) = h(y) = \{x\}$ so $h : B^n \to \{x\}$ and $h(x) = x$. Since $f$ is continuous, this also implies that $B^n$ is contractible. A continuous function such as $f : B^n \times [0, 1] \to B^n$, or more generally $f : Z \times [0, 1] \to Z$, is often called a *homotopy* and written $f_t : Z \to Z$ where $f_t(z) = f(z, t)$. The homotopy $f_t$ between the identity and the retraction $f_1$ of $Z$ on $Y$ means that $Z$ and $Y$ are "topologically" equivalent (in some sense). Thus the ball $B^n$ and the point $\{x\}$ are topologically equivalent. More generally, if $Y$ is contractible and there is a strong retraction of $Z$ on $Y$, then $Z$ is also contractible.

**Lemma 3.21** *Let $X$ be a topological space. If $Y$ is contractible and $Y \subset Z \subset X$ such that $Y$ is a deformation retract of $Z$, then $Z$ is contractible.*

*Proof* Let $g : Z \times [0, 1] \to Z$ be the strong retraction of $Z$ on $Y$, and let $f : Y \times [0, 1] \to Y$ be the contraction of $Y$ to $y \in Y$. Define

$$r : Z \times \left[0, \frac{1}{2}\right] \to Z \text{ by } r(x, t) = g(z, 2t)$$

$$r' : Z \times \left[\frac{1}{2}, 1\right] \to Z \text{ by } r'(z, t) = f\big(g(z, 1), 2t - 1\big).$$

To see this define a contraction $s$ of $Z$ onto $y_0$. Note that $r(z, \frac{1}{2}) = r'(z, \frac{1}{2})$, since $g(z, 1) = f(g(z, 1), 0)$. This follows because $g$ is a strong retraction and so $g(z, 1) \in Y$, $g(y, 1) = f(y, 0) = y$ if $y \in Y$. Clearly $s : Z \times [0, 1] \to Z$ (defined by $s(z, t) = r(z, t)$ if $t < \frac{1}{2}$, $s(z, t) = r'(z, t)$ if $t \geq \frac{1}{2}$) is continuous and is the identity at $t = 0$. Finally if $t = 1$, then $s(z, 1) = f(g(z, 1), 1) = y_0$. □

**Lemma 3.22** *If $Z$ is a (non-empty) compact, convex set in $\Re^n$, then it is contractible.*

*Proof* For any $x$ in the interior of $Z$ there exists some $\xi > 0$ such that $B^n = \mathrm{clos}(B_d(x, \xi))$ is contained in $Z$. (Indeed $\xi$ can be chosen so that $B^n$ is contained in the interior of $Z$.) As observed in Example 3.3, the closed ball, $B^n$, is both compact and strictly convex. By Lemma 3.17, the distance function $d(z, -) : B^n \to \Re$ is continuous for each $z \in Z$, and so there exists a point $\overline{y}(z)$ in $B^n$, say, such that $d(z, \overline{y}(z)) < d(z, y))\ \forall y \in B^n$. Then $d(z, \overline{y}(z)) = d(z, B^n)$, the distance between $z$ and $B^n$. Indeed $d(z, \overline{y}(z)) = 0$ iff $z \in B^n$. Moreover for each $z \in Z$, $\overline{y}(z)$ is unique. Define the function $f : Z \times [0, 1] \to Z$ by $f(z,t) = tz + (1 - t)\overline{y}(z)$. Since $\overline{y}(z) \in B^n \subset Z$ for each $z$, and $Z$ is convex, $f(z, t) \in Z$ for all $t \in (0, 1]$. Clearly if $z \in B^n$ then $f(z, t) = z$ for all $t \in [0, 1]$ and $f(-, 1) = h : Z \to B^n$ is a retraction. Thus $f$ is a strong retraction, and $B^n$ is a deformation retract of $Z$. By Lemma 3.21, $Z$ is contractible. □

Note that compactness of $Z$ is not strictly necessary for the validity of this lemma.

**Lemma 3.23** *If $Z$ is contractible to $z_0$ and $Y \subset Z$ is a retract of $Z$ by $h : Z \to Y$ then $Y$ is contractible to $h(z_0)$.*

*Proof* Let $f : Z \times [0, 1] \to Z$ be the contraction of $Z$ on $z_0$, and let $h : Z \to Y$ be the retraction. Clearly $h \circ f : Z \times [0, 1] \to Z \to Y$. If $y \in Y$, then $f(y, 0) = y$ and $h(y) = y$ (because $h$ is a retraction). Thus $h \circ f(y, 0) = y$. Moreover $f(z, 1) = z_0\ \forall \in Z$, so $h \circ f(z, 1) = h(z_0)$. □

Clearly being a deformation retract of $Z$ is a much stronger property than being a retract of $Z$. Both of these properties are useful in proving that any compact convex set has the fixed point property, and that the sphere is neither contractible nor has the fixed point property.

Remember that the sphere of radius $\xi$ in $\Re^n$, with center $x$, is

$$S^{n-1} = \mathrm{Boundary}\big(\mathrm{clos}\big(B_d(x, \xi)\big)\big) = \{y \in \Re^n : d(x, y) = \xi\}.$$

Now let $x_0 \in S^{n-1}$ be the north pole of the sphere. We shall give an intuitive argument why $D = S^{n-1}\setminus\{x_0\}$ is contractible, but $S^{n-1}$ is not contractible.

*Example 3.9* Let $D = S^{n-1}\setminus\{x_0\}$ and let $Z$ be a copy of $D$ which is flattened at the south pole. Let $D_0$ be the flattened disc round the South Pole, $x_S$. Clearly $D_0$ is homeomorphic to an $(n - 1)$ dimensional ball $B^{n-1}$ centered on $x_S$. Then $Z$ is homeomorphic to the object $D_0 \times [0, 1)$. There is obviously a strong retraction of $D$ onto $D_0$. This retraction may be thought of as the function that moves any point $z \in S^{n-1}\setminus\{x_0\}$ down the lines of longitude to $D_0$. Since $D_0$ is compact, convex it is contractible to $x_S$ and thus, by Lemma 3.20, there is a contraction $f : D \times [0, 1] \to D$ to $x_S$.

## 3.7 Choice on Compact Sets

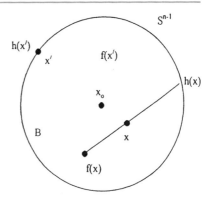

**Fig. 3.22** The retraction of $B^n$ on $S^{n-1}$

To indicate why $S^{n-1}$ cannot be contractible, let us suppose without loss of generality, that $g : S^{n-1} \times [0, 1] \to S^{n-1}$ is a contraction of $S^{n-1}$ to $x_S$, and that $g$ extends the contraction $f : D \times [0, 1] \to D$ (i.e., $g(z, t) = f(z, t)$ whenever $z \in D$). Now $f(y, t)$ maps each point $y \in D$ to a point further south on the longitudinal line through $y$. If we require $g(y, t) = f(y, t)$ for each $y \in \mathcal{D}$, and we require $g$ to be continuous at $(x_0, 0)$ then it is necessary for $g(x_0, t)$ to be an equatorial circle in $S^{n-1}$. In other words if $g$ is a function it must fail continuity at $(x_0, 0)$. While this is not a formal argument, the underlying idea is clear: The sphere $S^{n-1}$ contains a hole, and it is topologically different from the ball $B^n$.

**Brouwer's Theorem** *Any compact, convex set in $\Re^n$ has the fixed point property.*

*Proof* We prove first that the ball $B^n \equiv B$ has the fixed point property. Suppose otherwise: that is there exists a continuous function $f : B \to B$ with $x \neq f(x)$ for all $x \in B$.

Since $f(x) \neq x$, construct the arc from $f(x)$ to $x$ and extend this to the boundary of $B$. Now label the point where the arc and the boundary of $B$ intersect as $h(x)$. Since the boundary of $B$ is $S^{n-1}$, we have constructed a function $h : B \to S^{n-1}$. It is easy to see that $h$ is continuous (because $f$ is continuous). Moreover if $x' \in S^{n-1}$ then $h(x') \in S^{n-1}$. Since $S^{n-1} \subset B$, it is clear that $h : B \to S^{n-1}$ is a retraction. (See Fig. 3.22.) By Lemma 3.23, the contractibility of $B$ to its center, $x_0$ say, implies that $S^{n-1}$ is contractible to $h(x_0)$. But Example 3.9 indicates that $S^{n-1}$ is not contractible. The contradiction implies that any continuous function $f : B \to B$ has a fixed point.

Now let $Y$ be any compact convex set in $\Re^n$. Then there exists for some $\xi$ and $y_0 \in Y$, a closed $\xi$-ball, centered at $y_0$, such that $Y$ is contained in $B = \text{clos}(B_d(y_0, \xi))$. As in the proof of Lemma 3.22, there exists a strong retraction $g : B \times [0, 1] \to B$; so $Y$ is a deformation retract of $B$. (See Fig. 3.23.) In particular $g(-, 1) = h : B \to Y$ is a retraction. If $f : Y \to Y$ is continuous, then $f \circ h : B \to Y \to Y \subset B$ is continuous and has a fixed point. Since the image of $f \circ h$ is in $Y$, this fixed point, $y_1$, belongs to $Y$. Hence $f \circ h(y_1) = y_1$ for some $y_1 \in Y$. But $h = \text{id}$ (the identity) on $Y$, so $h(y_1) = y_1$ and thus $y_1 = f(y_1)$. Consequently $y_1 \in Y$ is a fixed point of $f$. Thus $Y$ has the fixed point property. □

**Fig. 3.23** The strong retraction of $B$ on $Y$

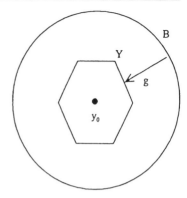

*Example 3.10* The standard compact, convex set in $\Re^n$ is the $(n-1)$-simplex $\Delta^{n-1}$ defined by

$$\Delta = \Delta^{n-1} = \left\{ x = (x_1, \ldots, x_n) \in \Re^n : \sum_{i=1}^n x_i = 1, \text{ and } x_i \geq 0 \; \forall \, i \right\}.$$

$\Delta$ has $n$ vertices $\{x_i^0\}$, where $x_i^0 = (0, \ldots, 1, \ldots, 0)$ with a 1 in the $i$th entry. An *edge* between $x_i^0$ and $x_j^0$ is the arc $\langle\langle x_i^0, x_j^0 \rangle\rangle$ or convex set of the form

$$\left\{ x \in \Re^n : x = \lambda x_i^0 + (1-\lambda) x_j^0 \text{ for } \lambda \in [0,1] \right\}.$$

An $s$-dimensional face of $\Delta$ is simply the convex hull of $(s+1)$ different vertices. Note in particular that there are $n$ different $(n-2)$ dimensional faces. Such a face is opposite the vertex $x_i^0$, so we may label this face $\Delta_i^{n-2}$. These $n$ different faces have empty intersection. However any subfamily, $\mathcal{F}$, of this family of $n$ faces (where $\mathcal{F}$ has cardinality at most $(n-1)$), does have a non-empty intersection. In fact if $\mathcal{F}$ has cardinality $(n-1)$ then the intersection is a vertex.

Brouwer's Theorem allows one to derive further results on the existence of choice.

**Lemma 3.24** *Let $Q : \Delta \to \Re^n$ be an LDC correspondence from the $(n-1)$ dimensional simplex to $\Re^n$, such that $Q(x)$ is both non-empty and convex, for each $x \in \Delta$. Then there exists a continuous selection, $f$, for $Q$, namely a continuous function $f : \Delta \to \Re^n$ such that $f(x) \in Q(x)$ for all $x \in \Delta$.*

*Proof* Since $Q(x) \neq \Phi$, $\forall x \in \Delta$, then for each $x \in \Delta, x \in Q^{-1}(y)$ for some $y \in \Re^n$. Hence $\{Q^{-1}(y) : y \in \Re^n\}$ is a cover for $\Delta$. Since $Q$ is LDC, $Q^{-1}(y)$ is open, $\forall y \in \Re^n$, and thus the cover is an open cover. $\Delta$ is compact. As in the proof of Lemma 3.9, there is a finite index set, $A = \{y_1, \ldots, y_k\}$ of points in $\Re^n$ such that $\{Q^{-1}(y_i) : y_i \in A\}$ covers $\Delta$. Define $\alpha_i : \Delta \to \Re$ by $\alpha_i(x) = d(x, \Delta \backslash Q^{-1}(y_i))$ for $i = 1, \ldots, k$, and let $g_i : \Delta \to \Re$ be given by $g_i(x) = \alpha_i(x) / \sum_{j=1}^k \alpha_j(x)$. As before,

$d$ is the distance operator, so $\alpha_i(x)$ is the distance from $x$ to $\Delta\setminus Q^{-1}(y_i)$. $\{g_i\}$ is known as a *partition of unity* for $Q$. Clearly $\sum g_i(x) = 1$ for all $x \in \Delta$, and $g_i(x) = 0$ iff $x \in \Delta\setminus Q^{-1}(y_i)$ (since $\Delta\setminus Q^{-1}(y_i)$ is closed and thus compact). Finally define $f : \Delta \to \Re^n$ by $f(x) = \sum_{i=1}^{k} g_i(x) y_i$. By the construction, $g_i(x) = 0$ iff $y_i \notin Q(x)$, thus $f(x)$ is a convex combination of points all in $Q(x)$. Since $Q(x)$ is convex, $f(x) \in Q(x)$. By Lemma 3.17, each $\alpha_i$ is continuous. Thus $f$ is continuous. □

**Lemma 3.25** *Let $P : \Delta \to \Delta$ be an LDC correspondence such that for all $x \in \Delta$, $P(x)$ is convex and $x \notin \mathrm{Con}\, P(x)$, the convex hull of $P(x)$. Then the choice $C_P(\Delta)$ is non empty.*

*Proof* Suppose $C_P(\Delta) = \Phi$. Then $P(x) \neq \Phi\ \forall x \in \Delta$. By Lemma 3.24, there exists a continuous function $f : \Delta \to \Delta$ such that $f(x) \in P(x)\ \forall x \in \Delta$. By Brouwer's Theorem, there exists a fixed point $x_0 \in \Delta$ such that $x_0 = f(x_0)$. This contradicts $x \notin \mathrm{Con}\, P(x), \forall x \in \Delta$. Thus $C_P(\Delta) \neq \Phi$. □

These two lemmas are stated for correspondences with domain the finite dimensional simplex, $\Delta$. Clearly they are valid for correspondences with domain a (finite dimensional) compact convex space. However both results can be extended to (infinite dimensional) topological vector spaces. The general version of Lemma 3.24 is known as Michael's Selection Theorem (Michael 1956). However it is necessary to impose conditions on the domain and codomain spaces. In particular it is necessary to be able to construct a partition of unity. For this purpose we can use a condition call "paracompactness" rather than compactness. Paracompactness of a space $X$ requires that there exist, at any point $x \in X$, an open set $U_x$ containing $x$, such that for any open cover $\{U_i\}$ of $X$, only finitely many of the open sets of the cover intersect $U_x$. To construct the continuous selection it is also necessary that the codomain $Y$ of the correspondence has a norm, and is complete (essentially this means that a limit of a convergent sequences of points is contained in $Y$). A complete normed topological vector space $Y$ is called a *Banach* space. We also need $Y$ to be "separable" (ie $Y$ contains a countable dense subset.) If $Y$ is a separable *Banach* space we say it is admissible.

Michael's Selection Theorem employs a property, called lower hemi-continuity.

**Definition 3.9** A correspondence $Q : X \to Y$ between the topological spaces, $X$ and $Y$, is *lower hemi-continuous* (LHC) if whenever $U$ is open in $Y$, then the set $\{x \in X : Q(x) \cap U \neq \Phi\}$ is open in $X$.

**Michael's Selection Theorem** *Suppose $Q : X \to Y$ is a lower hemi-continuous correspondence from a paracompact, Hausdorff topological space $X$ into the admissible space $Y$, such that $Q(x)$ is non-empty closed and convex, for all $x \in X$. Then there exists a continuous selection $f : X \to Y$ for $Q$.*

Lemma 3.24 also provides the proof for a useful intersection theorem known as the Knaster-Kuratowski-Mazurkiewicz (KKM) Theorem.

Before stating this theorem, consider an arbitrary collection $\{x_1, \ldots, x_k\}$ of distinct points in $\Re^n$. Then clearly the convex hull, $\Delta$, of these points can be identified with a $(k-1)$-dimensional simplex. Let $S \subset \{1, \ldots, k\}$ be any index set, and let $\Delta_S$ be the simplex generated by this collection of $s - 1$ points (where $s = |S|$).

**KKM Theorem** *Let $R : X \to Y$ be a correspondence between a convex set $X$ contained in a Hausdorff topological vector space $Y$ such that $R(x) \neq \Phi$ for all $x \in X$. Suppose that for at least one point $x_0 \in X$, $R(x_0)$ is compact. Suppose further that $R(x)$ is closed for all $x \in X$. Finally for any set $\{x_1, \ldots, x_k\}$ of points in $X$, suppose that*

$$\mathrm{Con}\{x_1, \ldots, x_k\} \subset \bigcup_{i=1}^{k} R(x_i).$$

*Then $\bigcap_{x \in X} R(x)$ is non empty.*

*Proof* By Lemma 3.8, since $R(x_0)$ is compact, $\bigcap_{x \in X} R(x)$ is non-empty iff $\bigcap_{i=1}^{k} R(x_i) \neq \Phi$ for any finite index set. So let $K = \{1, \ldots, k\}$ and let $\Delta$ be the $(k-1)$-dimensional simplex spanned by $\{x_1, \ldots, x_k\}$. Define $P : \Delta \to \Delta$ by $P(x) = \{y \in \Delta : x \in \Delta \backslash R(y)\}$ and define $Q : \Delta \to \Delta$ by $Q(x) = \mathrm{Con}\, P(x)$, the convex hull of $P(x)$.

But $P^{-1}(y) = \{x \in \Delta : y \in P(x)\} = \Delta \backslash R(y)$ is an open set, in $\Delta$, and so $P$ is LDC. Thus $Q$ is LDC. Now suppose that $\bigcap_{i \in K} R(x_i) = \Phi$.

Thus for each $x \in \Delta$ there exists $x_i (i \in K)$ such that $x \notin R(x_i)$. But then $x \in \Delta - R(x_i)$ and so $x \in P^{-1}(x_i)$. In particular, for each $x \in \Delta$, $P(x)$, and thus $Q(x)$, is non-empty. Moreover $\{Q^{-1}(x_i) : i \in K\}$ is an open cover for $\Delta$. As in the proof of Lemma 3.24, there is a partition of unity for $Q$. (We need $Y$ to be Hausdorff for this construction.) In particular there exists a continuous selection $f : \Delta \to \Delta$ for $Q$. By Brouwer's Theorem, $f$ has a fixed point $x_0 \in \Delta$. Thus $x_0 \in \mathrm{Con}\, P(x_0)$, and so $x_0 \in \mathrm{Con}\{y_1, \ldots, y_k\}$ where $y_i \in P(x_0)$ for $i \in K$. But then $x_0 \in \Delta \backslash R(y_i)$ for $i \in K$, and so $x_0 \notin R(y_i)$ for $i \in K$.

Hence

$$\mathrm{Con}\{y_1, \ldots, y_k\} \not\subset \bigcup_{i=1}^{k} R(y_i).$$

This contradicts the hypothesis of the Theorem. Consequently $\bigcap_{i \in K} R(x_i) \neq \Phi$ for any finite vertex set $K$. By compactness $\bigcap_{x \in X} R(x) \neq \Phi$. □

We can immediately use the KKM theorem to prove a fixed point theorem for a *correspondence P* from a compact convex set $X$ to a Hausdorff topological vector space, $Y$. In particular $X$ need not be finite dimensional.

**Browder Fixed Point Theorem** *Let $Q : X \to X$ be a correspondence where $X$ is a compact convex subset of the Hausdorff topological vector space, $Y$. Suppose further that $Q$ is LDC, and that $Q(x)$ is convex and non-empty for all $x \in X$. Then there exists $x_0 \in X$ such that $x_0 \in Q(x_0)$.*

*Proof* Suppose that $x \notin Q(x) \ \forall x \in X$. Define $R : X \to X$ by $R(x) = X \backslash Q^{-1}(x)$. Since $Q$ is LDC, $R(x)$ is closed and thus compact $\forall x \in X$. To use KKM, we seek to show that Con $\{x_1, \ldots, x_k\} \subset \bigcup_{i=1}^{k} R(x_i)$ for any finite index set, $K = \{1, \ldots, k\}$.

We proceed by contradiction. That is, suppose that there exists $x_0$ in $X$ with $x_0 \in$ Con $\{x_1, \ldots, x_k\}$ but $x_0 \notin R(x_i)$ for $i \in K$. Then $x_0 \in Q^{-1}(x_i)$, so $x_i \in Q(x_0), \forall i \in K$. But then $x_0 \in$ Con $Q(x_0)$. Since $Q(x)$ is convex $\forall \ x \in X$, this implies that $x_0 \in Q(x_0)$, a contradiction. Consequently $x \in R(x_i)$ for some $i \in K$. By the KKM Theorem, $\bigcap_{x \in X} R(x) \neq \Phi$.

Thus $\exists \ x_0 \in X$ with $x_0 \in R(x)$, and so $x_0 \in X \backslash Q^{-1}(x) \ \forall x \in X$. Thus $x_0 \notin Q^{-1}(x)$ and $x \notin Q(x_0) \ \forall x \in X$. This contradicts the assumption that $Q(x) \neq \Phi \ \forall x \in X$. Hence $\exists \ x_0 \in X$ with $x_0 \in Q(x_0)$. □

**Ky Fan Theorem** *Let $P : X \to X$ be an LDC correspondence where $X$ is a compact convex subset of the Hausdorff topological vector space, $Y$. If $x \notin$ Con $P(x)$, $\forall x \in X$, then the choice $C_P(X_0) \neq \Phi$ for any compact convex subset, $X_0$ of $X$.*

*Proof* Define $Q : X \to X$ by $Q(x) =$ Con $P(x)$. If $Q(x) \neq \Phi$ for all $x \in X$, then by the Browder fixed point theorem, $\exists \ x_0 \in X$ with $x_0 \in Q(x_0)$. This contradicts $x \notin$ Con $P(x) \ \forall \ x \in X$. Hence $Q(x_0) = \Phi$ for some $x_0 \in X$. Thus $C_P(X) = \{x \in X : P(x) = \Phi\}$ is non-empty. The same inference is valid for any compact, convex subset $X_0$ of $X$. □

## 3.8 Political and Economic Choice

The results outlined in the previous section are based on the intersection property of a family of closed sets. With compactness, this result can be extended to the case of a correspondence $R : X \to X$ to show that $\bigcap_{x \in X} R(x) \neq \Phi$. If we regard $R$ as derived from an LDC correspondence $P : X \to X$ by $R(x) = X \backslash P^{-1}(x)$ then $R(x)$ can be interpreted as the set of points "no worse than" or "at last as good as" $x$.

But then the choice $C_P(X) = \bigcap_{x \in X} R(x)$, since such a choice must be at least as good as any other point. The finite-dimensional version (Lemma 3.25) of the proof that the choice is non-empty is based simply on a fixed point argument using Brouwer's Theorem. To extend the result to an infinite dimensional topological vector space we reduced the problem to one on a finite dimensional simplex, $\Delta$, spanned by $\{x_1, \ldots, x_k : x_i \in X\}$ and then showed essentially that $\bigcap_{x \in \Delta} R(x)$ is non empty. By compactness then $\bigcap_{x \in X} R(x)$ is non-empty. There is, in fact, a related infinite dimensional version of Brouwer's Fixed Point Theorem, known as Schauder's fixed point theorem for a continuous function $f : X \to Y$, where $Y$ is a compact subset of the convex Banach space $X$.

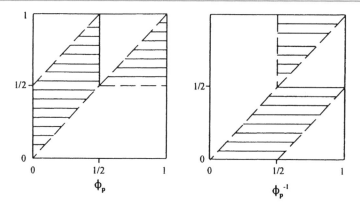

**Fig. 3.24** Illustration of the Ky Fan Theorem

One technique for proving existence of an equilibrium price vector in an exchange economy (as discussed in Sect. 3.5) is to construct a continuous function $f : \Delta \to \Delta$, where $\Delta$ is the price simplex of feasible price vectors, and show that $f$ has a fixed point (using either the Brouwer or Schauder fixed point theorems).

An alternative technique is to use the Ky Fan Theorem to prove existence of an equilibrium price vector. This technique permits a proof even when preferences are not representable by utility functions. More importantly, perhaps, it can be used in the infinite dimensional case.

*Example 3.11* To illustrate the Ky Fan Theorem with a simple example, consider Fig. 3.24, which reproduces Fig. 1.11 from Chap. 1.

It is evident that the inverse preference, $P^{-1}$, is not LDC: for example, $P^{-1}(\frac{3}{4}) = (\frac{1}{4}, \frac{1}{2}] \cup (\frac{3}{4}, 1]$ which is not open. As we saw earlier the choice of $P$ on the unit interval is empty. In fact, to ensure existence of a choice we can require simply that $P$ be lower hemi-continuous. This we can do by deleting the segment $(\frac{1}{2}, 1)$ above the point $\frac{1}{2}$. If the choice were indeed empty, then by Michael's Selection Theorem we could find a continuous selection $f : [0, 1] \to [0, 1]$ for $P$. By Brouwer's fixed point theorem $f$ has a fixed point, $x_0$, say, with $x_0 \in P(x_0)$. By inspection the fixed point must be $x_0 = \frac{1}{2}$. If we require $P$ to be irreflexive (since it is a strict preference relation) then this means that $\frac{1}{2} \notin P(\frac{1}{2})$ and so the choice must be $C_P([0, 1]) = \{\frac{1}{2}\}$. Notice that the preference displayed in Fig. 3.24 cannot be represented by a utility function. This follows because the implicit indifference relation is intransitive. See Fig. 3.25.

The Ky Fan Theorem gives a proof of the existence of a choice for a "spatial voting game". Remember a social choice procedure, $\sigma$ is *simple* iff it is defined by a family $\mathcal{D}$ of decisive coalitions, each a subset of the society $M$. In this case if $\pi = (P_1, \ldots, P_m)$ is a profile on the topological space $X$, then the social preference is given by

$$x\sigma(\pi)y \quad \text{iff } x \in \bigcup_{A \in \mathcal{D}} \bigcap_{i \in A} P_i(y) = P_{\mathcal{D}}(y).$$

**Fig. 3.25** The graph of indifference

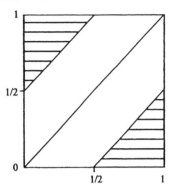

Here we use $P_i : X \to X$ to denote the preference correspondence of individual $i$. For coalition $A$, then $x \in P_A(y) = \bigcap_{i \in A} P_i(y)$ means every member of $A$ prefers $x$ to $y$.

Thus $x \in \bigcup_{A \in \mathcal{D}} P_A(y)$ means that for some coalition $A \in \mathcal{D}$, all members of $A$ prefer $x$ to $y$. Finally we write $x \in P_\mathcal{D}(y)$ for the condition that $x$ is socially preferred to $y$. The choice for $\sigma(\pi)$ on $X$ is then

$$C_{\sigma(\pi)}(X) = \{x : P_\mathcal{D}(x) = \Phi\}.$$

**Nakamura Theorem** *Suppose $X$ is a compact convex topological vector space of dimension $n$. Suppose that $\pi = (P_1, \ldots, P_m)$ is a profile on $X$ such that each preference $P_i : X \to X$ is (i) LDC; and (ii) semi-convex, in the sense that $x \notin \text{Con } P_i(x)$ for all $x \in X$. If $\sigma$ is simple and has Nakamura number $k(\sigma)$, and if $n \leq k(\sigma) - 2$ then the choice $C_{\sigma(\pi)}(X)$ is non-empty.*

*Proof* For any point $x$, $y \in P_\mathcal{D}^{-1}(x)$ means $x \in P_\mathcal{D}(y)$ and so $x \in P_i(y)$ $\forall i \in A$, some $A \in \mathcal{D}$. Thus $y \in P_i^{-1}(x)$ $\forall i \in A$ or $y \in \bigcap_{i \in A} P_i^{-1}(x)$ or $y \in \bigcup_\mathcal{D} \bigcap_A P_i^{-1}(x)$. But each $P_i$ is LDC and so $P_i^{-1}(x)$ is open, for all $x \in X$. Finite intersection of open sets is open, and so $P_\mathcal{D}$ is LDC.

Now suppose that $P_\mathcal{D}$ is not semi-convex (that is $x \in \text{Con } P_\mathcal{D}(x)$ for some $x \in X$). Since $X$ is $n$-dimensional and convex, this means it is possible to find a set of points $\{x_1, \ldots, x_{n+1}\}$ such that $x \in \text{Con}\{x_1, \ldots, x_{n+1}\}$ and such that $x_j \in P_\mathcal{D}(x)$ for each $j = 1, \ldots, n+1$. Without loss of generality this means there exists a subfamily $\mathcal{D}' = \{A_1, \ldots, A_{n+1}\}$ of $\mathcal{D}$ such that $x_j \in P_{A_j}(x)$. Now $n + 1 \leq k(\sigma) - 1$, and by the definition of the Nakamura number, the collegium $K(\mathcal{D}')$ is non-empty. In particular there is some individual $i \in A_j$, for $j = 1, \ldots, n+1$. Hence $x_j \in P_i(x)$ for $j = 1, \ldots, n+1$. But then $x \in \text{Con } P_i(x)$. This contradicts the semi-convexity of individual preference. Consequently $P_\mathcal{D}$ is semi-convex. The conditions of the Ky Fan Theorem are satisfied, and so $P_\mathcal{D}(\bar{x}) = \Phi$ for some $\bar{x} \in X$. Thus $C_{\sigma(\pi)}(x) \neq \Phi$. □

It is worth observing that in finite dimensional spaces the Ky Fan Theorem is valid with the continuity property weakened to lower hemi-continuity (LHC). Note first that if $P$ is LDC then it is LHC; this follows because $\{x \in X : P(x) \cap V \neq \Phi\} = \bigcup_{y \in V}(P^{-1}(y) \cap X)$ is the union of open sets and thus open.

Moreover (as suggested in Example 3.11) if $P$ is LHC and the choice is non empty, then the correspondence $x \to \operatorname{Con} P(x)$ has a continuous selection $f$ (by Michael's Selection Theorem). By the Brouwer Fixed Point Theorem, there is a fixed point $x_o$ such that $x_o \in \operatorname{Con} P(x_0)$. This violates semi-convexity of $P$. Thus the Nakamura Theorem is valid when preferences are LHC. The finite dimensional version of the Ky Fan Theorem can be used to show existence of a Nash equilibrium (Nash 1950).

**Definition 3.10**
(i) A *Game* $G = \{(P_i, X) : i \in M\}$ for society $M$ consists of a *strategy space*, $X_i$, and a strict preference correspondence $P_i : X \to X$ for each $i \in M$, where $X = \Pi_i X_i = X_1 \times \cdots \times X_m$ is the joint strategy space.
(ii) In a game $G$, the Nash improvement correspondence for individual $i$ is defined by

$$\hat{P}_i : X \to X \quad \text{where } y \in \hat{P}_i(x) \text{ iff } y \in P_i(x) \text{ and}$$
$$y = (x_1, \ldots, x_{i-1}, x_i^*, \ldots x_m),$$
$$x = (x_1, \ldots, x_{i-1}, x_i, \ldots, x_m).$$

(iii) The overall Nash improvement correspondence is

$$\hat{P} = \bigcup_{i \in M} P_i : X \to X.$$

(iv) A point $\overline{x} \in X$ is a Nash Equilibrium for the game $G$ iff $\hat{P}(\overline{x}) = \Phi$.

Bergstrom (1975, 1992) showed the following.

**Bergstrom's Theorem** *Let $G = \{(P_i, X)\}$ be a game, and suppose each $X_i \subset \Re^n$ is a non-empty compact, convex subset of $\Re^n$. Suppose further that for all $i \in M$, $\hat{P}_i$ is both semi-convex and LHC. Then there exists a Nash equilibrium for $G$.*

*Proof* Since each $\hat{P}_i$ is LHC, it follows easily that $\hat{P} : X \to X$ is LHC. To see that $\hat{P}$ is semi-convex, suppose that $y \in \operatorname{Con} \hat{P}(x)$, then $y = \sum \lambda_i y^i$, $\sum_{i \in N} \lambda_i = 1$, $\lambda_i \geq 0$, $\forall i \in N$ and $y^i \in \hat{P}_i(x)$. By the definition of $\hat{P}_j$, $y - x = \sum_{i \in M} \lambda_i z^i$ where $z^i = y^i - x$.

This follows because $y^i$ and $x$ only differ on the $i$th coordinate, and so $z^i = (0, \ldots, z_0^i, \ldots)$ where $z_0^i \in X_i$. Moreover, $\lambda_i \neq 0$ iff $z^i \neq 0$ because $\hat{P}_i$ is semi-convex.

**Fig. 3.26** Failure of semi-convexity

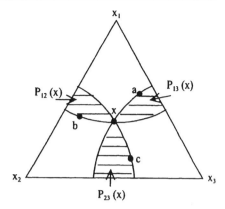

Clearly $\{z^i : i \in M, \lambda_i \neq 0\}$ is a linearly independent set, so $y = x$ iff $\lambda_i = 0$ $\forall i \in M$. But then $y^i = x$ $\forall i \in M$, which again violates semi-convexity of $P_i$. Thus $y \neq x$ and so $\hat{P}$ is semi-convex. By the Ky Fan Theorem, for $X$ finite dimensional, the choice of $\hat{P}$ on $X$ is non-empty. Thus the Nash Equilibrium is non-empty. □

Although the Nakamura Theorem guarantees existence of a social choice for a social choice rule, $\sigma$, for any semi-convex and LDC profile in dimension at most $k(\sigma) - 2$, it is easy to construct situations in higher dimensions with empty choice. The example we now present also describes a game with no Nash equilibrium.

*Example 3.12* Consider a simply voting procedure with $M = \{1, 2, 3\}$ and let $\mathcal{D}$ consist of any coalition with at least two voters. Let $X$ be a compact convex set in $\Re^2$, and construct preferences for each $i$ in $M$ as follows. Each $i$ has a "bliss point" $x_i \in X$ and a preference $P_i$ on $X$ such that for $y, x \in P_i(y)$ iff $\|x - x_i\| < \|y - y_i\|$. The preference is clearly LDC and semi-convex (since $P_i(x)$ is a convex set and $x \notin P(x)$ for all $x \in X$). Now let $\Delta = \text{Con}\{x_1, x_2, x_3\}$ the 2-dimensional simplex in $X$ (for convenience suppose all $x_i$ are distinct and in the interior of $X$). For each $A \subset M$ let $P_A(x) = \bigcap_{i \in A} P_i(x)$ as before.

In particular the choice for the strict Pareto rule is $C_{P_M}(X) = C_M(X) = \Delta$. This can be seen by noting that if $x \in C_M(X)$ iff there is no point $y \in M$ such that $\|y - x_i\| < \|x - x_i\|$ $\forall i \in M$. Clearly this condition holds exactly at those points in $\Delta$. For this reason preferences of this kind are called "Euclidean preferences".

Now consider a point in the interior of $X$. At $x$ the preferred sets for the three coalitions ($\mathcal{D}' = \{1, 2\}, \{1, 3\}, \{2, 3\}$) do not satisfy the semi-convexity property. Figure 3.26 shows that

$$x \in \text{Con}\{P_{12}(x), P_{13}(x), P_{23}(x)\}.$$

While $P_{\mathcal{D}'}$ is LDC, it violates semi-convexity. Thus the Ky Fan Theorem cannot be used to guarantee existence of a choice. To illustrate the connection with Walker's Theorem, note also that there is a voting cycle. That is 1 prefers $a$ to $b$ to $c$, 2 prefers $b$ to $c$ to $a$, and 3 prefers $c$ to $a$ to $b$. The reason the cycle can be constructed is that

**Fig. 3.27** Empty Nash equilibrium

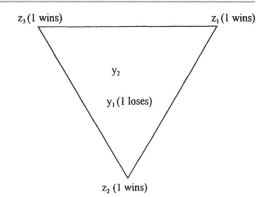

the Nakamura number is 3, and dimension $(X) = 2$, thus $n = k(\sigma) - 1$. In fact there is no social choice in $X$. This voting cycle also defines a game with no Nash equilibrium.

Let $\{1, 2\}$ be the two parties, and let the "strategy spaces" be $X_1 = X_2 = \Delta$. Say party $i$ adopts a winning position $y_i \in X_i$ over party $j'$s position, $y_j \in X_j$ iff $y_i \in P_{\mathcal{D}'}(y_j)$, so party $i$ gains a majority over party $j$. This induces a Nash improvement correspondence $\hat{P}_j$, for $j = 1, 2$.

For example $\hat{P}_1 : Y \to Y$ where $Y = X_1 \times X_2$ is defined by $(y_1^*, y_2) \in \hat{P}_1(y_1, y_2)$ iff $y_1^* \in P_{\mathcal{D}'}(y_2)$, so $y_1^*$ is a winning position over $y_2$ but $y_1 \notin P_{\mathcal{D}'}(y_2)$, so $y_1$ is not a winning position. In other words party 1 prefers to move from $y_1$ to $y_1^*$, so as to win. It is evident if we choose $y_1, y_2$ and the points $\{z_1, z_2, z_3\}$ as in Fig. 3.27, then $z_i \in P_{\mathcal{D}'}(y_2)$ so $(z_i, y_2) \in \hat{P}_1(y_1, y_2)$ for $i = 1, 2, 3$ and $y_1 \in \text{Con}\{z_1, z_2, z_3\}$.

Thus $(y_1, y_2) \in \text{Con}\, \hat{P}_1(y_1, y_2)$. Because of the failure of semi-convexity of both $\hat{P}_1$ and $\hat{P}_2$, a Nash equilibrium cannot be guaranteed. In fact the Nash equilibrium is empty.

We now briefly indicate how the Ky Fan Theorem can be used to show existence of an equilibrium price vector in an exchange economy. First of all each individual $i$ initially holds a vector of endowments $e_i \in \mathfrak{R}^n$. A price vector $p \in \Delta^{n-1}$ belongs to the $(n-1)$-dimensional simplex: that is $p = (p_1, \ldots, p_n)$ such that $\sum_{i=1}^{n} p_i = 1$.

An *allocation* $x \in X = \prod_{i \in N} X_i \subset (\mathfrak{R}^n)^m$ where $X_i$ is $i$'s consumption set in $\mathfrak{R}_+^n$ (here $x_i \in \mathfrak{R}_+^n$ iff $x_{ij} \geq 0, j = 1, \ldots, n$). At the price vector $p$, the $i$th budget set is

$$B_i(p) = \{x_i \in \mathfrak{R}_+^n : \langle p, x \rangle \leq \langle p, e_i \rangle\}.$$

At price $p$, the *demand vector* $\overline{x} = (\overline{x}_1, \ldots, \overline{x}_m)$ satisfies the *optimality condition* $\hat{P}_i(\overline{x}) \cap \{x \in X : x_i \in B_i(p)\} = \Phi$ for each $i$.

As before $\hat{P}_i : X \to X$ is the Nash improvement correspondence (as in Definition 3.10).

As we discussed earlier in Sect. 3.5.1, an equilibrium price vector $\overline{p}$ is a price vector $\overline{p} = (p_1, \ldots, p_n)$ such that the demand vector $\overline{x}$ satisfies the optimality condition at $\overline{p}$ and such that total demand does not exceed supply. This latter condition

## 3.8 Political and Economic Choice

requires that $\sum_{i \in M} \overline{x}_i \leq \sum_{i \in M} e_i$ (the two terms are both vectors in $\Re^n$). That is, if we use the suffix of $x_j$ to denote commodity $j$, then $\sum_{i \in M}(\overline{x}_{ij} - e_{ij}) \leq 0$ for each $j = 1, \ldots, n$.

Note also that a transformation $p > \lambda p$, for a real number $\lambda > 0$, does not change the budget set.

This follows because $B_i(p) = \{x_i \in \Re_+^n : \langle p, x \rangle \leq \langle p, e_i \rangle\} = B_i(\lambda p)$.

Consequently if $\overline{p}$ is an equilibrium price vector, then so is $\lambda \overline{p}$. Without loss of generality, then, we can normalize the price vector, $p$, so that $\|p\| = 1$ for some norm on $\Re^n$. We may do this for example by assuming that $\sum_{j=1}^n p_j = 1$ and that $p_j \geq 0 \, \forall j$.

For this reason we let $\Delta^{n-1}$ represent the set of all price vectors.

A further point is worth making. Since we assume that $p_j \geq 0 \, \forall j$ it is possible for commodity $j$ to have zero price. But then the good must be in excess supply. To ensure this, we require the equilibrium price vector $\overline{p}$ and $i$'s demand $\overline{x}_i$ at $\overline{p}$ to satisfy the condition $\langle \overline{p}, \overline{x}_i \rangle = \langle \overline{p}, \overline{e}_i \rangle$ As we noted in Sect. 3.5.1, this can be ensured by assuming that preference is locally non-satiated in $X$. That is if we let $\overline{P}_i(x) \subset X_i$ be the projection of $\hat{P}_i$ onto $X_i$ at $x$, then for any neighborhood $U$ of $x_i$ in $X_i$ there exists $x_i' \in U$ such that $x_i' \in \overline{P}_i(x)$.

To show existence of a price equilibrium we need to define a price adjustment mechanism.

To this end define:

$$\hat{P}_0 : \Delta \times X \to \Delta \quad \text{by}$$

$$\hat{P}_0(p, x) = \left\{ p' \in \Delta : \left\langle p' - p, \sum_{i \in M}(x_i - e_i) \right\rangle > 0 \right\}. \quad (**)$$

Now let $X^* = \Delta \times X$ and define $P_0^* : X^* \to X^*$ by $(p', x) \in P_0^*(p, y)$ iff $x = y$ and $p' \in \hat{P}_0(p, x)$.

In the same way for each $i \in M$ extend $\hat{P}_i : X \to X$ to $P_i^* : X^* \to X^*$ by letting $(p', x) \in P_i^*(p, y)$ iff $p' = p$ and $x \in \hat{P}_i(y)$. This defines an *exchange game* $G_e = \{(P_i^*, X^*) : i = 0, \ldots, m\}$.

Bergstrom (1992) has shown (under the appropriate conditions of semi-convexity, LHC and local monotonicity for each $\hat{P}_i$) that there is a Nash equilibrium to $G_e$. Note that $e \in (\Re^n)^m$ is the initial vector of endowments. We can show that the Nash equilibria comprise a non-empty set $\{(\overline{p}, \overline{x}) \in \Delta \times X\}$ where $\overline{x} = (\overline{x}_1, \ldots, \overline{x}_m)$ is a vector of final demands for the members of $M$, and $\overline{p}$ is an equilibrium price vector.

A Nash equilibrium $(\overline{p}, \overline{x}) \in \Delta \times X$ satisfies the following properties:

(i) Since $P_i^*(\overline{p}, \overline{x}) = \Phi$ for each $i \in N$, it follows that $\overline{x}_i$ is $i$'s demand. Moreover by local monotonicity we have $\langle \overline{p}, \overline{x}_i \rangle = \langle \overline{p}, e_i \rangle$, $\forall i \in M$. Thus

$$\sum_{i \in M} \langle \overline{p}, \overline{x}_i - e_i \rangle = 0. \quad (*)$$

(ii) Now $P_0^*(\overline{p}, \overline{x}) = \Phi$ so $\langle p - \overline{p}, \sum_{i \in M}(\overline{x}_i - e_i) \rangle \leq 0 \, \forall p \in \Delta$ (See (**)).

Suppose that $\sum_{i \in M}(\bar{x}_i - e_i) > 0$. Then it is clearly possible to find $p \in \Delta$ such that $p_j > 0$, for some $j$, with $\langle p, \sum_{i \in M}(x_i - e_i) \rangle > 0$. But this violates (*) and (**). Consequently $\sum_{i \in M}(\bar{x}_{ij} - e_{ij}) \leq 0$ for $j = 1, \ldots, n$.

Thus $\bar{x} \in (\Re^n)^m$ satisfies the feasibility constraint $\sum_{i \in M} \bar{x}_i \leq \sum_{i \in M} e_i$.

Hence $(\bar{p}, \bar{x})$ is a *free-disposal* equilibrium, in the sense that total demand may be less than total supply. Bergstrom (1992) demonstrates how additional assumptions on individual preference are sufficient to guarantee equality of demand and supply. Section 4.4, below, discusses this more fully.

## Further Reading

The reference for the Kuhn-Tucker Theorems is:

Kuhn, H. W., & Tucker, A. W. (1950). Non-linear programming. In *Proceedings: 2nd Berkeley symposium on mathematical statistics and probability*, Berkeley: University of California Press.

A useful further reference for economic applications is:

Heal, E. M. (1973). *Theory of economic planning*. Amsterdam: North Holland.

The classic references on fixed point theorems and the various corollaries of the theorems are:

Brouwer, L. E. J. (1912). Uber Abbildung von Mannigfaltikeiten. *Mathematische Annalen, 71*, 97–115.
Browder, F. E. (1967). A new generalization of the Schauder fixed point theorem. *Mathematische Annalen, 174*, 285–290.
Browder, F. E. (1968). The fixed point theory of multivalued mappings in topological vector spaces. *Mathematische Annalen, 177*, 283–301.
Fan, K. (1961). A generalization of Tychonoff's fixed point theorem. *Mathematische Annalen, 142*, 305–310.
Knaster, B., Kuratowski, K., & Mazerkiewicz, S. (1929). Ein Beweis des Fixpunktsatze fur $n$-dimensionale Simplexe. *Fundamenta Mathematicae, 14*, 132–137.
Michael, E. (1956). Continuous selections I. *Annals of Mathematics, 63*, 361–382.
Nash, J. F. (1950). Equilibrium points in $n$-person games. *Proceedings of the National Academy of Sciences of the United States of America, 36*, 48–49.
Schauder, J. (1930). Der Fixpunktsatze in Funktionalräumn. *Studia Mathematica, 2*, 171–180.

A useful discussion of the relationships between the theorems can be found in:

Border, K. (1985). *Fixed point theorems with applications to economics and game theory*. Cambridge: Cambridge University Press.
Hildenbrand, W., & Kirman, A. P. (1976). *Introduction to equilibrium analysis*. Amsterdam: Elsevier.

The proof of the Browder fixed point theorem by KKM is given in:

Yannelis, N., & Prabhakar, N. (1983). Existence of maximal elements and equilibria in linear topological spaces. *Journal of Mathematical Economics, 12*, 233–245.

Applications of the Fan Theorem to show existence of a price equilibrium can be found in:

Aliprantis, C., & Brown, D. (1983). Equilibria in markets with a Riesz space of commodities. *Journal of Mathematical Economics, 11*, 189–207.
Bergstrom, T. (1975). *The existence of maximal elements and equilibria in the absence of transitivity* (Typescript). Washington University in St. Louis.
Bergstrom, T. (1992). When non-transitive relations take maxima and competitive equilibrium can't be beat. In W. Neuefeind & R. Riezman (Eds.), *Economic theory and international trade*, Berlin: Springer.
Shafer, W. (1976). Equilibrium in economies without ordered preferences or first disposal. *Journal of Mathematical Economics, 3*, 135–137.
Shafer, W., & Sonnenschein, H. (1975). Equilibrium in abstract economies without ordered preferences. *Journal of Mathematical Economics, 2*, 345–348.

The application of the idea of the Nakamura number to existence of a voting equilibrium can be found in:

Greenberg, J. (1979). Consistent majority rules over compact sets of alternatives. *Econometrica, 41*, 285–297.
Schofield, N. (1984). Social equilibrium and cycles on compact sets. *Journal of Economic Theory, 33*, 59–71.
Strnad, J. (1985). The structure of continuous-valued neutral monotonic social functions. *Social Choice and Welfare, 2*, 181–195.

Finally there are results on existence of a joint political economic equilibrium, namely an outcome $(\overline{p}, \overline{t}, \overline{x}, \overline{y})$, where $(\overline{p}, \overline{x})$ is a market equilibrium, $\overline{t}$ is an equilibrium tax schedule voted on under a rule $\mathcal{D}$ and $\overline{y}$ is an allocation of publicly provided goods.

Konishi, H. (1996). Equilibrium in abstract political economies: with an application to a public good economy with voting. *Social Choice and Welfare, 13*, 43–50.

# Differential Calculus and Smooth Optimisation

Under certain conditions a continuous function $f : \Re^n \to \Re^m$ can be approximated at each point $x$ in $\Re^n$ by a linear function $df(x) : \Re^n \to \Re^m$, known as the differential of $f$ at $x$. In the same way the differential $df$ may be approximated by a bilinear map $d^2 f(x)$. When all differentials are continuous then $f$ is called smooth. For a smooth function $f$, Taylor's Theorem gives a relationship between the differentials at a point $x$ and the value of $f$ in a neighbourhood of a point. This in turn allows us to characterise maximum points of the function by features of the first and second differential. For a real-valued function whose preference correspondence is convex we can virtually identify critical points (where $df(x) = 0$) with maxima.

In the maximisation problem for a smooth function on a "smooth" constraint set, we seek critical points of the Lagrangian, introduced in the previous chapter. In particular in economic situations with exogenous prices we may characterise optimum points for consumers and producers to be points where the differential of the utility or production function is given by the price vector. Finally we use these results to show that for a society the set of Pareto optimal points belongs to a set of generalised critical points of a function which describes the preferences of the society.

## 4.1 Differential of a Function

A function $f : \Re \to \Re$ is *differentiable* at $x \in \Re$ if $\lim_{h \to 0} \frac{f(x+h) - f(x)}{h}$ exists (and is neither $+\infty$ nor $-\infty$). When this limit exists we shall write it as $\frac{df}{dx}|_x$. Another way of writing this is that as $(x_n) \to x$ then

$$\frac{f(x_n) - f(x)}{x_n - x} \to \frac{df}{dx}\bigg|_x,$$

the *derivative* of $f$ at $x$.

This means that there is a real number $\lambda(x) = \frac{df}{dx}|_x \in \Re$ such that $f(x) = \lambda(x)h + \epsilon|h|$, where $\epsilon \to 0$ as $h \to 0$.

**Fig. 4.1** The differential

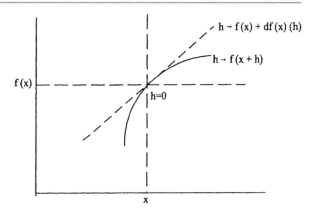

Let $df(x)$ be the linear function $\Re \to \Re$ given by $df(x)(h) = \lambda(x)h$. Then the map $\Re \to \Re$ given by

$$h \to f(x) + df(x)(h) = g(x+h)$$

is a "first order approximation" to the map

$$h \to f(x+h).$$

In other words the maps $h \to g(x+h)$ and $h \to f(x+h)$ are "tangent" to one another where "tangent" means that

$$\frac{|f(x+h) - g(x+h)|}{|h|}$$

approaches 0 as $h \to 0$. Note that the map $h \to f(x) + df(x)(h) = g(x+h)$ has a straight line graph, and so $df(x)$ is a "linear approximation" to the function $f$ at $x$. See Fig. 4.1.

*Example 4.1*
1. Suppose $f : \Re \to \Re : x \to x^2$. Then $\lim_{h \to 0} \frac{f(x+h) - f(x)}{h} = \lim_{h \to 0} \frac{(x+h)^2 - x^2}{h} = \lim_{h \to 0} \frac{2hx + h^2}{h} = 2x + \lim_{h \to 0} h = 2x$. Similarly if $f : \Re \to \Re : x \to x^r$ then $df(x) = rx^{r-1}$.
2. Suppose $f : \Re \to \Re : x \to \sin x$. Then

$$\lim_{h \to 0} \left( \frac{\sin(x+h) - \sin x}{h} \right)$$

$$= \lim_{h \to 0} \left( \frac{\sin x (\cos h - 1) + \cos x \sin h}{h} \right)$$

$$= \lim_{h \to 0} \frac{\sin x}{h} \left( \frac{-h^2}{2} \right) + \lim_{h \to 0} \frac{\cos x}{h} (h)$$

$$= \cos x.$$

## 4.1 Differential of a Function

3. $f : \Re \to \Re : x \to e^x$.

$$\lim_{h \to 0} \frac{e^{x+h} - e^x}{h} = \lim_{h \to 0} \frac{e^x}{h}\left[1 + h + \frac{h^2}{2} \cdots - 1\right] = e^x.$$

4. $f : \Re \to \Re : x \to x^4$ if $x \geq 0$, $x^2$ if $x < 0$. Consider the limit as $h$ approaches 0 from above (*i.e.*, $h \to 0_+$). Then

$$\lim_{h \to 0_+} \frac{f(0+h) - f(0)}{h} = \frac{h^4 - 0}{h} = h^3 = 0.$$

The limit as $h$ approaches 0 from below is

$$\lim_{h \to 0_-} \frac{f(0+h) - f(0)}{h} = \frac{h^3 - 0}{h} = h^2 = 0.$$

Thus $df(0)$ is defined and equal to 0.

5. $f : \Re \to \Re$, by

$$x \to -x^2 \quad x \leq 0$$
$$x \to (x-1)^2 - 1 \quad 0 < x \leq 1$$
$$x \to -x \quad x > 1$$

$$\lim_{x \to 0} f(x) = 0$$
$$\lim_{x \to 0_+} f(x) = 0.$$

Thus $f$ is continuous at $x = 0$.

$$\lim_{x \to 1_-} f(x) = -1$$
$$\lim_{x \to 1_+} f(x) = -1.$$

Thus $f$ is continuous at $x = 1$.

$$\lim_{x \to 0_-} df(x) = \lim_{x \to 0_-} (-2x) = 0$$
$$\lim_{x \to 0_+} df(x) = \lim_{x \to 0_+} 2(x-1) = -2$$
$$\lim_{x \to 1_-} df(x) = \lim_{x \to 1_-} 2(x-1) = 0$$
$$\lim_{x \to 1_+} df(x) = \lim_{x \to 1_+} (-1) = -1.$$

Hence $df(x)$ is not continuous at $x = 0$ and $x = 1$. See Fig. 4.2.
To extend the definition to the higher dimension case, we proceed as follows:

**Fig. 4.2** Example 4.1.5

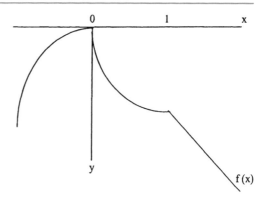

**Definition 4.1** Let $X, Y$ be two normed vector spaces, with norms $\|\ \|_X, \|\ \|_Y$, and suppose $f, g : X \to Y$. Then say $f$ and $g$ are *tangent* at $x \in X$ iff

$$\lim_{\|h\|_X \to 0} \frac{\|f(x+h) - g(x+h)\|_Y}{\|h\|_X} = 0.$$

If there exists a linear map $df(x) : X \to Y$ such that the function $g : X \to Y$ given by

$$g(x+h) = f(x) + df(x)(h)$$

is tangent to $f$ at $x$, then $f$ is said to be differentiable at $x$, and $df(x)$ is called the *differential* of $f$ at $x$.

In other words $df(x)$ is the differential of $f$ at $x$ iff there is a linear approximation $df(x)$ to $f$ at $x$, in the sense that

$$f(x+h) - f(x) = df(x)(h) + \|h\|_X \mu(h)$$

where $\mu : X \to Y$ and $\|\mu(h)\|_Y \to 0$ as $\|h\|_X \to 0$.

Note that since $df(x)$ is a linear map from $X$ to $Y$, then its image is a vector subspace of $Y$, and $df(x)(\underline{0})$ is the origin, $\underline{0}$, in $Y$.

Suppose now that $f$ is defined on an open ball $U$ in $X$.

For some $x \in U$, consider an open neighborhood $V$ of $x$ in $U$. The image of the map

$$h \to g(x+h) \quad \text{for each } h \in U$$

will be of the form $f(x) + df(x)(h)$, which is to say a linear subspace of $Y$, but translated by the vector $f(x)$ from the origin.

If $f$ is differentiable at $x$, then we can regard $df(x)$ as a linear map from $X$ to $Y$, so $df(x) \in L(X, Y)$, the set of linear maps from $X$ to $Y$. As we have shown in Sect. 3.2 of Chap. 3, $L(X, Y)$ is a normed vector space, when $X$ is finite dimensional. For example, for $k \in L(X, Y)$ we can define $\|k\|$ by

$$\|k\| = \sup\{\|k(x)\|_Y : x \in X \text{ s.t. } \|x\|_X = 1\}.$$

Let $\mathcal{L}(X, Y)$ be $L(X, Y)$ with the topology induced from this norm.

## 4.1 Differential of a Function

When $f : U \subset X \to Y$ is continuous we shall call $f$ a $C^0$-map. If $f$ is $C^0$, and $df(x)$ is defined at $x$, then $df(x)$ will be linear and thus continuous, in the sense that $df(x) \in \mathcal{L}(X, Y)$.

Hence we can regard $df$ as a map

$$df : U \to \mathcal{L}(X, Y).$$

It is important to note here that though the map $df(x)$ may be continuous, the map $df : U \to \mathcal{L}(X, Y)$ need not be continuous at $x$. However when $f$ is $C^0$, and the map

$$df : U \to \mathcal{L}(X, Y)$$

is continuous for all $x \in U$, then we shall say that $f$ is a $C^1$-differentiable map on $U$. Let $C_0(U, Y)$ be the set of maps from $U$ to $Y$ which are continuous on $U$, and let $C_1(U, Y)$ be the set of maps which are $C^1$-differentiable on $U$. Clearly $C_1(U, Y) \subset C_0(U, Y)$. If $f$ is a differentiable map, then $df(x)$, since it is linear, can be represented by a matrix. Suppose therefore that $f : \Re^n \to \Re$, and let $\{e_1, \ldots, e_n\}$ be the standard basis for $\Re^n$. Then for any $h \in \Re^n, h = \sum_{i=1}^{n} h_i e_i$ and so $df(x)(h) = \sum_{i=1}^{n} h_i df(x)(e_i) = \sum_{i=1}^{n} h_i \alpha_i$ say.

Consider the vector $(0, \ldots, h_i, \ldots, 0) \in \Re^n$.

Then by the definitions

$$\alpha_i = df(x)(0, \ldots, e_i, \ldots, 0)$$
$$= \underset{h_i \to o}{\text{Lim}} \left\{ \frac{f(x_1, \ldots, x_i + h_i, \ldots,) - f(x_1, \ldots, x_i, \ldots, x_n)}{h_i} \right\} = \left. \frac{\partial f}{\partial x_i} \right|_x,$$

where $\left. \frac{\partial f}{\partial x_i} \right|_x$ is called the *partial derivative* of $f$ at $x$ with respect to the $i$th coordinate, $x_i$.

Thus the linear function $df(x) : \Re \to \Re$ can be represented by a "row vector" or matrix

$$Df(x) = \left( \left.\frac{\partial f_j}{\partial x_1}\right|_x, \ldots, \left.\frac{\partial f}{\partial x_n}\right|_x \right).$$

Note that this representation is dependent on the particular choice of the basis for $\Re^n$. This matrix $Df(x)$ can also be regarded as a vector in $\Re^n$, and is then called the *direction gradient* of $f$ at $x$. The $i$th coordinate of $Df(x)$ is the partial derivative of $f$ with respect to $x_i$ at $x$.

If $h$ is a vector in $\Re^n$ with coordinates $(h_1, \ldots, h_n)$ with respect to the standard basis, then

$$df(x)(h) = \sum_{i=1}^{n} h_i \left.\frac{\partial f}{\partial x_i}\right|_x = \langle Df(x), h_i \rangle$$

where $\langle Df(x), h \rangle$ is the scalar product of $h$ and the direction gradient $Df(x)$.

In the same way if $f : \Re^n \to \Re^m$ and $f$ is differentiable at $x$, then $df(x)$ can be represented by the $n \times m$ matrix

$$Df(x) = \left(\frac{\partial f_j}{\partial x_i}\right)_x, \quad i = 1, \ldots, n; \; j = 1, \ldots, m$$

where $f(x) = f(x_1, \ldots, x_n) = (f_1(x), \ldots, f_j(x), \ldots, f_m(x))$. This matrix is called the *Jacobian* of $f$ at $x$. We may define the *norm* of $Df(x)$ to be

$$\|Df(x)\| = \sup\left\{\left|\frac{\partial f_j}{\partial x_i}\right|_x : i = 1, \ldots, n; \; j = 1, \ldots, m\right\}.$$

When $f$ has domain $U \subset \Re^n$, then continuity of $Df : U \to M(n, m)$, where $M(n, m)$ is the set of $n \times m$ matrices, implies the continuity of each partial derivative

$$U \to \Re : x \to \left.\frac{\partial f_j}{\partial x_i}\right|_x.$$

Note that when $f : \Re \to \Re$ then $\frac{\partial f}{\partial x}|_x$ is written simply as $\frac{df}{dx}|_x$ and is a real number.

Then the linear function $df(x) : \Re \to \Re$ is given by $df(x)(h) = (\frac{df}{dx}|_x)h$.

To simplify notation we shall not distinguish between the linear function $df(x)$ and the real number $\frac{df}{dx}|_x$ when $f : \Re \to \Re$.

Suppose now that $f : U \subset X \to Y$ and $g : V \subset Y \to Z$ such that $g \circ f : U \subset X \to Z$ exists. If $f$ is differentiable at $x$, and $g$ is differentiable at $f(x)$ then $g \circ f$ is differentiable at $x$ and is given by

$$d(g \circ f)(x) = dg(f(x)) \circ df(x).$$

In terms of Jacobian matrices this is $D(g \circ f)(x) = Dg(f(x)) \circ Df(x)$, or $\frac{\partial g_k}{\partial x_i} = \sum_{j=1}^{m} \frac{\partial g_k}{\partial f_j} \frac{\partial f_j}{\partial x_i}$, i.e.,

$$k\text{th row} \begin{pmatrix} \frac{\partial g_k}{\partial f_1} & \cdots & \frac{\partial g_k}{\partial f_m} \end{pmatrix} \begin{pmatrix} \frac{\partial f_1}{\partial x_i} \\ \vdots \\ \frac{\partial f_m}{\partial x_i} \end{pmatrix}$$
$i$th column

This is also known as the *chain-rule*.

If $\text{Id} : \Re^n \to \Re^n$ is the identity map then clearly the Jacobian matrix of $\text{Id}$ must be the identity matrix.

Suppose now that $f : U \subset \Re^n \to V \subset \Re^n$ is differentiable at $x$, and has an inverse $g = f^{-1}$ which is differentiable. Then $g \circ f = \text{Id}$ and so $\text{Id} = D(g \circ f)(x) = Dg(f(x)) \circ Df(x)$. Thus $D(f^{-1})(f(x)) = [Df(x)]^{-1}$.

In particular, for this to be the case $Df(x)$ must be an $n \times n$ matrix of rank $n$. When this is so, $f$ is called a *diffeomorphism*.

## 4.1 Differential of a Function

**Fig. 4.3** Example 4.2

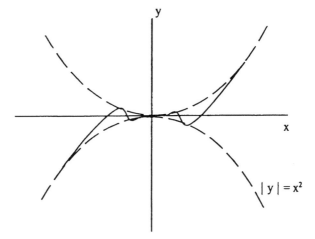

On the other hand suppose $f : X \to \Re$ and $g : Y \to \Re$, where $f$ is differentiable at $x \in X$ and $g$ is differentiable at $y \in Y$.

Let $fg : X \times Y \to \Re : (x, y) \to f(x)g(y)$. From the chain rule, $d(fg)(x, y) \times (h, k) = f(x) \, dg(y)(k) + g(y) \, df(x)(h)$, and so $d(fg)(x, y) = f(x)dg(y) + g(y)df(x)$.

Hence $fg$ is differentiable at the point $(x, y) \in X \times Y$.

When $X = Y = \Re$, and $(fg)(x) = f(x)g(x)$ then $d(fg)(x) = f dg(x) + g df(x)$, where this is called the *product rule*.

*Example 4.2* Consider the function $f : \Re \to \Re$ given by $x \to x^2 \sin \frac{1}{x}$ if $x \neq 0$; 0 if $x = 0$. See Fig. 4.3.

We first of all verify that $f$ is continuous. Let $g(x) = x^2$, $h(x) = \sin \frac{1}{x} = \rho[m(x)]$ where $m(x) = \frac{1}{x}$ and $\rho(y) = \sin(y)$. Since $m$ is continuous at any non zero point, both $h$ and $g$ are continuous. Thus $f$ is continuous at $x \neq 0$. (Compare with Example 3.1.)

Now $\lim_{x \to 0} x^2 \sin \frac{1}{x} = \lim_{y \to +\infty} \frac{\sin y}{y^2}$. But $-1 \leq \sin y \leq 1$, and so $\lim_{y \to +\infty} \frac{\sin y}{y^2} = 0$.

Hence $x_n \to 0$ implies $f(x_n) \to 0 = f(0)$. Thus $f$ is also continuous at $x = 0$. Consider now the differential of $f$. By the product rule, since $f = gh$,

$$d(gh)(x) = x^2 dh(x) + \left(\sin \frac{1}{x}\right) dg(x).$$

Since $h(x) = \rho[m(x)]$, by the chain rule,

$$dh(x) = d\rho(m(x)) \cdot dm(x)$$
$$= \cos\left[(m(x))\left(-\frac{1}{x^2}\right)\right]$$
$$= -\frac{1}{x^2} \cos \frac{1}{x}.$$

Thus

$$df(x) = d(gh)(x)$$
$$= x^2 \left[ -\frac{1}{x^2} \cos \frac{1}{x} + 2x \sin \frac{1}{x} \right]$$
$$= -\cos \frac{1}{x} + 2x \sin \frac{1}{x},$$

for any $x \neq 0$.

Clearly $df(x)$ is defined and continuous at any $x \neq 0$. To determine if $df(0)$ is defined, let $k = \frac{1}{h}$. Then

$$\lim_{h \to 0_+} \frac{f(0+h) - f(0)}{h} = \lim_{h \to 0_+} \frac{h^2 \sin \frac{1}{h}}{h}$$
$$= \lim_{h \to 0_+} h \sin \frac{1}{h}$$
$$= \lim_{k \to +\infty} \frac{\sin k}{k}.$$

But again $-1 \leq \sin k \leq 1$ for all $k$, and so $\lim_{k \to +\infty} \frac{\sin k}{k} = 0$. In the same way $\lim_{h \to 0_-} \frac{h^2 \sin \frac{1}{h}}{h} = 0$. Thus $\lim_{h \to 0} \frac{f(0+h) - f(0)}{h} = 0$, and so $df(0) = 0$. Hence $df(0)$ is defined and equal to zero.

On the other hand consider $(x_n) \to 0_+$. We show that $\lim_{x_n \to 0_+} df(x_n)$ does not exist. By the above $\lim_{x \to 0_+} df(x) = \lim_{x \to 0_+} [2x \sin \frac{1}{x} - \cos \frac{1}{x}]$. While $\lim_{x \to 0_+} 2x \sin \frac{1}{x} = 0$, there is no limit for $\cos \frac{1}{x}$ as $x \to 0_+$ (see Example 3.1). Thus the function $df : \mathfrak{R} \to \mathcal{L}(\mathfrak{R}, \mathfrak{R})$ is not continuous at the point $x = 0$.

The reason for the discontinuity of the function $df$ at $x = 0$ is that in any neighbourhood $U$ of the origin, there exist an "infinite" number of non-zero points, $x'$, such that $df(x') = 0$. We return to this below.

## 4.2 $C^r$-Differentiable Functions

### 4.2.1 The Hessian

Suppose that $f : X \to Y$ is a $C^1$-differentiable map, with domain $U \subset X$. Then as we have seen $df : U \to \mathcal{L}(X, Y)$ where $\mathcal{L}(X, Y)$ is the topological vector space of linear maps from $X$ to $Y$ with the norm

$$\|k\| = \sup \{ \|k(x)\|_Y : x \in X \text{ s.t. } \|x\|_X = 1 \}.$$

Since both $U$ and $\mathcal{L}(X, Y)$ are normed vector spaces, and $df$ is continuous, $df$ may itself be differentiable at a point $x \in U$. If $df$ is differentiable, then its derivative at $x$ is written $d^2 f(x)$, and will itself be a linear approximation of the map $df$ from $X$ to $\mathcal{L}(X, Y)$. If $df$ is $C^1$, then $df$ will be continuous, and $d^2 f(x)$ will also be a continuous map. Thus $d^2 f(x) \in \mathcal{L}(X, \mathcal{L}(X, Y))$.

## 4.2 $C^r$-Differentiable Functions

When $d^2 f : U \to \mathcal{L}(X, \mathcal{L}(X,Y))$ is itself continuous, and $f$ is $C^1$-differentiable, then say $f$ is $C^2$-differentiable. Let $C_2(U, Y)$ be the set of $C^2$-differentiable maps on $U$. In precisely the same way say that $f$ is $C^r$-differentiable iff $f$ is $C^{r-1}$-differentiable, and the $r$th derivative $df : U \to \mathcal{L}(X, \mathcal{L}(X, \mathcal{L}(X, \ldots)))$ is continuous.

The map is called *smooth* or $C^\infty$ if $d^r f$ is continuous for all $r$.

Now the second derivative $d^2 f(x)$ satisfies $d^2 f(x)(h)(k) \in Y$ for vectors $h, k \in X$. Moreover $d^2 f(x)(h)$ is a linear map from $X$ to $Y$ and $d^2 f(x)$ is a linear map from $X$ to $\mathcal{L}(X, Y)$.

Thus $d^2 f(x)$ is linear in both factors $h$ and $k$. Hence $d^2 f(x)$ may be regarded as a map

$$H(x) : X \times X \to Y$$

where $H(x)(h, k) = d^2 f(x)(h)(k) \in Y$.

Moreover $d^2 f(x)$ is linear in both $h$ and $k$, and so $H(x)$ is linear in both $h$ and $k$. As in Sect. 2.3.3, we say $H(x)$ is bilinear.

Let $L^2(X; Y)$ be the set of *bilinear maps* $X \times X \to Y$. Thus $H \in L^2(X; Y)$ iff

$$H(\alpha_1 h_1 + \alpha_2 h_2, k) = \alpha_1 H(h_1, k) + \alpha_2 H(h_2, k)$$
$$H(h, \beta_1 k_1 + \beta_2 k_2) = \beta_1 H(h, k_1) + \beta_2 H(h, k_2)$$

for any $\alpha_1, \alpha_2, \beta_1, \beta_2 \in Re, h, h_1, h_2, k, k_1, k_2 \in X$.

Since $X$ is a finite-dimensional normed vector space, so is $X \times X$, and thus the set of bilinear maps $L^2(X; Y)$ has a norm topology. Write $\mathcal{L}^2(X, Y)$ when the set of bilinear maps has this topology. The continuity of the second differential $d^2 f : U \to \mathcal{L}(X, \mathcal{L}(X, Y))$ is equivalent to the continuity of the map $H : U \to \mathcal{L}^2(X; Y)$, and we may therefore regard $d^2 f$ as a map $d^2 f : U \to \mathcal{L}^2(X; Y)$. In the same way we may regard $d^r f$ as a map $d^r f : U \to \mathcal{L}^r(X; Y)$ where $\mathcal{L}^r(X; Y)$ is the set of maps $X^r \to Y$ which are linear in each component, and is endowed with the norm topology.

Suppose now that $f : \Re^n \to \Re$ is a $C^2$-map, and consider a point $x = (x_1, \ldots, x_n)$ where the coordinates are chosen with respect to the standard basis. As we have seen the differential $df : U \to \mathcal{L}(\Re^n, \Re)$ can be represented by a continuous function

$$Df : x \to \left( \left.\frac{\partial f}{\partial x_1}\right|_x, \ldots, \left.\frac{\partial f}{\partial x_n}\right|_x \right).$$

Now let $\partial f_j : U \to \Re$ be the continuous function $x \to \left.\frac{\partial f}{\partial x_j}\right|_x$. Clearly the differential of $\partial f_j$ will be

$$x \to \left( \left.\frac{\partial}{\partial x_1}(\partial f_j)\right|_x, \ldots, \left.\frac{\partial}{\partial x_n}(\partial f_j)\right|_x \right);$$

write $\left.\frac{\partial}{\partial x_i}(\partial f_j)\right|_x = \partial f_{ji} = \left.\frac{\partial}{\partial x_i}\left(\frac{\partial f}{\partial x_j}\right)\right|_x$.

Then the differential $d^2 f(x)$ can be represented by the matrix array

$$(\partial f_{ji})_x = \begin{pmatrix} \frac{\partial}{\partial x_1}(\frac{\partial f}{\partial x_1}) & \cdots & \frac{\partial}{\partial x_1}(\frac{\partial f}{\partial x_n}) \\ \vdots & \vdots & \\ \frac{\partial}{\partial x_n}(\frac{\partial f}{\partial x_1}) & \cdots & \frac{\partial}{\partial x_n}(\frac{\partial f}{\partial x_n}) \end{pmatrix}_x.$$

This $n \times n$ matrix we shall also write as $Hf(x)$ and call the *Hessian matrix* of $f$ at $x$. Note that $Hf(x)$ is dependent on the particular basis, or coordinate system for $X$.

From elementary calculus it is known that

$$\frac{\partial}{\partial x_i}\left(\frac{\partial f}{\partial x_j}\right)\bigg|_x = \frac{\partial}{\partial x_j}\left(\frac{\partial f}{\partial x_i}\right)\bigg|_x$$

and so the matrix $Hf(x)$ is *symmetric*.

Consequently, as in Sect. 2.3.3, $Hf(x)$ may be regarded as a *quadratic form* given by

$$D^2 f(x)(h, k) = \langle h, Hf(x)(k) \rangle$$

$$= (h_1, \ldots, h_n)\left(\frac{\partial}{\partial x_i}\left(\frac{\partial f}{\partial x_j}\right)\begin{pmatrix} k_1 \\ k_n \end{pmatrix}\right)$$

$$= \sum_{i=1}^n \sum_{j=1}^n h_i \frac{\partial}{\partial x_i}\left(\frac{\partial f}{\partial x_j}\right) k_j.$$

As an illustration if $f : \Re^2 \to \Re$ is $C^2$ then $D^2 f(x) : \Re^2 \times \Re^2 \to \Re$ is given by

$$D^2 f(x)(h, h) = \begin{pmatrix} h_1 & h_2 \end{pmatrix} \begin{pmatrix} \frac{\partial^2 f}{\partial x_1^2} & \frac{\partial^2 f}{\partial x_1^2 \partial x_2} \\ \frac{\partial^2 f}{\partial x_1^2 \partial x_1} & \frac{\partial^2 f}{\partial x_2^2} \end{pmatrix} \begin{pmatrix} h_1 \\ h_2 \end{pmatrix}$$

$$= \left(h_1^2 \frac{\partial^2 f}{\partial x_1^2} + 2h_1 h_2 \frac{\partial^2 f}{\partial x_1 \partial x_2} + h_2^2 \frac{\partial^2 f}{\partial x_2^2}\right)\bigg|_x.$$

In the case that $f : \Re \to \Re$ is $C^2$, then $\frac{\partial^2 f}{\partial x^2}|_x$ is simply written as $\frac{d^2 f}{dx^2}|_x$, a real number. Consequently the second differential $D^2 f(x)$ is given by

$$D^2 f(x)(h, h) = h\left(\frac{d^2 f}{dx^2}\bigg|_x\right)h$$

$$= h^2 \frac{d^2 f}{dx^2}\bigg|_x.$$

We shall not distinguish in this case between the linear map $D^2 f(x) : \Re^2 \to \Re$ and the real number $\frac{d^2 f}{dx^2}|_x$.

**Fig. 4.4** Rolle's Theorem

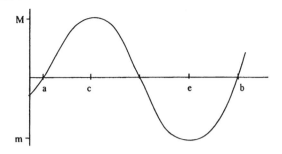

### 4.2.2 Taylor's Theorem

From the definition of the derivative of a function $f : X \to Y$, $df(x)$ is the linear approximation to $f$ in the sense that $f(x+h) - f(x) = df(x)(h) + \|h\|_x \mu(h)$ where the "error" $\|h\|_x \mu(h)$ approaches zero as $h$ approaches zero. Taylor's Theorem is concerned with the "accuracy" of this approximation for a small vector $h$, by using the higher order derivatives. Our principal tool in this is the following. If $f : U \subset X \to \Re$ and the convex hull $[x, x+h]$ of the points $x$ and $x+h$ belongs to $U$, then there is some point $z \in [x, x+h]$ such that $f(x+h) = f(x) + df(z)(h)$.

To prove this result we proceed as follows.

**Lemma 4.1** (Rolle's Theorem) *Let $f : U \to \Re$ where $U$ is an open set in $\Re$ containing the compact interval $I = [a, b]$, and $a < b$. Suppose that $f$ is continuous and differentiable on $U$, and that $f(a) = f(b)$. Then there exists a point $c \in (a, b)$ such that $df(c) = 0$.*

*Proof* From the Weierstrass Theorem (and Lemma 3.16) $f$ attains its upper and lower bounds on the compact interval, $I$. Thus there exists finite $m, M \in \Re$ such that $m \leq f(x) \leq M$ for all $x \in I$.

If $f$ is constant on $I$, so $m = f(x) = M, \forall x \in I$, then clearly $df(x) = 0$ for all $x \in I$.

Then there exists a point $c$ in the interior $(a, b)$ of $I$ such that $df(c) = 0$. Suppose that $f$ is not constant. Since $f$ is continuous and $I$ is compact, there exist points $c, e \in I$ such that $f(c) = M$ and $f(e) = m$. Suppose that neither $c$ nor $e$ belong to $(a, b)$. In this case we obtain $a = e$ and $b = c$, say. But then $M = m$ and so $f$ is the constant function. When $f$ is not the constant function either $c$ or $e$ belongs to the interior $(a, b)$ of $I$.

1. Suppose $c \in (a, b)$. Clearly $f(c) - f(x) \geq 0$ for all $x \in I$. Since $c \in (a, b)$ there exists $x \in I$ s.t. $x > c$, in which case $\frac{f(x) - f(c)}{x - c} \leq 0$. On the other hand there exists $x \in I$ s.t. $x < c$ and $\frac{f(x) - f(c)}{x - c} \geq 0$. By the continuity of $df$ at $x$, $df(c) = \text{Lim}_{x \to c_+} \frac{f(x) - f(c)}{x - c} = \text{Lim}_{x \to c_-} \frac{f(x) - f(c)}{x - c} = 0$. Since $c \in (a, b)$ and $df(x) = 0$ we obtain the result. See Fig. 4.4.

2. If $e \in (a, b)$ then we proceed in precisely the same way to show $df(e) = 0$. Thus there exists some point $c \in (a, b)$, say, such that $df(c) = 0$. □

**Fig. 4.5** Lemma 4.2

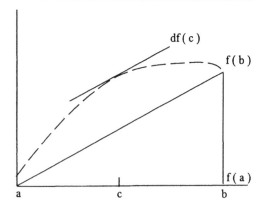

Note that when both $c$ and $e$ belong to the interior of $I$, then these maximum and minimum points for the function $f$ are critical points in the sense that the derivative is zero.

**Lemma 4.2** *Let $f : \Re \to \Re$ where $f$ is continuous on the interval $I = [a, b]$ and $df$ is continuous on $(a, b)$. Then there exists a point $c \in (a, b)$ such that $df(c) = \frac{f(b)-f(a)}{b-a}$.*

*Proof* Let $g(x) = f(b) - f(x) - k(b - x)$ and $k = \frac{f(b)-f(a)}{b-a}$. Clearly, $g(a) = g(b) = 0$. By Rolle's Theorem, there exists some point $c \in (a, b)$ such that $dg(c) = 0$. But $dg(c) = k - df(c)$. Thus $df(c) = \frac{f(b)-f(a)}{b-a}$. See Fig. 4.5. □

**Lemma 4.3** *Let $f : \Re \to \Re$ be continuous and differentiable on an open set containing the interval $[a, a + h]$. Then there exists a number $t \in (0, 1)$ such that*

$$f(a + h) = f(a) + df(a + th)(h).$$

*Proof* Put $b = a + h$. By the previous lemma there exists $c \in (a, a + h)$ such that

$$df(c) = \frac{f(b) - f(a)}{b - a}.$$

Let $t = \frac{c-a}{b-a}$. Clearly $t \in (0, 1)$ and $c = a + th$. But then $df(a + th) : \Re \to \Re$ is the linear map given by $df(a + th)(h) = f(b) - f(a)$, and so $f(a + h) = f(a) + df(a + th)(h)$. □

**Mean Value Theorem** *Let $f : U \subset X \to \Re$ be a differentiable function on $U$, where $U$ is an open set in the normed vector space $X$. Suppose that the line segment*

$$[x, x + h] = \{z : z = x + th \text{ where } t \in [0, 1]\}$$

## 4.2 $C^r$-Differentiable Functions

belongs to $U$. Then there is some number $t \in (0, 1)$ such that

$$f(x + h) = f(x) + df(x + th)(h).$$

*Proof* Define $g : [0, 1] \to \Re$ by $g(t) = f(x + th)$. Now $g$ is the composition of the function

$$\rho : [0, 1] \to U : t \to x + th$$

with $f : [x, x + h] \to \Re$.

Since both $\rho$ and $f$ are differentiable, so is $g$. By the chain rule,

$$dg(t) = df\bigl(\rho(t)\bigr) \circ d\rho(t)$$
$$= df(x + th)(h).$$

By Lemma 4.3, there exists $t \in (0, 1)$ such that $dg(t) = \frac{g(1)-g(0)}{1-0}$. But $g(1) = f(x+h)$ and $g(0) = f(x)$. Hence $df(x+th)(h) = f(x+h) - f(x)$. □

**Lemma 4.4** *Suppose $g : U \to \Re$ is a $C^2$-map on an open set $U$ in $\Re$ containing the interval $[0, 1]$. Then there exists $\theta \in (0, 1)$ such that*

$$g(1) = g(0) + dg(0) + \frac{1}{2}d^2g(\theta).$$

*Proof* (Note here that we regard $dg(t)$ and $d^2g(t)$ as real numbers.) Now define $k(t) = g(t) - g(0) - t\,dg(0) - t^2[g(1) - g(0) - dg(0)]$.

Clearly $k(0) = k(1) = 0$, and so by Rolle's Theorem, there exists $\theta \in (0, 1)$ such that $dk(\theta) = 0$. But $dk(t) = dg(t) - dg(0) - 2t[g(1) - g(0) - dg(0)]$. Hence $dk(0) = 0$.

Again by Rolle's Theorem, there exists $\theta' \in (0, \theta)$ such that $d^2k(\theta') = 0$. But $d^2k(\theta') = d^2g(\theta') - 2[g(1) - g(0) - dg(0)]$.

Hence $g(1) - g(0) - dg(0) = \frac{1}{2}d^2g(\theta')$ for some $\theta' \in (0, 1)$. □

**Lemma 4.5** *Let $f : U \subset X \to \Re$ be a $C^2$-function on an open set $U$ in the normed vector space $X$. If the line segment $[x, x + h]$ belongs to $U$, then there exists $z \in (x, x + h)$ such that*

$$f(x + h) = f(x) + df(x)(h) + \frac{1}{2}d^2 f(z)(h, h).$$

*Proof* Let $g : [0, 1] \to \Re$ by $g(t) = f(x + th)$. As in the mean value theorem, $dg(t) = df(x + th)(h)$. Moreover $d^2g(t) = d^2 f(x + th)(h, h)$.

By Lemma 4.4, $g(1) = g(0) + dg(0) + \frac{1}{2}d^2g(\theta')$ for some $\theta' \in (0, 1)$.

Let $z = x + \theta'h$. Then $f(x + h) = f(x) + df(x)(h) + \frac{1}{2}d^2 f(h, h)$. □

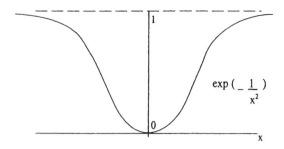

**Fig. 4.6** The flat function

**Taylor's Theorem** *Let $f : U \subset X \to \Re$ be a smooth (or $C^\infty$-) function on an open set $U$ in the normed vector space $X$. If the line segment $[x, x+h]$ belongs to $U$, then $f(x+h) = f(x) + \sum_{r=1}^{n} \frac{1}{r!} d^r f(x)(h, \ldots, h) + R_n(h)$ where the error term $R_n(h) = \frac{1}{(n+1)!} d^{n+1} f(z)(h, \ldots, h)$ and $z \in (x, x+h)$.*

*Proof* By induction on Lemma 4.5, using the mean value theorem. □

The *Taylor series of $f$ at $x$ to order $k$* is

$$[f(x)]_k = f(x) + \sum_{r=1}^{k} \frac{1}{r!} d^r f(x)(h, \ldots, h).$$

When $f$ is $C^\infty$, then $[f(x)]_k$ exists for all $k$. In the case when $X = \Re^n$ and the error term $R_k(h)$ approaches zero as $k \to \infty$, then the Taylor series $[f(x)]_k$ will converge to $f(x+h)$.

In general however $[f(x)]_k$ need not converge, or if it does converge then it need not converge to $f(x+h)$.

*Example 4.3* To illustrate this, consider the *flat* function $f : \Re \to \Re$ given by

$$f(x) = \exp\left(-\frac{1}{x^2}\right), \quad x \neq 0,$$
$$= 0 \quad x = 0.$$

Now $Df(x) = -\frac{2}{x^3} \exp(-\frac{1}{x^2})$ for $x \neq 0$. Since $y^{\frac{3}{2}} e^{-y} \to 0$ as $y \to \infty$, we obtain $Df(x) \to 0$ as $x \to 0$. However

$$Df(0) = \lim_{h \to 0} \frac{f(0+h) - f(0)}{h} = \lim_{h \to 0} \frac{1}{h} \exp\left(-\frac{1}{h^2}\right) = 0.$$

Thus $f$ is both continuous and $C^1$ at $x = 0$. In the same way $f$ is $C^r$ and $D^r f(0) = 0$ for all $r > 1$. Thus the Taylor series is $[f(0)]_k = 0$. However for small $h > 0$ it is evident that $f(0+h) \neq 0$. Hence the Taylor series cannot converge to $f$.

These remarks lead directly to classification theory in differential topology, and are beyond the scope of this work. The interested reader may refer to Chillingsworth (1976) for further discussion.

## 4.2.3  Critical Points of a Function

Suppose now that $f : U \subset \Re^n \to \Re$ is a $C^2$-map. Once a coordinate basis is chosen for $\Re^n$, then $D^2 f(x)$ may be regarded as a quadratic form. In matrix notation this means that

$$D^2 f(x)(h, h) = h^t H f(x) h.$$

As we have seen in Chap. 2 if the Hessian matrix $Hf(x) = (\partial f_{ji})_x$ has all its eigenvalues positive, then $D^2 f(x)(h, h) > 0$ for all $h \in \Re^n$, and so $Hf(x)$ will be *positive definite*.

Conversely $Hf(x)$ is negative definite iff all its eigenvalues are strictly negative.

**Lemma 4.6** *If $f : U \subset \Re^n \to \Re$ is a $C^2$ map on $U$, and the Hessian matrix $Hf(x)$ is positive (negative) definite at $x$, then there is a neighbourhood $V$ of $x$ in $U$ s.t. $Hf(y)$ is positive (negative) definite for all $y \in V$.*

*Proof* If $Hf(x)$ is positive definite, then as we have seen there are $n$ different algebraic relationships between the partial derivatives $\partial f_{ji}(x)$ for $j = 1, \ldots, n$ and $i = 1, \ldots, n$, which characterise the roots $\lambda_1(x), \ldots, \lambda_n(x)$ of the characteristic equation

$$\left| Hf(x) - \lambda(x) I \right| = 0.$$

But since $f$ is $C^2$, the map $D^2 f : U \to L^2(\Re^n; \Re)$ is continuous. In particular this implies that for each $i, j$ the map $x \to \frac{\partial}{\partial x_i}(\frac{\partial f}{\partial x_j})|_x = \partial f_{ji}(x)$ is continuous. Thus if $\partial f_{ji}(x) > 0$ then there is a neighbourhood $V$ of $x$ in $U$ such that $\partial f_{ji}(y) > 0$ for all $y \in V$. Moreover if

$$C(x) = C\big(\partial f_{ji}(x) : i = 1, \ldots, n; j = 1, \ldots, n\big)$$

is an algebraic sentence in $\partial f_{ji}(x)$ such that $C(x) > 0$ then again there is a neighbourhood $V$ of $x$ in $U$ such that $C(y) > 0$ for all $y \in V$.

Thus there is a neighborhood $V$ of $x$ in $U$ such that $\lambda_i(x) > 0$ for $i = 1, \ldots, n$ implies $\lambda_i(y) > 0$ for $i = 1, \ldots, n$, and all $y \in V$. Hence $Hf(x)$ is positive definite at $x \in U$ implies that $Hf(y)$ is positive definite for all $y$ in some neighborhood of $x$ in $U$. The same argument holds if $Hf(x)$ is negative definite at $x$. □

**Definition 4.2** Let $f : U \subset \Re^n \to \Re$ where $U$ is an open set in $\Re^n$. A point $x$ in $U$ is called
1. a *local strict maximum* of $f$ in $U$ iff there exists a neighbourhood $V$ of $x$ in $U$ such that $f(y) < f(x)$ for all $y \in V$ with $y \neq x$;
2. a *local strict minimum* of $f$ in $U$ iff there exists a neighbourhood $V$ of $x$ in $U$ such that $f(y) > f(x)$ for all $y \in V$ with $y \neq x$;
3. a *local maximum* of $f$ in $U$ iff there exists a neighbourhood $V$ of $x$ in $U$ such that $f(y) \leq f(x)$ for $y \in V$;

4. a *local minimum* of $f$ in $U$ iff there exists a neighbourhood $V$ of $x$ in $U$ such that $f(y) \geq f(x)$ for all $y \in V$.
5. Similarly a *global* (strict) *maximum* (or minimum) on $U$ is defined by requiring $f(y) < (\leq, >, \geq) f(x)$ respectively on $U$.
6. If $f$ is $C^1$-differentiable then $x$ is called a *critical point* iff $df(x) = 0$, the zero map from $\Re^n$ to $\Re$.

**Lemma 4.7** *Suppose that* $f : U \subset \Re^n \to \Re$ *is a* $C^2$*-function on an open set* $U$ *in* $\Re^n$. *Then* $f$ *has a local strict maximum (minimum) at* $x$ *if*
1. $x$ *is a critical point of* $f$ *and*
2. *the Hessian* $Hf(x)$ *is negative (positive) definite.*

*Proof* Suppose that $x$ is a critical point and $Hf(x)$ is negative definite. By Lemma 4.5

$$f(y) = f(x) + df(x)(h) + \frac{1}{2}d^2 f(z)(h, h)$$

whenever the line segment $[x, y] \in U$, $h = y - x$ and $z = x + \theta h$ for some $\theta \in (0, 1)$.

Now by the assumption there is a coordinate base for $\Re^n$ such that $Hf(x)$ is negative definite. By Lemma 4.6, there is a neighbourhood $V$ of $x$ in $U$ such that $Hf(y)$ is negative definite for all $y$ in $V$. Let $N_\epsilon(x) = \{x + h : \|h\| < \epsilon\}$ be an $\epsilon$-neighborhood in $V$ of $x$. Let $S_{\frac{\epsilon}{2}}(\underline{0}) = \{h \in \Re^n : \|h\| = \frac{1}{2}\epsilon\}$.

Clearly any vector $x + h$, where $h \in S_{\frac{\epsilon}{2}}(\underline{0})$ belongs to $N_\epsilon(x)$, and thus $V$. Hence $Hf(z)$ is negative definite for any $z = x + \theta h$, where $h \in S_{\frac{\epsilon}{2}}(\underline{0})$, and $\theta \in (0, 1)$. Thus $Hf(z)(h, h) < 0$, and any $z \in [x, x + h]$.

But also by assumption $df(x) = 0$ and so $df(x)(h) = 0$ for all $h \in \Re^n$. Hence $f(x + h) = f(x) + \frac{1}{2}d^2 f(z)(h, h)$ and so $f(x + h) < f(x)$ for $h \in S_{\frac{\epsilon}{2}}(\underline{0})$. But the same argument is true for any $h$ satisfying $\|h\| < \frac{\epsilon}{2}$. Thus $f(y) < f(x)$ for all $y$ in the open ball of radius $\frac{\epsilon}{2}$ about $x$. Hence $x$ is a local strict maximum. The same argument when $Hf(x)$ is positive definite shows that $x$ must be a local strict minimum. □

In Sect. 2.3 we defined a quadratic form $A^* : \Re^n \times \Re^n \to \Re$ to be non-degenerate iff the *nullity* of $A^*$, namely $\{x : A^*(x, x) = 0\}$, is $\{\underline{0}\}$. If $x$ is a critical point of a $C^2$-function $f : U \subset \Re^n \to \Re$ such that $d^2 f(x)$ is non-degenerate (when regarded as a quadratic form), then call $x$ a *non-degenerate critical point*.

The dimension $(s)$ of the subspace of $\Re^n$ on which $d^2 f(x)$ is negative definite is called the *index* of $f$ at $x$, and $x$ is called a critical point of index $s$.

If $x$ is a non-degenerate critical point, then when any coordinate system for $\Re^n$ is chosen, the Hessian $Hf(x)$ will have $s$ eigenvalues which are negative, and $n - s$ which are positive.

For example if $f : \Re \to \Re$, then only three cases can occur at a critical point
1. $d^2 f(x) > 0 : x$ is a local minimum;

## 4.2 $C^r$-Differentiable Functions

2. $d^2 f(x) < 0 : x$ is a local maximum;
3. $d^2 f(x) = 0 : x$ is a degenerate critical point.

If $f : \Re^2 \to \Re$ then a number of different cases can occur. There are three non-degenerate cases:

1. $Hf(x) = \begin{bmatrix} 1 & 0 \\ 0 & 1 \end{bmatrix}$, say, with respect to a suitable basis; $x$ is a local minimum since both eigenvalues are positive. Index $= 0$.
2. $Hf(x) = \begin{bmatrix} -1 & 0 \\ 0 & -1 \end{bmatrix}$; $x$ is a local maximum, both eigenvalues are negative. Index $= 2$.
3. $Hf(x) = \begin{bmatrix} 1 & 0 \\ 0 & -1 \end{bmatrix}$; $x$ is a *saddle point* or index 1 non-degenerate critical point.

In the degenerate cases, one eigenvalue is zero and so $\det(Hf(x)) = 0$.

*Example 4.4* Let $f : \Re^2 \to \Re : (x, y) \to xy$. The differential at $(x, y)$ is $Df(x, y) = (y, x)$. Thus $H = Hf(x, y) = \begin{pmatrix} 0 & 1 \\ 1 & 0 \end{pmatrix}$. Clearly $(0, 0)$ is the critical point. Moreover $|H| = -1$ and so $(0, 0)$ is non-degenerate. The eigenvalues $\lambda_1, \lambda_2$ of the Hessian satisfy $\lambda_1 + \lambda_2 = 0, \lambda_1 \lambda_2 = -1$. Thus $\lambda_1 = 1, \lambda_2 = -1$. Eigenvectors of $Hf(x, y)$ are $v_1 = (1, 1)$ and $v_2 = (1, -1)$ respectively. Let $P = \frac{1}{\sqrt{2}} \begin{pmatrix} 1 & 1 \\ 1 & -1 \end{pmatrix}$ be the normalized eigenvector (basis change) matrix, so $P^{-1} = P$. Then

$$\wedge = \frac{1}{2} \begin{pmatrix} 1 & 1 \\ -1 & 1 \end{pmatrix} \begin{pmatrix} 0 & 1 \\ 1 & 0 \end{pmatrix} \begin{pmatrix} 1 & 1 \\ -1 & 0 \end{pmatrix} = \begin{pmatrix} 1 & 0 \\ 0 & -1 \end{pmatrix}.$$

Consider a vector $h = (h_1, h_2) \in \Re^2$. In matrix notation, $h^t H h = h^t P \wedge P^{-1} h$. Now

$$P(h) = \frac{1}{\sqrt{2}} \begin{pmatrix} 1 & 1 \\ 1 & -1 \end{pmatrix} \begin{pmatrix} h_1 \\ h_2 \end{pmatrix} = \frac{1}{\sqrt{2}} \begin{pmatrix} h_1 + h_2 \\ h_1 - h_2 \end{pmatrix}.$$

Thus $h^t H h = \frac{1}{2}[(h_1 + h_2)^2 - (h_1 - h_2)^2] = 2h_1 h_2$.

It is clear that $D^3 f(0, 0) = 0$. Hence from Taylor's Theorem,

$$f(0 + h_1, 0 + h_2) = f(0) + Df(0)(h) + \frac{1}{2} D^2 f(0)(h, h)$$

and so $f(h_1, h_2) = \frac{1}{2} h^t H h = h_1 h_2$.

Suppose we make the basis change represented by $P$. Then with respect to the new basis $\{v_1, v_2\}$ the point $(x, y)$ has coordinates $(\frac{1}{\sqrt{2}}(h_1 + h_2), \frac{1}{\sqrt{2}}(h_1 - h_2))$.

Thus $f$ can be represented in a neighbourhood of the origin as

$$(h_1, h_2) \to \frac{1}{\sqrt{2}}(h_1 + h_2) \frac{1}{\sqrt{2}}(h_1 - h_2) = \frac{1}{2}(h_1^2 - h_2^2).$$

Notice that with respect to this new coordinate system

**Fig. 4.7** The flat function

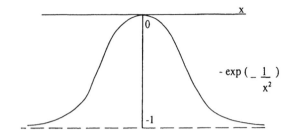

$$Df(h_1, h_2) = (h_1, -h_2), \quad \text{and}$$

$$Hf(h_1, h_2) = \begin{pmatrix} 1 & 0 \\ 0 & -1 \end{pmatrix}.$$

In the eigenspace $E_1 = \{(x, y) \in \Re^2 : x = y\}$ the Hessian has eigenvalue 1 and so $f$ has a local minimum at 0, when restricted to $E_1$.

Conversely in the eigenspace $E_{-1} = \{(x, y) \in \Re^2 : x + y = 0\}$, $f$ has a local maximum at 0.

More generally when $f : \Re^2 \to \Re$ is a quadratic function in $x, y$, then at a critical point $f$ can be represented, after a suitable coordinate change, either as
1. $(x, y) \to x^2 + y^2$ an index 0, minimum
2. $(x, y) \to -x^2 - y^2$ an index 2, maximum
3. $(x, y) \to x^2 - y^2$ an index 1 saddle point.

*Example 4.5* Let $f : \Re^3 \to \Re : (x, y, z) \to x^2 + 2y^2 + 3z^2 + xy + xz$. Therefore $Df(x, y, z) = (2x + y + z, 4y + x, 6z + x)$.

Clearly $(0, 0, 0)$ is the only critical point.

$$H = Hf(x, y, z) = \begin{pmatrix} 2 & 1 & 1 \\ 1 & 4 & 0 \\ 1 & 0 & 6 \end{pmatrix}$$

$|H| = 38$ and so $(0, 0, 0)$ is non-degenerate. It can be shown that the eigenvalues of the matrix are strictly positive, and so $H$ is positive definite and $(0, 0, 0)$ is a minimum of the function. Thus $f$ can be written in the form $(u, v, w) \to au^2 + bv^2 + cw^2$, where $a, b, c > 0$ and $(u, v, w)$ are the new coordinates after a (linear) basis change.

Notice that Lemma 4.7 does not assert that a local strict maximum (or minimum) of a $C^1$-function must be a critical point where the Hessian is negative (respectively positive) definite.

For example consider the "flat" function $f : \Re \to \Re$ given by $f(x) = -\exp(-\frac{1}{x^2})$, and $f(0) = 0$. As we showed in Example 4.3, $df(0) = d^2 f(0) = 0$.

Yet clearly $0 > -\exp(-\frac{1}{a^2})$ for any $a \neq 0$. Thus 0 is a global strict maximum of $f$ on $\Re$, although $d^2 f$ is not negative definite. See Fig. 4.7.

## 4.2 $C^r$-Differentiable Functions

A local maximum or minimum must however be a critical point. If the point is a local maximum for example then the Hessian can have no positive eigenvalues. Consequently $D^2 f(x)(h, h) \leq 0$ for all $h$, and so the Hessian must be negative semi-definite. As the flat function indicates, the Hessian may be identically zero at the local maximum.

**Lemma 4.8** *Suppose that $f : U \subset \Re^n \to \Re$ is a $C^2$-function on an open set $u$ in $\Re^n$. Then $f$ has a local maximum or minimum at $x$ only if $x$ is a critical point of $f$.*

*Proof* Suppose that $df(x) \neq 0$. Then we seek to show that $x$ can be neither a local maximum nor minimum at $x$.

Since $df(x)$ is a linear map from $\Re^n$ to $\Re$ it is possible to find some vector $h \in \Re^n$ such that $df(x)(h) > 0$.

Choose $h$ sufficiently small so that the line segment $[x, x + h]$ belongs to $U$.

Now $f$ is $C^1$, and so $df : U \to \mathcal{L}(\Re^n, \Re)$ is continuous. In particular since $df(x) \neq 0$ then for some neighbourhood $V$ of $x$ in $U$, $df(y) \neq 0$ for all $y \in V$. Thus for all $y \in V$, $df(y)(h) > 0$ (see Lemma 4.18 for more discussion of this phenomenon).

By the mean value theorem there exists $t \in (0, 1)$ such that $f(x + h) = f(x) + df(x + th)(h)$. By choosing $h$ sufficiently small, the vector $x + th \in V$. Hence $f(x + h) > f(x)$. Consequently $x$ cannot be a local maximum.

But in precisely the same way if $df(x) \neq 0$ then it is possible to find $h$ such that $df(x)(h) < 0$. A similar argument can then be used to show that $f(x + h) < f(x)$ and so $x$ cannot be a local minimum. Hence if $x$ is either a local maximum or minimum of $f$ then it must be a critical point of $f$. □

**Lemma 4.9** *Suppose that $f : U \subset \Re^n \to \Re$ is a $C^2$-function on an open set $U$ in $\Re^n$. If $f$ has a local maximum at $x$ then the Hessian $d^2 f$ at the critical point must be negative semi-definite (i.e., $d^2 f(x)(h, h) \leq 0$ for all $h \in \Re^n$).*

*Proof* We may suppose that $x$ is a critical point. Suppose further that for some co-ordinate basis at $x$, and vector $h \in \Re^n$, $d^2 f(x)(h, h) > 0$. From Lemma 4.6, by the continuity of $d^2 f$ there is a neighbourhood $V$ of $x$ in $U$ such that $d^2 f(x')(h, h) > 0$ for all $x' \in V$.

Choose an $\epsilon$-neighbourhood of $x$ in $V$, and choose $\alpha > 0$ such that $\|\alpha h\| = \frac{1}{2}\epsilon$. Clearly $x + \alpha h \in V$. By Taylor's Theorem, there exists $z = x + \theta \alpha h$, $\theta \in (0, 1)$, such that $f(x + \alpha h) = f(x) + df(x)(h) + \frac{1}{2} d^2 f(z)(\alpha h, \alpha h)$. But $d^2 f(z)$ is bilinear, so $d^2 f(z)(\alpha h, \alpha h) = \alpha^2 d^2 f(z)(h, h) > 0$ since $z \in V$. Moreover $df(x)(h) = 0$. Thus $f(x + \alpha h) > f(x)$.

Moreover for *any* neighbourhood $U$ of $x$ it is possible to choose $\epsilon$ sufficiently small so that $x + \alpha h$ belongs to $U$. Thus $x$ cannot be a local maximum. □

Similarly if $f$ has a local minimum at $x$ then $x$ must be a critical point with positive semi-definite Hessian.

*Example 4.6* Let $f : \Re^2 \to \Re : (x, y) \to x^2y^2 - 4yx^2 + 4x^2$; $Df(x, y) = (2xy^2 - 8xy + 8x, 2x^2y - 4x^2)$. Thus $(x, y)$ is a critical point when
1. $x(2y^2 - 8y + 8) = 0$ and
2. $2x^2(y - 2) = 0$.

Now $2y^2 - 8y + 8 = 2(y - 2)^2$. Thus $(x, y)$ is a critical point *either* when $y = 2$ or $x = 0$.

Let $S(f)$ be the set of critical points. Then $S(f) = V_1 \cup V_2$ where

$$V_1 = \{(x, y) \in \Re^2 : x = 0\}$$
$$V_2 = \{(x, y) \in \Re^2 : y = 2\}.$$

Now

$$Hf(x, y) = \begin{pmatrix} 2(y-2)^2 & -4x(2-y) \\ -4x(2-y) & 2x^2 \end{pmatrix}$$

and so when $(x, y) \in V_1$ then $Hf(x, y) = \begin{pmatrix} \mu^2 & 0 \\ 0 & 0 \end{pmatrix}$, and when $(x, y) \in V_2$, then $Hf(x, y) = \begin{pmatrix} 0 & 0 \\ 0 & \tau^2 \end{pmatrix}$.

For suitable $\mu$ and $\tau$, any point in $S(f)$ is degenerate. On $V_1 \setminus \{(0, 0)\}$ clearly a critical point is not negative semi-definite, and so such a point cannot be a local maximum. The same is true for a point on $V_2 \setminus \{(0, 0)\}$.

Now $(0, 0) \in V_1 \cap V_2$, and $Hf(0, 0) = (0)$. Lemma 4.9 does not rule out $(0, 0)$ as a local maximum. However it should be obvious that the origin is a local minimum.

Unlike Examples 4.4 and 4.5 no linear change of coordinate bases transforms the function into a quadratic canonical form.

To find a local maximum point we therefore seek all critical points. Those which have negative definite Hessian must be local maxima. Those points remaining which do not have a negative semi-definite Hessian cannot be local maxima, and may be rejected. The remaining critical points must then be examined.

A $C^2$-function $f : \Re^n \to \Re$ with a non-degenerate Hessian at every critical point is called a *Morse function*. Below we shall show that any Morse function can be represented in a *canonical* form such as we found in Example 4.4. For such a function, local maxima are precisely those critical points of index $n$. Moreover, any smooth function with a degenerate critical point can be "approximated" by a Morse function.

Suppose now that we wish to maximise a $C^2$-function on a compact set $K$. As we know from the Weierstrass theorem, there does exist a maximum. However, Lemmas 4.8 and 4.9 are now no longer valid and it is possible that a point on the boundary of $K$ will be a local or global maximum but not a critical point. However, Lemma 4.7 will still be valid, and a negative definite critical point will certainly be a local maximum.

A further difficulty arises since a local maximum need not be a global maximum. However, for concave functions, local and global maxima coincide. We discuss maximisation by smooth optimisation on compact convex sets in the next section.

## 4.3 Constrained Optimisation

### 4.3.1 Concave and Quasi-concave Functions

In the previous section we obtained necessary and sufficient conditions for a critical point to be a local maximum on an open set $U$. When the set is not open, then a local maximum need not be a critical point. Previously we have defined the differential of a function only on an open set. Suppose now that $Y \subset \Re^n$ is compact and therefore closed, and has a boundary $\partial Y$. If $df$ is continuous at each point in the interior, Int $(Y)$ of $Y$, then we may extend $df$ over $Y$ by defining $df(x)$, at each point $x$ in the boundary, $\partial Y$ of $Y$, to be the limit $df(x_k)$ for any sequence, $(x_k)$, of points in Int $(Y)$, which converge to $x$. More generally we shall say a function $f : Y \subset \Re^n \to \Re$ is $C^1$ on the *admissible* set $Y$ if $df : Y \to \mathcal{L}(\Re^n, \Re)$ is defined and continuous in the above sense at each $x \in Y$. We now give an alternative definition of the differential of a $C^1$-function $f : Y \to \Re$. Suppose that $Y$ is convex and both $x$ and $x + h$ belong to $Y$. Then the arc $[x, x + h] = \{z \in \Re^n : z = x + \lambda h, \lambda \in [0, 1]\}$ belongs to $Y$.

Now $df(x)(h) = \lim_{\lambda \to 0_+} \frac{f(x+\lambda h)-f(x)}{\lambda}$ and thus $df(x)(h)$ is often called the *directional derivative* of $f$ at $x$ in the direction $h$.

Finding maxima of a function becomes comparatively simple when $f$ is a concave or quasi-concave function (see Sect. 3.4 for definitions of these terms). Our first result shows that if $f$ is a concave function then we may relate $df(x)(y - x)$ to $f(y)$ and $f(x)$.

**Lemma 4.10** *If $f : Y \subset \Re^n \to \Re$ is a concave $C^1$-function on a convex admissible set $Y$ then*

$$df(x)(y - x) \geq f(y) - f(x).$$

*Proof* Since $f$ is concave

$$f(\lambda y + (1 - \lambda)x) \geq \lambda f(y) + (1 - \lambda) f(x)$$

for any $\lambda \in [0, 1]$ whenever $x, y \in Y$. But then $f(x + \lambda(y - x) - f(x)) \geq \lambda[f(y) - f(x)]$, and so

$$df(x)(y - x) = \lim_{\lambda \to 0_+} \frac{f(x + \lambda(y - x)) - f(x)}{\lambda}$$

$$\geq f(y) - f(x). \qquad \square$$

This enables us to show that for a concave function, $f$, a critical point of $f$ must be a global maximum when $Y$ is open.

First of all call a function $f : Y \subset \Re^n \to \Re$ *strictly quasi-concave* iff $Y$ is convex and for all $x, y \in Y$

$$f(\lambda y + (1 - \lambda)x) > \min(f(x), f(y)) \quad \text{for all } \lambda \in (0, 1).$$

Remember that $f$ is quasi-concave if

$$f(\lambda y + (1-\lambda)x) \geq \min(f(x), f(y)) \quad \text{for all } \lambda \in [0, 1].$$

As above let $P(x; Y) = \{y \in Y : f(y) > f(x)\}$ be the preferred set of a function $f$ on the set $Y$. A point $x \in Y$ is a global maximum of $f$ on $Y$ iff $P(x; Y) = \Phi$. When there is no chance of misunderstanding we shall write $P(x)$ for $P(x; Y)$. As shown in Sect. 3.4, if $f$ is (strictly) quasi-concave then, $\forall x \in Y$, the preferred set, $P(x)$, is (strictly) convex.

**Lemma 4.11**
1. If $f : Y \subset \Re^n \to \Re$ is a concave or strictly quasi-concave function on a convex admissible set, then any point which is a local maximum of $f$ is also a global maximum.
2. If $f : U \subset \Re^n \to \Re$ is a concave $C^1$-function where $U$ is open and convex, then any critical point of $f$ is a global maximum on $U$.

*Proof*
1. Suppose that $f$ is concave or strictly quasi-concave, and that $x$ is a local maximum but not a global maximum on $Y$. Then there exists $y \in Y$ such that $f(y) > f(x)$.

    Since $Y$ is convex, the line segment $[x, y]$ belongs to $Y$. For any neighbourhood $U$ of $x$ in $Y$ there exists some $\lambda^* \in (0, 1)$ such that, for $\lambda \in (0, \lambda^*)$, $z = \lambda y + (1-\lambda)x \in U$. But by concavity

    $$f(z) \geq \lambda f(y) + (1-\lambda) f(x) > f(x).$$

    Hence in any neighbourhood $U$ of $x$ in $Y$ there exists a point $z$ such that $f(z) > f(x)$. Hence $x$ cannot be a local maximum. Similarly by strict quasi-concavity

    $$f(z) > \min(f(x), f(y)) = f(x),$$

    and so, again, $x$ cannot be a local maximum. By contradiction a local maximum must be a global maximum.
2. If $f$ is $C^1$ and $U$ is open then by Lemma 4.8, a local maximum must be a critical point. By Lemma 4.10, $df(x)(y - x) \geq f(y) - f(x)$. Thus $df(x) = 0$ implies that $f(y) - f(x) \leq 0$ for all $y \in Y$. Hence $x$ is a global maximum of $f$ on $Y$. □

Clearly if $x$ were a critical point of a concave function on an open set then the Hessian $d^2 f(x)$ must be negative semi-definite. To see this, note that by Lemma 4.11, the critical point must be a global maximum, and thus a local maximum. By Lemma 4.9, $d^2 f(x)$ must be negative semi-definite. The same is true if $f$ is quasi-concave.

**Lemma 4.12** *If $f : U \subset \Re^n \to \Re$ is a quasi-concave $C^2$-function on an open set, then at any critical point, $x$, $d^2 f(x)$ is negative semi-definite.*

## 4.3 Constrained Optimisation

*Proof* Suppose on the contrary that $df(x) = 0$ and $d^2 f(x)(h, h) > 0$ for some $h \in \Re^n$. As in Lemma 4.6, there is a neighbourhood $V$ of $x$ in $U$ such that $d^2 f(z)(h, h) > 0$ for all $z$ in $V$.

Thus there is some $\lambda^* \in (0, 1)$ such that, for all $\lambda \in (0, \lambda^*)$, there is some $z$ in $V$ such that

$$f(x + \lambda h) = f(x) + df(x)(\lambda h) + \lambda^2 d^2 f(z)(h, h), \quad \text{and}$$

$$f(x - \lambda h) = f(x) + df(x)(-\lambda h) + (-\lambda h)^2 d^2 f(z)(h, h),$$

where $[x - \lambda h, x + \lambda h]$ belongs to $U$. Then $f(x + \lambda h) > f(x)$ and $f(x - \lambda h) > f(x)$. Now $x \in [x - \lambda h, x + \lambda h]$ and so by quasi-concavity,

$$f(x) \geq \min\bigl(f(x + \lambda h), f(x - \lambda h)\bigr).$$

By contradiction $d^2 f(x)(h, h) \leq 0$ for all $h \in \Re^n$. □

For a concave function, $f$, on a convex set $Y$, $d^2 f(x)$ is negative semi-definite not just at critical points, but at every point in the interior of $Y$.

**Lemma 4.13**
1. If $f : U \subset \Re^n \to \Re$ is a concave $C^2$-function on an open convex set $U$, then $d^2 f(x)$ is negative semi-definite for all $x \in U$.
2. If $f : Y \subset \Re^n \to \Re$ is a $C^2$-function on an admissible convex set $Y$ and $d^2 f(x)$ is negative semi-definite for all $x \in Y$, then $f$ is concave.

*Proof*
1. Suppose there exists $x \in U$ and $h \in \Re^n$ such that $d^2 f(x)(h, h) > 0$. By the continuity of $d^2 f$, there is a neighbourhood $V$ of $x$ in $U$ such that $d^2 f(z)(h, h) > 0$, for $z \in V$. Choose $\theta \in (0, 1)$ such that $x + \theta h \in V$. Then by Taylor's theorem there exists $z \in (x, x + \theta h)$ such that

$$f(x + \theta h) = f(x) + df(x)(\theta h) + \frac{1}{2} d^2 f(z)(\theta h, \theta h)$$
$$> f(x) + df(x)(\theta h).$$

   But then $df(x)(\theta h) < f(x + \theta h) - f(x)$. This contradicts $df(x)(y - x) \geq f(y) - f(x)$, $\forall x, y$ in $U$. Thus $d^2 f(x)$ is negative semi-definite.

2. If $x, y \in Y$ and $Y$ is convex, then the arc $[x, y] \subset Y$. Hence there is some $z = \lambda x + (1 - \lambda) y$, where $\lambda \in (0, 1)$, such that

$$f(y) = f(x) + df(x)(y - x) + d^2 f(z)(y - x, y - x)$$
$$\leq f(x) + df(x)(y - x).$$

   But in the same way $f(x) - f(z) \leq df(z)(x - z)$ and $f(y) - f(z) \leq df(z)(y - z)$. Hence

$$f(z) \geq \lambda \bigl[f(x) - df(z)(x - z)\bigr] + (1 - \lambda)\bigl[f(y) - df(z)(y - z)\bigr].$$

Now $\lambda df(z)(x-z) + (1-\lambda)df(z)(y-z) = df(z)[\lambda x + (1-\lambda)y - z] = df(z)(0) = 0$, since $df(z)$ is linear.
Hence $f(z) \geq \lambda f(x) + (1-\lambda)f(y)$ for any $\lambda \in [0,1]$ and so $f$ is concave. $\square$

We now extend the analysis to a quasi-concave function and characterise the preferred set $P(x; Y)$.

**Lemma 4.14** *Suppose $f : Y \subset \Re^n \to \Re$ is a quasi-concave $C^1$-function on the convex admissible set $Y$.*
1. *If $f(y) \geq f(x)$ then $df(x)(y-x) \geq 0$.*
2. *If $df(x)(y-x) > 0$, then there exists some $\lambda^* \in (0,1)$ such that $f(z) > f(x)$ for any*

$$z = \lambda y + (1-\lambda)x \quad \text{where } \lambda \in (0, \lambda^*).$$

*Proof*
1. By the definition of quasi-concavity $f(\lambda y + (1-\lambda)x) > f(x)$ for all $\lambda \in [0,1]$. But then, as in the analysis of a concave function,

$$f(x + \lambda(y-x)) - f(x) \geq 0$$

and so $df(x)(y-x) = \text{Lim}_{\lambda \to 0_+} \frac{f(x+\lambda(y-x)) - f(x)}{\lambda} \geq 0$.
2. Now suppose $f(x) \geq f(z)$ for all $z$ in the line segment $[x, y]$. Then $df(x)(y-x) = \text{Lim}_{\lambda \to 0_+} \frac{f(x+\lambda)(y-x) - f(x)}{\lambda} \leq 0$, contradicting $df(x)(y-x) > 0$. Thus there exists $z^* = \lambda^* y + (1-\lambda^*)x$ such that $f(z^*) > f(x)$. But then for all $z \in (x, \lambda^* y + (1-\lambda^*)x)$, $f(z) > f(x)$. $\square$

The property that $f(y) \geq f(x) \Rightarrow df(x)(y-x) \geq 0$ is often called *pseudo-concavity*.

We shall also say that $f : Y \subset \Re^n \to \Re$ is *strictly pseudo-concave* iff for any $x, y$ in $Y$, with $y \neq x$ then $f(y) \geq f(x)$ implies that $df(x)(y-x) > 0$. (Note we do not require $Y$ to be convex, but we do require it to be admissible.)

**Lemma 4.15** *Suppose $f : Y \subset \Re^n \to \Re$ is a strictly pseudo-concave function on an admissible set $Y$.*
1. *Then $f$ is strictly quasi-concave when $Y$ is convex.*
2. *If $x$ is a critical point, then it is a global strict maximum.*

*Proof*
1. Suppose that $f$ is not strictly quasi-concave. Then for some $x, y \in Y$, $f(\lambda^* y + (1-\lambda^*)x) \leq \min(f(x), f(y))$ for some $\lambda^* \in (0,1)$. Without loss of generality suppose $f(x) \leq f(y)$, and $f(\lambda y + (1-\lambda)x) \leq f(x)$ for all $\lambda \in (0,1)$. Then $df(x)(y-x) = \text{Lim}_{\lambda \to 0_+} \frac{f(\lambda y + (1-\lambda)x) - f(x)}{\lambda} \leq 0$. But by strict pseudo-concavity, we require that $df(x)(y-x) \geq 0$. Thus $f$ must be strictly quasi-concave.

## 4.3 Constrained Optimisation

2. If $df(x) = 0$ then $df(x)(y - x) = 0$ for all $y \in U$. Hence $f(y) < f(x)$ for all $y \in U, y \neq x$. Thus $x$ is a global strict maximum. □

As we have observed, when $f$ is a quasi-concave function on $Y$, the preferred set $P(x; Y)$ is a convex set in $Y$. Clearly when $f$ is continuous then $P(x; Y)$ is open in $Y$. As we might expect from the Separating Hyperplane Theorem, $P(x; Y)$ will then belong to an "open half space". To see this note that Lemma 4.14 establishes (for a quasi-concave $C^1$ function $f$) that the weakly preferred set

$$R(x; Y) = \{y \in Y : f(y) \geq f(x)\}$$

belongs to the closed half-space

$$H(x; Y) = \{y \in Y : df(x)(y - x) \geq 0\}.$$

When $Y$ is open and convex the boundary of $H(x; Y)$ is the hyperplane $\{y \in Y : df(x)(y - x) = 0\}$ and $H(x; Y)$ has relative interior $\overset{0}{H}(x; Y) = \{y \in Y : df(x)(y - x) > 0\}$. Write $H(x), R(x), P(x)$ for $H(x; Y), R(x; Y), P(x; Y)$, etc., when $Y$ is understood.

**Lemma 4.16** *Suppose $f : U \subset \Re^n \to \Re$ is $C^1$, and $U$ is open and convex.*

1. *If $f$ is quasi-concave, with $df(x) \neq 0$ then $P(x) \subset \overset{0}{H}(x)$,*
2. *If $f$ is concave or strictly pseudo-concave then $P(x) \subset \overset{0}{H}(x)$ for all $x \in U$. In particular if $x$ is a critical point, then $P(x) = \overset{0}{H}(x) = \Phi$.*

*Proof*

1. Suppose that $df(x) \neq 0$ but that $P(x) \not\subset \overset{0}{H}(x)$. However both $P(x)$ and $\overset{0}{H}(x)$ are open sets in $U$. By Lemma 4.14, $R(x) \subset H(x)$, and thus the closure of $P(x)$ belongs to the closure of $\overset{0}{H}(x)$ in $U$. Consequently there must exist a point $y$ which belongs to $P(x)$ yet $df(x)(y - x) = 0$, so $y$ belongs to the boundary of $\overset{0}{H}(x)$. Since $P(x)$ is open there exists a neighbourhood $V$ of $y$ in $P(x)$, and thus in $R(x)$. Since $y$ is a boundary point of $\overset{0}{H}(x)$ in any neighbourhood $V$ of $y$ there exists $z$ such that $z \notin H(x)$. But this contradicts $R(x) \subset H(x)$. Hence $P(x) \subset \overset{0}{H}(x)$.
2. Since a concave or strictly pseudo-concave function is a quasi-concave function, (1) establishes that $P(x) \subset \overset{0}{H}(x)$ for all $x \in U$ such that $df(x) \neq 0$. By Lemmas 4.11 and 4.15, if $df(x) = 0$ then $x$ is a global maximum. Hence $P(x) = \overset{0}{H}(x) = \Phi$. □

Lemma 4.8 shows that if $\overset{0}{H}(x) \neq \Phi$, then $x$ cannot be a global maximum on the open set $U$.

**Fig. 4.8** Illustration of Lemma 4.16

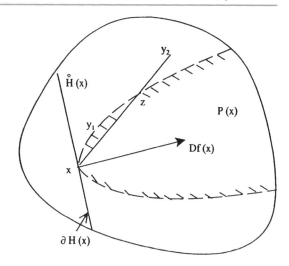

Moreover for a concave or strictly pseudo-concave function $P(x)$ is empty if $\overset{0}{H}(x)$ is empty (*i.e.*, when $df(x) = 0$).

Figure 4.8 illustrates these observations. Let $P(x)$ be the preferred set of a quasi-concave function $f$ (at a non-critical point $x$).

Point $y_1$ satisfies $f(y_1) = f(x)$ and thus belongs to $R(x)$ and hence $H(x)$. Point $y_2 \in H(x) \setminus P(x)$ but there exists an open interval $(x, z)$ belonging to $[x, y_2]$ and to $P(x)$.

We may identify the linear map $df(x) : \Re^n \to \Re$ with a vector $Df(x) \in \Re^n$ where $df(x)(h) = \langle Df(x), h \rangle$ the scalar product of $Df(x)$ with $h$. $Df(x)$ is the *direction gradient*, and is *normal* to the indifference surface at $x$, and therefore to the hyperplane $\partial H(x) = \{y \in Y : df(x)(y - x) = 0\}$.

To see this intuitively, note that the indifference surface $I(x) = \{y \in Y : f(y) = f(x)\}$ through $x$, and the hyperplane $\partial H(x)$ are tangent at $x$. Just as $df(x)$ is an approximation to the function $f$, so is the hyperplane $\partial H(x)$ an approximation to $I(x)$, near to $x$.

As we shall see in Example 4.7, a quasi-concave function, $f$, may have a critical point $x$ (so $\overset{0}{H}(x) = \Phi$) yet $P(x) \neq \Phi$. For example, if $Y$ is the unit interval, then $f$ may have a degenerate critical point with $P(x) \neq \Phi$. Lemma 4.16 establishes that this cannot happen when $f$ is concave and $Y$ is an open set. The final Lemma of this section extends Lemma 4.16 to the case when $Y$ is admissible.

**Lemma 4.17** *Let $f : Y \subset \Re^n \to \Re$ be $C^1$, and let $Y$ be a convex admissible set.*

1. *If $f$ is quasi-concave on $Y$, then $\forall x \in Y$, $\overset{0}{H}(x; Y) \neq \Phi$ implies $P(x; y) \neq \Phi$. If $x$ is a local maximum of $f$, then $\overset{0}{H}(x; Y) = \Phi$.*
2. *If $f$ is a strictly pseudo-concave function on an admissible set $Y$, and $x$ is a local maximum, then it is a global strict maximum.*

**Fig. 4.9** Illustration of Lemma 4.17

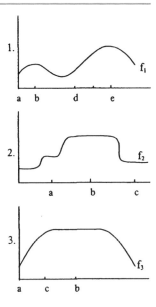

3.  If $f$ is concave $C^1$ or strictly pseudo-concave on $Y$, then $P(x;Y) \subset \overset{0}{H}(x : Y) \; \forall x \in Y$. Hence $\overset{0}{H}(x;Y) = \Phi \Rightarrow P(x;Y) = \Phi$.

*Proof*

1.  If $\overset{0}{H}(x;Y) \neq \Phi$ then $df(x)(y - x) > 0$ for some $y \in Y$. Then by Lemma 4.14(2), in any neighbourhood $U$ of $x$ in $Y$ there exists $z$ such that $f(z) > f(x)$. Hence $x$ cannot be a local maximum, and indeed $P(x) \neq \Phi$.
2.  By Lemma 4.15, $f$ must be quasi-concave. By (1), if $x$ is a local maximum then $df(x)(y-x) \leq 0$ for all $y \in Y$. By definition this implies that $f(y) < f(x)$ for all $y \in Y$ such that $y \neq x$. Thus $x$ is a global strict maximum.
3.  If $f$ is concave and $y \in P(x;Y)$, then $f(y) > f(x)$. By Lemma 4.10, $df(x)(y - x) \geq f(y) - f(x) > 0$. Thus $y \in \overset{0}{H}(x)$.
4.  If $f$ is strictly pseudo-concave then $f(y) > f(x)$ implies $df(x)(y - x) > 0$, and so $P(x;Y) \subset \overset{0}{H}(x;Y)$. □

*Example 4.7* These results are illustrated in Fig. 4.9.
1.  For the general function $f_1$, $b$ is a critical point and local maximum, but not a global maximum ($e$). On the compact interval $[a, d]$, $d$ is a local maximum but not a global maximum.
2.  For the quasi-concave function $f_2$, $a$ is a degenerate critical point but neither a local nor global maximum, while $b$ is a degenerate critical point which is also a global maximum.
    Point $c$ is a critical point which is also a local maximum. However on $[b, c]$, $c$ is not a global maximum.

3. For the concave function $f_3$, clearly $b$ is a degenerate (but negative semi-definite) critical point, which is also a local and global maximum. Moreover on the interval $[a, c]$, $c$ is the local and global maximum, even though it is not a critical point. Note that $df_3(c)(a - c) < 0$.

Lemma 4.17 suggests that we call any point $x$ in an admissible set $Y$ a *generalized critical point in $Y$* iff $\overset{0}{H}(x; Y) = \Phi$, of course if $df(x) = 0$, then $\overset{0}{H}(x; Y) = \Phi$, but the converse is not true when $x$ is a boundary point.

Lemma 4.17 shows that (i) for a quasi-concave $C^1$-function, a global maximum is a local maximum is a generalised critical point; (ii) for a concave $C^1$-or strictly pseudo-concave function a critical point is a generalised critical point is a local maximum is a global maximum.

### 4.3.2 Economic Optimisation with Exogenous Prices

Suppose now that we wish to find the maximum of a quasi-concave $C^1$-function $f : Y \to \Re$ subject to a constraint $g(x) \geq 0$ where $g$ is also a quasi-concave $C^1$-function $g : Y \to \Re$.

As we know from the previous section, when $P_f(x) = \{y \in Y : f(y) > f(x)\}$ and $df(x) \neq 0$, then $P_f(x) \subset H_f(x) = \{y \in Y : df(x)(x - y) \geq 0\}$.

Suppose now that $H_g(x) = \{y \in Y : dg(x)(x - y) \geq 0\}$ has the property that $H_g(x) \cap \overset{0}{H}_f(x) = \Phi$, and $x$ satisfies $g(x) = 0$.

In this case, there exists no point $y$ such that $g(y) \geq 0$ and $f(y) > f(x)$.

A condition that is sufficient for the disjointness of the two half-spaces $\overset{0}{H}_f(x)$ and $H_g(x)$ is clearly that $\lambda dg(x) + df(x) = 0$ for some $\lambda > 0$.

In this case if $df(x)(v) > 0$, then $dg(x)(v) < 0$, for any $v \in \Re^n$.

Now let $L = L_\lambda(f, g)$ be the Lagrangian $f + \lambda g : Y \to \Re$. A *sufficient* condition for $x$ to be a solution to the optimisation problem is that $dL(x) = 0$.

Note however that this is not a necessary condition. As we know from the previous section it might well be the case for some point $x$ on the boundary of the admissible set $Y$ that $dL(x) \neq 0$ yet there exists no $y \in Y$ such that

$$dg(x)(x - y) \geq 0 \quad \text{and} \quad df(x)(x - y) > 0.$$

Figure 4.10 illustrates such a case when

$$Y = \{(x, y) \in \Re^2 : x \geq 0, y \geq 0\}.$$

We shall refer to this possibility as the *boundary problem*.

As we know we may represent the linear maps $df(x), dg(x)$ by the direction gradients, or vectors normal to the indifference surfaces, labelled $Df(x), Dg(x)$.

When the maps $df(x), dg(x)$ satisfy $df(x) + \lambda dg(x) = 0, \lambda > 0$, then the direction gradients are *positively dependent* and satisfy $Df(x) + \lambda Dg(x) = 0$.

## 4.3 Constrained Optimisation

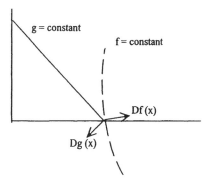

**Fig. 4.10** Optimisation of $f$ on a constraint, $g$

In the example $Df(x)$ and $Dg(x)$ are not positively dependent, yet $x$ is a solution to the optimisation problem.

It is often common to make some boundary assumption so that the solution does not belong to the boundary of the feasible or admissible set $Y$.

In the more general optimisation problem $(f, g) : Y \to \Re^{m+1}$, the Kuhn Tucker theorem implies that a global saddle point $(x^*, \lambda^*)$ to the Lagrangian $L_\lambda(f, g) = f + \sum \lambda_i g_i$ gives a solution $x^*$ to the optimisation problem. Aside from the boundary problem, we may find the global maxima of $L_\lambda(f, g)$ by finding the critical points of $L_\lambda(f, g)$.

Thus we must choose $x^*$ such that

$$df(x^*) + \sum_{i=1}^{m} \lambda_i \, dg_i(x^*) = 0.$$

Once a coordinate system is chosen this is equivalent to finding $x^*$, and coefficients $\lambda_1, \ldots, \lambda_m$ all non-negative such that

$$Df(x^*) + \sum_{i=1}^{m} \lambda_i Dg_i(x^*) = 0.$$

The Kuhn Tucker Theorem also showed that if $x^*$ is such that $g_i(x^*) > 0$, then $\lambda_i = 0$ and if $g_i(x^*) = 0$ then $\lambda_i > 0$.

*Example 4.8* Maximise the function $f : \Re \to \Re$:

$$x \to x^2 : x \geq 0$$

$$x \to 0 : x < 0$$

subject to $g_1(x) = x \geq 0$ and $g_2(x) = 1 - x \geq 0$. Now $L_\lambda(x) = x^2 + \lambda_1 x + \lambda_2(1 - x)$; $\frac{\partial L}{\partial x} = 2x + \lambda_1 - \lambda_2 = 0$, $\frac{\partial L}{\partial \lambda_1} = x = \frac{\partial L}{\partial \lambda_2} = 1 - x = 0$.

Clearly these equations have no solution. By inspection the solution cannot satisfy $g_1(x) = 0$. Hence choose $\lambda_1 = 0$ and solve

$$L_\lambda(x) = x^2 + \lambda(1 - x).$$

**Fig. 4.11** The Lagrangian $L(f, g)$

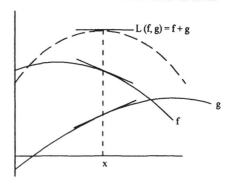

Then $\frac{\partial L}{\partial x} = 2x - \lambda$, $\frac{\partial L}{\partial \lambda} = 1 - x = 0$. Thus $x = 1$ and $\lambda = 2$ is a solution.

Suppose now that $f, g_1, \ldots, g_m$ are all concave functions on the convex admissible set $Y = \{x \in \Re^n : x_i \geq 0, i = 1, \ldots, n\}$. Obviously if $z = \alpha y + (1 - \alpha)x$, then

$$L_\lambda(f, g)(z) = f(z) + \sum_{i=1}^{m} \lambda_i g_i(z)$$

$$\geq \alpha f(y) + (1 - \alpha)f(x) + \sum_{i=1}^{m} \lambda_i \left[\alpha g_i(y) + (1 - \alpha)g_i(x)\right]$$

$$= \alpha L_\lambda(f, g)(y) + (1 - \alpha)L_\lambda(f, g)(x).$$

Thus $L_\lambda(f, g)$ is a concave function. By Lemma 4.11, $x^*$ is a global maximum of $L_\lambda(f, g)$ iff $dL(f, g)(x^*) = 0$ (aside from the boundary problem).

For more general functions, to find the global maximum of the Lagrangian $L_\lambda(f, g)$, and thus the optimum to the problem $(f, g)$, we find the critical points of $L_\lambda(f, g)$. Those critical points which have negative definite Hessian will then be local maxima of $L_\lambda(f, g)$. However we still have to examine the local maxima when the Hessian of the Lagrangian is negative semi-definite to find the global maxima. Even in this general case, any solution $x^*$ to the problem $(f, g)$ must be a global maximum for a suitably chosen Lagrangian $L_\lambda(f, g)$, and thus must satisfy the first order condition

$$Df(x^*) + \sum_{i=1}^{m} \lambda_i Dg_i(x^*) = 0$$

(again, subject to the boundary problem).

*Example 4.9* Maximise $f : \Re^2 \to \Re : (x, y) \to xy$ subject to the constraint $g(x, y) = 1 - x^2 - y^2 \geq 0$. We seek a solution to the first order condition:

$$DL(x, y) = Df(x, y) + \lambda Dg(x, y) = 0.$$

Thus $(y, x) + \lambda(-2x, -2y) = 0$ or $\lambda = \frac{y}{2x} = \frac{x}{2y}$ so $x^2 = y^2$.

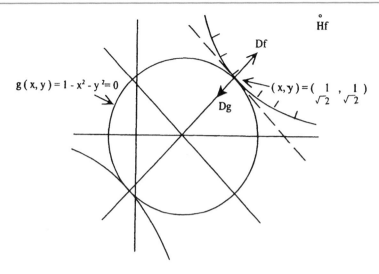

**Fig. 4.12** Example 4.9

For $x = -y$, $\lambda < 0$ and so $Df(x, y) = |\lambda| Dg(x, y)$, (corresponding to a minimum of $f$ on the feasible set $g(x, y) \geq 0$).

Thus we choose $x = y$ and $\lambda = \frac{1}{2}$. For $(x, y)$ on the boundary of the constraint set we require $1 - x^2 - y^2 = 0$. Hence $x = y = \pm \frac{1}{\sqrt{2}}$.

The Lagrangian is therefore $L = xy + \frac{1}{2}(1 - x^2 - y^2)$ with differential (with respect to $x, y$)

$$DL(x, y) = (y - x, x - y) \quad \text{and Hessian}$$

$$HL(x, y) = \begin{pmatrix} -1 & 1 \\ 1 & -1 \end{pmatrix}.$$

The eigenvalues of $HL$ are $-2$, $0$ corresponding to eigenvectors $(1, -1)$ and $(1, 1)$ respectively. Hence $HL$ is negative semi-definite, and so for example the point $(\frac{1}{\sqrt{2}}, \frac{1}{\sqrt{2}})$ is a local maximum for the Lagrangian. See Fig. 4.12.

As we have observed in Example 3.4, the function $f(x, y) = xy$ is not quasi-concave on $\Re^2$, and hence it is not the case that $P_f \subset H_f$. However on $\Re_+^2 = \{(x, y) \in \Re^2 : x \geq 0, y \geq 0\}$, $f$ is quasi-concave, and so the optimality condition $H_g(x, y) \cap H_f(x, y) = \Phi$ is sufficient for an optimum.

Note also that $Df(x, y) = (y, x)$ and so the origin $(0, 0)$ is a critical point of $f$. However setting $DL(x, y) = 0$ at $(x, y) = (0, 0)$ requires $\lambda = 0$. In this case however

$$HL(x, y) = \begin{pmatrix} 0 & 1 \\ 1 & 0 \end{pmatrix}$$

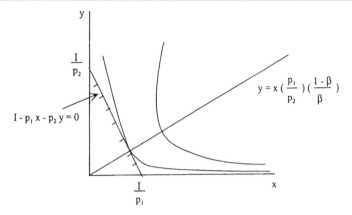

**Fig. 4.13** Optimising on a budget constraint

and as in Example 4.4, $HL$ is non-degenerate with eigenvalues $+1, -1$. Hence $HL$ is not negative semi-definite, and so $(0, 0)$ cannot be a local maximum for the Lagrangian.

However if we were to maximise $f(x, y) = -xy$ on the feasible set $\Re_+^2$ subject to the same constraint then $L$ would be maximized at $(0, 0)$ with $\lambda = 0$.

*Example 4.10* In Example 3.7 we examined the maximisation of a convex preference correspondence of a *consumer* subject to a budget constraint of the form $B(p) = \{x \in \Re_+^n : \langle p, x \rangle \leq \langle p, e \rangle = I\}$, given by an exogenous price vector $p \in \Re_+^n$, and initial endowment vector $e \in \Re_+^n$.

Suppose now that the preference correspondence is given by a utility function:
$$f : \Re_+^2 \to \Re : (x, y) \to \beta \log x + (1 - \beta) \log y, \quad 0 < \beta < 1.$$

Clearly $Df(x, y) = (\frac{\beta}{x}, \frac{1-\beta}{y})$, and so
$$Hf(x, y) = \begin{pmatrix} \frac{-\beta}{x^2} & 0 \\ 0 & \frac{-(1-\beta)}{y^2} \end{pmatrix}$$

is negative definite. Thus $f$ is concave on $\Re^2$. The budget constraint is
$$g(x, y) = I - p_1 x - p_2 y \geq 0$$

where $p_1, p_2$ are the given prices of commodities $x, y$. The first order condition on the Lagrangian is: $Df(x, y) + \lambda Dg(x, y) = 0$, i.e., $(\frac{\beta}{x}, \frac{1-\beta}{y}) + \lambda(-p_1, -p_2) = 0$ and $\lambda > 0$. Hence $\frac{p_1}{p_2} = \frac{\beta}{x} \cdot \frac{y}{1-\beta}$. See Fig. 4.13.

Now $f$ is concave and has no critical point within the constraint set. Thus $(x, y)$ maximises $L_\lambda$ iff
$$y = x \left(\frac{p_1}{p_2}\right)\left(\frac{1-\beta}{\beta}\right) \quad \text{and} \quad g(x, y) = 0.$$

## 4.3 Constrained Optimisation

Thus $y = \frac{I-p_1x}{p_2}$ and so $x = \frac{I\beta}{p_1}$, $y = \frac{I(1-\beta)}{p_2}$, and $\lambda = \frac{1}{I}$, is the marginal utility of income.

Now consider a situation where prices vary. Then optimal consumption $(x^*, y^*) = (d_1(p_1, p_2), d_2(p_1, p_2))$ where $d_i(p_1, p_2)$ is the *demand* for commodity $x$ or $y$. As we have just shown, $d_1(p_1, p_2) = \frac{I\beta}{p_1}$, and $d_2(p_1, p_2) = \frac{I(1-\beta)}{p_2}$.

Suppose that all prices are increased by the same proportion i.e., $(p_1', p_2') = \alpha(p_1, p_2), \alpha > 0$.

In this exchange situation $I' = p_1'e_1 + p_2'e_2 = \alpha I = \alpha(p_1e_1 + p_2e_2)$.

Thus $x' = \frac{I'\beta}{p_1'} = \frac{\alpha I \beta}{\alpha p_1} = x'$, and $y' = y$.

Hence $d_i(\alpha p_1, \alpha p_2) = d_i(p_1, p_2)$, for $i = 1, 2$. The demand function is said to be *homogeneous* in prices.

Suppose now that income is obtained from supplying labor at a wage rate $w$ say. Let the supply of labor by the consumer be $e = 1 - x_3$, where $x_3$ is leisure time and enters into the utility function.

Then $f : \Re^3 \to \Re : (x_1 x_2 x_3) \to \sum_{i=1}^{3} a_i \log x_i$ and the budget constraint is $p_1 x_1 + p_2 x_2 \leq (1 - x_3) w$, or $g(x_1, x_2, x_3) = w - (p_1 x_1 + p_2 x_2 + w x_3) \geq 0$. The first order condition is $(\frac{a_1}{x_1}, \frac{a_2}{x_2}, \frac{a_3}{x_3}) = \lambda(p_1, p_2, w), \lambda > 0$.

Clearly the demand function will again be homogeneous, since $d(p_1, p_2, w) = d(\alpha p_1, \alpha p_2, \alpha w)$.

For the general consumer optimisation problem, we therefore *normalise* the price vector. In general, in an $n$-commodity exchange economy let

$$\Delta = \{p \in \Re^n_+ : \|p\| = 1\}$$

be the *price simplex*. Here $\| \ \|$ is a convenient norm on $\Re^n$.

If $f : \Re^n_+ \to \Re$ is the utility function, let

$$D^* f(x) = \frac{Df(x)}{\|Df(x)\|} \in \Delta.$$

Suppose then that $x^* \in \Re^n$ is a maximum of $f : \Re^n_+ \to \Re$ subject to the budget constraint $\langle p, x \rangle \leq I$.

As we have seen the first order condition is

$$Df(x) + \lambda Dg(x) = 0,$$

where $Dg(x) = -p = (-p_1, \ldots, -p_n)$, $p \in \Delta$, and

$$Df(x) = \left(\frac{\partial f}{\partial x_1}, \ldots, \frac{\partial f}{\partial x_n}\right).$$

Thus $Df(x) = \lambda(p_1, \ldots, p_n) = \lambda p \in \Re^n_+$. But then $D^* f(x) = \frac{p}{\|p\|} \in \Delta$.

Subject to boundary problems, a *necessary* condition for optimal consumer behavior is that $D^* f(x) = \frac{p}{\|p\|}$.

As we have seen the optimality condition is that $\frac{\partial f}{\partial x_i} / \frac{\partial f}{\partial x_j} = \frac{p_i}{p_j}$, for the $i$th and $j$th commodity, where $\frac{\partial f}{\partial x_i}$ is often called the *marginal utility* of the $i$th commodity.

Now any point $y$ on the boundary of the budget set satisfies

$$\langle p, y \rangle = I = \frac{1}{\lambda}\langle Df(x^*), x^* \rangle.$$

Hence $y \in H(p, I)$, the hyperplane separating the budget set from the preferred set at the optimum $x^*$, iff $\langle Df(x^*), y - x^* \rangle = 0$.

Consider now the problem of maximisation of a profit function by a *producer*

$$\pi(x_1, \ldots, x_m, x_{m+1}, \ldots, x_n) = \sum_{j=1}^{n-m} p_{m+j} x_{m+j} - \sum_{j=1}^{m} p_j x_j,$$

where $(x_1, \ldots, x_m) \in \Re$ are inputs, $(x_{m+1}, \ldots, x_n)$ are outputs and $p \in \Re_+^n$ is a non-negative price vector.

As in Example 3.6, the set of *feasible* input-output combinations is given by the production set $G = \{x \in \Re_+^n : F(x) \geq 0\}$ where $F : \Re_+^n \to \Re$ is a smooth function and $F(x) = 0$ when $x$ is on the upper boundary or *frontier* of the production set $G$.

At a point $x$ on the boundary, the vector which is normal to the surface $\{x : F(x) = 0\}$ is

$$DF(x) = \left(\frac{\partial F}{\partial x_1}, \ldots, \frac{\partial F}{\partial x_n}\right)_x.$$

The first order condition for the Lagrangian is that

$$D\pi(x) + \lambda DF(x) = 0$$

or $(-p_1, \ldots, -p_m, p_{m+1}, \ldots, p_n) + \lambda(\frac{\lambda F}{\lambda x_1}, \ldots, \frac{\partial F}{\partial x_n}) = 0$.

For example with two inputs ($x_1$ and $x_2$) and one output ($x_3$) we might express maximum possible output $y$ in terms of $x_1$ and $x_2$, i.e., $y = g(x_1, x_2)$. Then the feasible set is

$$\{x \in \Re_+^3 : F(x_1, x_2, x_3) = g(x_1, x_2) - x_3 \geq 0\}.$$

Then $(-p_1, -p_2, p_3) + \lambda(\frac{\partial g}{\partial x_1}, \frac{\partial g}{\partial x_2}, -1) = 0$ and so

$$p_1 = p_3 \frac{\partial g}{\partial x_1}, \qquad p_2 = p_3 \frac{\partial g}{\partial x_2} \qquad \text{or} \qquad \frac{p_1}{p_2} = \frac{\partial g}{\partial x_1} \bigg/ \frac{\partial g}{\partial x_2}.$$

Here $\frac{\partial g}{\partial x_j}$ is called the *marginal product* (with respect to commodity $j$ for $j = 1, 2$).

For fixed $\bar{x} = (\bar{x}_1, \bar{x}_2)$ consider the locus of points in $\Re_+^2$ such that $y = g(\bar{x})$ is a constant. If $(\frac{\partial g}{\partial x_1}, \frac{\partial g}{\partial x_2})_{\bar{x}} \neq 0$ at $\bar{x}$, then by the implicit function theorem (discussed in the next chapter) we can express $x_2$ as a function $x_2(x_1)$ of $x_1$, only, near $\bar{x}$.

In this case $\frac{\partial g}{\partial x_1} + \frac{dx_2}{dx_1} \frac{\partial g}{\partial x_2} = 0$ and so $\frac{\partial g}{\partial x_1} \big/ \frac{\partial g}{\partial x_2}|_{\bar{x}} = \frac{dx_2}{dx_1}|_x = \frac{p_1}{p_2}$.

The ratio $\frac{\partial g}{\partial x_1} / \frac{\partial g}{\partial x_2}|_{\bar{x}}$ is called the *marginal rate of technical substitution* of $x_2$ for $x_1$ at the point $(\bar{x}_1, \bar{x}_2)$.

## 4.3 Constrained Optimisation

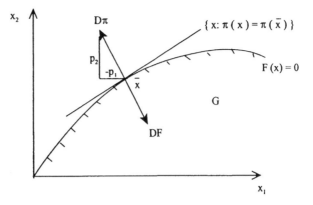

**Fig. 4.14** Maximising profit on a production set

*Example 4.11* There are two inputs $K$ (capital) and $L$ (labor), and one output, $Y$, say.

Let $g(K, L) = [dK^{-\rho} + (1-d)L^{-\rho}]^{\frac{-1}{\rho}}$. The feasibility constraint is

$$F(K, L, Y) = g(K, L) - Y \geq 0.$$

Let $-v, -w, p$ be the prices of capital, labor and the output. For optimality we obtain:

$$(-v, -w, p) + \lambda \left( \frac{\partial F}{\partial K}, \frac{\partial F}{\partial L}, \frac{\partial F}{\partial Y} \right) = 0.$$

On the production frontier, $g(K, L) = Y$ and so $p = -\lambda \frac{\partial F}{\partial Y} = \lambda$ since $\frac{\partial F}{\partial Y} = -1$.
Now let $X = [dk^{-\rho} + (1-d)L^{-\rho}]$.
Then $\frac{\partial F}{\partial K} = (-\frac{1}{\rho})[-\rho dk^{-\rho-1}] X^{-\frac{1}{\rho}-1}$.
Now $Y = X^{-\frac{1}{\rho}}$ so $Y^{1+\rho} = X^{-\frac{1}{\rho}-1}$. Thus $\frac{\partial F}{\partial K} = d(\frac{Y}{K})^{1+\rho}$.
Similarly $\frac{\partial F}{\partial L} = (1-d)(\frac{Y}{L})^{1+\rho}$. Thus $\frac{r}{w} = \frac{\partial F}{\partial K} / \frac{\partial F}{\partial L} = \frac{d}{1-d}(\frac{L}{K})^{1+\rho}$.

In the case just of a single output, where the production frontier is given by a function

$$x_{n+1} = g(x_1, \ldots, x_n) \quad \text{and} \quad (x_1, \ldots, x_n) \in \Re^n_+$$

is the input vector, then clearly the constraint set will be a convex set if and only if $g$ is a concave function. (See Example 3.3.) In this case the solution to the Lagrangian will give an optimal solution. See Fig. 4.14. However when the constraint set is not convex, then some solutions to the Lagrangian problem may be local minima. See Fig. 4.15 for an illustration.

As with the consumer, the optimum point on the production frontier is unchanged if all prices are multiplied by a positive number.

For a general consumer let $d : \Re^n_+ \to \Re^n_+$ be the demand map where

$$d(p_1, \ldots, p_n) = (x_1^*(p), \ldots, x_n^*(p)) = x^*(p)$$

**Fig. 4.15** A non-convex production set

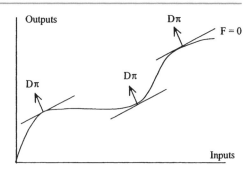

and $x^*(p)$ is any solution to the maximisation problem on the budget set

$$B(p_1, \ldots, p_n) = \{x \in \Re_+^n : \langle p, x \rangle \leq 1\}.$$

In general it need not be the case that $d$ is single-valued, and so it need not be a function.

As we have seen, $d$ is homogeneous in prices and so we may regard $d$ as a correspondence

$$d : \Delta \to \Re_+^n.$$

Similarly for a producer let $s : \Delta \to \Re_+^n$ where $s(p_1, \ldots, p_n) = (-x_1^*(p), \ldots, -x_m^*(p), x_{m+1}^*(p) \ldots)$ be the supply correspondence. (Here the first $m$ values are negative because these are inputs.)

Now consider a society $\{1, \ldots, i, \ldots, m\}$ and commodities named $\{1 \ldots j \ldots n\}$. Let $e_i \in \Re_+^n$ be the initial endowment vector of agent $i$, and $e = \sum_{i=1}^m e_i$ the total endowment of the society. Then a price vector $p^* \in \Delta$ is a *market-clearing price equilibrium* when $e + \sum_{i=1}^m s_i(p^*) = \sum_{i=1}^m d_i(p^*)$ where $s_i(p^*) \in \Re_+^n$ belongs to the set of optimal input-output vectors at price $p^*$ for agent $i$, and $d_i(p^*)$ is an optimal demand vector for consumer $i$ at price vector $p^*$.

As an illustration, consider a two person, two good exchange economy (without production) and let $e_{ij}$ be the initial endowment of good $j$ to agent $i$. Let $(f_1, f_2) : \Re_+^2 \to \Re^2$ be the $C^1$-utility functions of the two players.

At $(p_1, p_2) \in \Delta$, for optimality we have

$$\left( \frac{\partial f_i}{\partial x_{i1}}, \frac{\partial f_i}{\partial x_{i2}} \right) = \lambda_i (p_1, p_2), \quad \lambda_i > 0.$$

But $x_{1j} + x_{2j} = e_{1j} + e_{2j}$ for $j = 1$ or $2$, in market equilibrium. Thus $\frac{\partial f_i}{\partial x_{ij}} = -\frac{\partial f_i}{\partial x_{kj}}$ when $i \neq k$. Hence $\frac{1}{\lambda_1}(-\frac{\partial f_1}{\partial x_{11}}, \frac{\partial f_1}{\partial x_{12}}) = (p_1, p_2) = \frac{1}{\lambda_2}(\frac{\partial f_2}{\partial x_{11}}, -\frac{\partial f_2}{\partial x_{12}})$ or $(\frac{\partial f_1}{\partial x_{11}}, \frac{\partial f_1}{\partial x_{12}}) + \lambda(\frac{\partial f_2}{\partial x_{11}}, \frac{\partial f_2}{\partial x_{12}}) = 0$, for some $\lambda > 0$. See Fig. 4.16.

As we shall see in the next section this implies that the result $(x_{11}, x_{12}, x_{21}, x_{22})$ of optimal individual behaviour at the market-clearing price equilibrium is a *Pareto optimal outcome* under certain conditions.

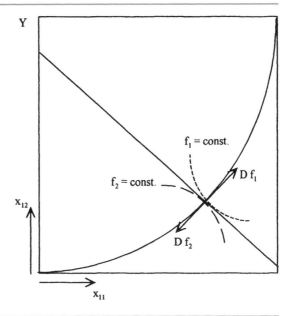

**Fig. 4.16** Two person, two good exchange economy

## 4.4 The Pareto Set and Price Equilibria

### 4.4.1 The Welfare and Core Theorems

Consider a society $M = \{1, \ldots, m\}$ of $m$ individuals where the preference of the $i$th individual in $M$ on a convex admissible set $Y$ in $\Re^n$ is given by a $C^1$-function $u_i : Y \subset \Re^n \to \Re$. Then $u = (u_1, \ldots, u_m) : Y \subset \Re^n \to \Re^m$ is called a $C^1$-*profile* for the society.

A point $y \in Y$ is said to be *Pareto preferred* (for the society $M$) to $x \in Y$ iff $u_i(y) > u_i(x)$ for all $i \in M$. In this case write $y \in P_M(x)$ and call $P_M : Y \to Y$ the *Pareto correspondence*. The (global) Pareto set for $M$ on $Y$ is the choice $P(u_1, \ldots, u_m) = \{y \in Y : P_M(y) = \Phi\}$. We seek to characterise this set.

In the same way as before we shall let

$$H_i(x) = \{y \in Y : du_i(x)(y - x) > 0\}$$

where $H_i : Y \to Y$ for each $i = 1, \ldots, m$. (Notice that $H_i(x)$ is open $\forall x \in Y$.)

Given a correspondence $P : Y \to Y$ the inverse correspondence $P^{-1} : Y \to Y$ is defined by

$$P^{-1}(x) = \{y \in Y : x \in P(y)\}.$$

In Sect. 3.3 we said a correspondence $P : Y \to Y$ (where $Y$ is a topological space) is *lower demi-continuous* (LDC) iff for all $x \in Y$, $P^{-1}(x)$ is open in $Y$.

Clearly if $u_i : W \to \Re$ is continuous then the preference correspondence $P_i : Y \to Y$ given by $P_i(x) = \{y : u_i(y) > u_i(x)\}$ is LDC, since

$$P_i^{-1}(y) = \{x \in Y : u_i(y) > u(x)\}$$

is open.

We show now that when $u_i : Y \to \Re$ is $C^1$ then $H_i : Y \to Y$ is LDC and as a consequence if $H_i(x) \neq \Phi$ then $P_i(x) \neq \Phi$.

This implies that if $x$ is a global maximum of $u_i$ on $Y$ (so $P_i(x) = \Phi$) then $H_i(x) = \Phi$ (so $x$ is a generalised critical point).

**Lemma 4.18** *If $u_i : Y \to \Re$ is a $C^1$-function on the convex admissible set $Y$, then $H_i : Y \to Y$ is lower demi-continuous and if $H_i(x) \neq \Phi$ then $P_i(x) \neq \Phi$.*

*Proof* Suppose that $H_i(x) \neq \Phi$. Then there exists $y \in Y$ such that $du_i(x)(y - x) > 0$. Let $h = y - x$.

By the continuity of $du_i : Y \to \mathcal{L}(\Re^n, \Re)$ there exists a neighbourhood $U$ of $x$ in $Y$, and a neighbourhood $V$ of $h$ in $\Re^n$ such that $du_i(z)(h') > 0$ for all $z \in U$, for all $h' \in V$.

Since $y \in H_i(x)$, we have $x \in H_i^{-1}(y)$. Now $h = y - x$. Let

$$U' = \{x' \in U : y - x' \in V\}.$$

For all $x' \in U'$, $du_i(x')(y - x') > 0$. Thus $U' \subset H_i^{-1}(y)$. Hence $H_i^{-1}(y)$ is open. This is true at each $y \in Y$, and so $H_i$ is $LDC$.

Suppose that $H_i(x) \neq \Phi$ and $h \in H_i(x)$. Since $H_i$ is LDC it is possible to choose $\lambda \in (0, 1)$, by Taylor's Theorem, such that

$$u_i(x + \lambda h) = u_i(x) + du_i(z)(\lambda h),$$

where $du_i(z)(h) > 0$, and $z \in (x, x + \lambda h)$. Thus $u_i(x + \lambda h) > u_i(x)$ and so $P_i(x) \neq \Phi$. □

When $u = (u_1, \ldots, u_m) : Y \to \Re^m$ is a $C^1$-profile then define the correspondence $H_M : Y \to Y$ by $H_M(x) = \bigcap_{i \in M} H_i(x)$ i.e., $y \in H_M(x)$ iff $du_i(x)(y - x) > 0$ for all $i \in M$.

**Lemma 4.19** *If $(u_1, \ldots, u_m) : Y \to \Re^m$ is a $C^1$-profile, then $H_M : Y \to Y$ is lower demi-continuous. If $H_M(x) \neq \Phi$ then $P_M(x) \neq \Phi$.*

*Proof* Suppose that $H_M(x) \neq \Phi$. Then there exists $y \in H_i(x)$ for each $i \in M$. Thus $x \in H_i^{-1}(y)$ for all $i \in M$. But each $H_i^{-1}(y)$ is open; hence $\exists$ an open neighbourhood $U_i$ of $x$ in $H_i^{-1}(y)$; let $U = \bigcap_{i \in M} U_i$. Then $x' \in U$ implies that $x' \in H_M^{-1}(y)$. Thus $H_M$ is LDC. As in the proof of Lemma 4.18 it is then possible to choose $h \in \Re^n$ such that, for all $i$ in $M$,

$$u_i(x + h) = u_i(x) + du_i(z)(h)$$

where $z$ belongs to $U$, and $du_i(z)(h) > 0$. Thus $x + h \in P_M(x)$ and so $P_M(x) \neq \Phi$. □

## 4.4 The Pareto Set and Price Equilibria

The set $\{x : H_M(x) = \Phi\}$ is called the *critical Pareto set*, and is often written as $\Theta_M$, or $\Theta(u_1, \ldots, u_m)$. By Lemma 4.19, $\Theta(u_1, \ldots, u_m)$ *contains* the Pareto set $P(u_1, \ldots, u_m)$.

Moreover we can see that $\Theta_M$ must be closed in $Y$. To see this suppose that $H_M(x) \neq \Theta$, and $y \in H_M(x) \neq \Phi$. Thus $x \in H_M^{-1}(y)$. But $H_M$ is LDC and so there is a neighbourhood $U$ of $x$ in $Y$ such that $x' \in H_M^{-1}(y)$ for all $x' \in U$. Then $y \in H_M(x')$ for all $x' \in U$, and so $H_M(x') \neq \Phi$ for all $x' \in U$.

Hence the set $\{x \in Y : H_M(x) \neq \Phi\}$ is open and so the critical Pareto set is closed.

In the same way, the Pareto correspondence $P_M : Y \to Y$ is given by $P_M(x) = \bigcap_{i \in M} P_i(x)$ where $P_i(x) = \{y : u_i(y) > u_i(x)\}$ for each $i \in M$. Since each $P_i$ is LDC, so must be $P_M$, and thus the Pareto set $P(u_1, \ldots, u_m)$ must also be closed.

Suppose now that $u_1, \ldots, u_m$ are all concave $C^1$-or strictly pseudo-concave functions on the convex set $Y$.

By Lemma 4.17, for each $i \in M$, $P_i(x) \subset H_i(x)$ at each $x \in Y$.

If $x \in \Theta(u_1, \ldots, u_m)$ then

$$\bigcap_{i \in M} P_i(x) \subset \bigcap_{i \in M} H_i(x) = \Phi$$

and so $x$ must also belong to the (global) Pareto set. Thus if $u = (u_1, \ldots, u_m)$ with each $u_i$ concave $C^1$ or strictly pseudo-concave, then the global Pareto set $P(u)$ and the critical Pareto set $\Theta(u)$ coincide. In this case we may more briefly say the preference profile represented by $u$ is *strictly convex*.

A point in $P(u_1, \ldots, u_m)$ is the precise analogue, in the case of a family of functions, of a maximum point for a single function, while a point in $\Theta(u_1, \ldots, u_m)$ is the analogue of a critical point of a single function $u : Y \to \Re$. In the case of a family or profile of functions, a point $x$ belongs to the critical Pareto set $\Theta_M(u)$, when a generalised Lagrangian $L(u_1, \ldots, u_m)$ has differential $dL(x) = 0$.

This allows us to define a Hessian for the family and determine which critical Pareto points are global Pareto points.

Suppose then that $u = (u_1, \ldots, u_m) : Y \to \Re^m$ where $Y$ is a convex admissible set in $\Re^n$ and each $u_i : Y \to \Re$ is a $C^1$-function.

A generalised Lagrangian $L(\lambda, u)$ for $u$ is a *semipositive* combination $\sum_{i=1}^m \lambda_i u_i$ where each $\lambda_i \geq 0$ but not all $\lambda_i = 0$.

For convenience let us write

$$\Re_+^m = \{x \in \Re^m : x_i \geq 0 \text{ for } i \in M\}$$
$$\overset{0}{\Re_+^m} = \{x \in \Re^m : x_i > 0 \text{ for } i \in M\}, \quad \text{and}$$
$$\overline{\Re_+^m} = \Re_+^m \setminus \{0\}.$$

Thus $\lambda \in \overline{\Re_+^m}$ iff each $\lambda_i \geq 0$ but not all $\lambda_i = 0$. Since each $u_i : Y \to \Re$ is a $C^1$-function, the differential at $x$ is a linear map $du_i(x) : \Re^n \to \Re$. Once a coordinate basis for $\Re^n$ is chosen, $du_i(x)$ may be represented by the row vector

$$Du_i(x) = \left(\left.\frac{\partial u_i}{\partial x}\right|_x, \ldots, \left.\frac{\partial u_i}{\partial x_n}\right|_x\right).$$

Similarly the profile $u : Y \to \Re^m$ has differential at $x$ represented by the $(n \times m)$ Jacobian matrix

$$Du(x) = \begin{pmatrix} Du_1(x) \\ \vdots \\ Du_m(x) \end{pmatrix} : \Re^n \to \Re^m.$$

Suppose now that $\lambda \in \Re^m$. Then define $\lambda \cdot Du(x) : \Re^n \to \Re$ by

$$(\lambda \cdot Du(x))(v) = \langle \lambda, Du(x)(v) \rangle$$

where $\langle \lambda, Du(x)(v) \rangle$ is the scalar product of the two vectors $\lambda, Du(x)(v)$ in $\Re^m$.

**Lemma 4.20** *The gradient vectors $\{Du_i(x) : i \in M\}$ are linearly dependent and satisfy the equation*

$$\sum_{i=1}^m \lambda_i Du_i(x) = 0$$

*iff $[\text{Im } Du(x)]^\perp$ is the subspace of $\Re^m$ spanned by $\lambda = (\lambda_1, \ldots, \lambda_m)$. Here $\lambda \in [\text{Im } Du(x)]^\perp$ iff $\langle \lambda, w \rangle = 0$ for all $w \in \text{Im } Du(x)$.*

*Proof*

$$\begin{aligned} \lambda \in \left[\text{Im } Du(x)\right]^\perp &\Leftrightarrow \langle \lambda, w \rangle = 0 \quad \forall\, w \in \text{Im } Du(x) \\ &\Leftrightarrow \langle \lambda, Du(x)(v) \rangle = 0 \quad \forall\, v \in \Re^n \\ &\Leftrightarrow (\lambda \cdot Du(x))(v) = 0 \quad \forall\, v \in \Re^n \\ &\Leftrightarrow \lambda \cdot Du(x) = 0. \end{aligned}$$

But $\lambda \cdot Du(x) = 0 \Leftrightarrow \sum_{i=1}^m \lambda_i Du_i(x) = 0$, where $\lambda = (\lambda_1, \ldots, \lambda_m)$. $\square$

**Theorem 4.21** *If $u : Y \to \Re^m$ is a $C^1$-profile on an admissible convex set and $x$ belongs to the interior of $Y$, then $x \in \Theta(u_1, \ldots, u_m)$ iff there exists $\lambda \in \overline{\Re_+^m}$ such that $dL(\lambda, u)(x) = 0$.*

*If $x$ belongs to the boundary of $Y$ and $dL(\lambda, u)(x) = 0$, for $\lambda \in \overline{\Re_+^m}$, then $x \in \Theta(u_1, \ldots, u_m)$.*

### 4.4 The Pareto Set and Price Equilibria

*Proof* Pick a coordinate basis for $\Re^n$. Suppose that there exists $\lambda \in \overline{\Re^m_+}$ such that

$$L(\lambda, u)(x) = \sum_{i=1}^{m} \lambda_i u_i(x) \in \Re,$$

satisfies $\sum_{i=1}^{m} \lambda_i Du_i(x) = 0$ (that is to say $DL(\lambda, u)(x) = 0$).

By Lemma 4.20 this implies that

$$\lambda \in \left[\mathrm{Im}(Du(x))\right]^{\perp}.$$

However suppose $x \notin \Theta(u_1, \ldots, u_m)$. Then there exists $v \in \Re^n$ such that $Du(x)(v) = w \in \overset{0}{\Re^m_+}$, i.e., $\langle Du_i(x), v \rangle = w_i > 0$ for all $i \in M$, where $w = (w_1, \ldots, w_m)$. But $w \in \mathrm{Im}\, Du(x)$ and $w \in \overset{0}{\Re^m_+}$.

Moreover $\lambda \in \overline{\Re^m_+}$ and so $\langle \lambda, w \rangle > 0$ (since not all $\lambda_i = 0$).

This contradicts $\lambda \in [\mathrm{Im}(Du(x))]^{\perp}$, since $\langle \lambda, w \rangle \neq 0$. Hence $x \in \Theta(u_1, \ldots, u_m)$. Thus we have shown that for any $x \in Y$, if $DL(\lambda, u)(x) = 0$ for some $\lambda \in \overline{\Re^m_+}$, then $x \in \Theta(u_1, \ldots, u_m)$. Clearly $DL(\lambda, u)(x) = 0$ iff $DL(\lambda, u)(x) = 0$, so we have proved sufficiency.

To show necessity, suppose that $\{Du_i(x) : i \in M\}$ are linearly independent. If $x$ belongs to the interior of $Y$ then for a vector $h \in \Re^n$ there exists a vector $y = x + \theta h$, for $\theta$ sufficiently small, so that $y \in Y$ and $\forall i \in M, \langle Du_i(x), h \rangle > 0$. Thus $x \notin \Theta(u_1, \ldots, u_m)$.

So suppose that $DL(\lambda, u)(x) = 0$ where $\lambda \neq 0$ but $\lambda \notin \overset{0}{\Re^m_+}$. Then for at least one $i$, $\lambda_i < 0$. But then there exists a vector $w \in \overset{0}{\Re^m_+}$ where $w = (w_1, \ldots, w_m)$ and $w_i > 0$ for each $i \in M$, such that $\langle \lambda, w \rangle = 0$. By Lemma 4.20, $w \in \mathrm{Im}(Du(x))$. Hence there exists a vector $h \in \Re^n$ such that $Du(x)(h) = w$.

But $w \in \overset{0}{\Re^m_+}$, and so $\langle Du_i(x), h \rangle > 0$ for all $i \in M$. Since $x$ belongs to the interior of $Y$, there exists a point $y = x + \alpha h$ such that $y \in H_i(x)$ for all $i \in M$. Hence $x \notin \Theta(u_1, \ldots, u_m)$.

Consequently if $x$ is an interior point of $Y$ then $x \in \Theta(u_1, \ldots, u_m)$ implies that $dL(\lambda, u)(x) = 0$ for some semipositive $\lambda$ in $\overline{\Re^m_+}$. □

*Example 4.12* To illustrate, we compute the Pareto set in $\Re^2$ when the utility functions are

$$u_1(x_1, x_2) = x_1^{\alpha} x_2 \quad \text{where } \alpha \in (0, 1), \quad \text{and}$$
$$u_2(x_1, x_2) = 1 - x_1^2 - x_2^2.$$

We maximise $u_1$ subject to the constraint $u_2(x_1, x_2) \geq 0$.

As in Example 4.9, the first order condition is

$$\left(\alpha x_1^{\alpha-1} x_2, x_1^{\alpha}\right) + \lambda(-2x_1, -2x_2) = 0.$$

**Fig. 4.17** Example 4.12

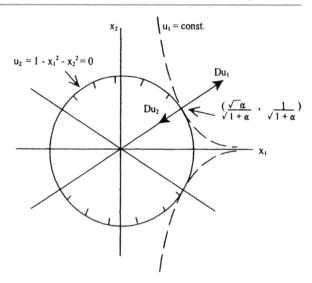

Hence $\lambda = \frac{\alpha x_1^{\alpha-1} x_2}{2x_1} = \frac{x_1^\alpha}{2x_2}$, so $\alpha x_2^2 = x_1^2$, or $x_1 = \pm\sqrt{\alpha} x_2$.

If $x_1 = -\sqrt{\alpha} x_2$ then $\lambda = \frac{x_1^\alpha}{\sqrt{\alpha}} 2(-x_1) < 0$, and so such a point does not belong to the critical Pareto set. Thus $(x_1, x_2) \in \Theta(u_1, u_2)$ iff $x_1 = x_2\sqrt{\alpha}$. Note that if $x_1 = x_2 = 0$ then the Lagrangian may be written as

$$L(\lambda, u)(0,0) = \lambda_1 u_1(0,0) + \lambda_2 u_2(0,0)$$

where $\lambda_1 = 0$ and $\lambda_2$ is any positive number. In the positive quadrant $\Re_+^2$, the critical Pareto set and global Pareto set coincide.

Finally to maximise $u_1$ on the set $\{(x_1, x_2) : u_2(x_1, x_2) \geq 0\}$ we simply choose $\lambda$ such that $u_2(x_1, x_2) = 0$.

Thus $x_1^2 + x_2^2 = \alpha x_2^2 + x_2^2 = 1$ or $x_2 = \frac{1}{\sqrt{1+\alpha}}$ and so $(x_1, x_2) = (\frac{\sqrt{\alpha}}{\sqrt{1+\alpha}}, \frac{1}{\sqrt{1+\alpha}})$.

In the next chapter we shall examine the critical Pareto set $\Theta(u_1, \ldots, u_m)$ and demonstrate that the set belongs to a singularity "manifold" which can be topologically characterised to be of "dimension" $m - 1$. This allows us then to examine the price equilibria of an exchange economy.

Note that in the example of a two person exchange economy studied in Sect. 4.3.2, we showed that the result of individual optimising behaviour led to an outcome, $x$, such that

$$Du_1(x) + \lambda Du_2(x) = 0$$

where $\lambda > 0$. As we have shown here this "market clearing equilibrium" must belong to the critical Pareto set $\Theta(u_1, u_2)$. Moreover, when both $u_1$ and $u_2$ represent strictly convex preferences, then this outcome belongs to the global Pareto set. We develop this in the following theorem.

## 4.4 The Pareto Set and Price Equilibria

**The Welfare Theorem for an Exchange Economy** *Consider an exchange economy where each individual in the society $M = \{1, \ldots, i, \ldots, m\}$ has initial endowment $e_i \in \Re_+^n$.*

1. *Suppose that the demand $x_i^*(p) \in \Re_+^n$ of agent $i$ at each price $p \in \Delta$ is such that $x_i^*(p)$ maximises the $C^1$-utility function $u_i : \Re^n \to \Re$ on the budget set $B_i(p) = \{x_i \in \Re_+^n : \langle p, x_i \rangle \leq \langle p, e_i \rangle\}$ and satisfies*
    (a) $Du_i(x_i^*(p)) = \lambda_i p, \lambda_i > 0$,
    (b) $\langle p, x_i^*(p) \rangle = \langle p, e_i \rangle$.
2. *Suppose further that $p^*$ is a market clearing price equilibrium in the sense that*
    $$\sum_{i=1}^m x_i^*(p^*) = \sum_{i=1}^m e_i \in \Re^n.$$
*Then $x^* = (x_1^*(p^*), \ldots, x_m^*(p)) \in \Theta(u_1, \ldots, u_m)$.*

*Moreover if each $u_i$ is either concave, or strictly pseudo-concave then $x^*$ belongs to the Pareto set, $P(u_1, \ldots, u_m)$.*

*Proof* We need to define the set of outcomes first of all. An outcome, $x$, is a vector

$$x = (x_1, \ldots, x_i, \ldots, x_m) \in \left(\Re_+^n\right)^m = \Re_+^{nm},$$

where $x_i = (x_{i1}, \ldots, x_{in}) \in \Re_+^n$ is an *allocation* for agent $i$. However there are $n$ resource constraints

$$\sum_{i=1}^m x_{ij} = \sum_{i=1}^m e_{ij} = e_{\cdot j},$$

for $j = 1, \ldots, n$, where $e_i = (e_{i1}, \ldots, e_{in}) \in \Re_+^n$ is the initial endowment of agent $i$.

Thus the set, $Y$, of feasible allocations to the members of $M$ is a hyperplane of dimension $n(m-1)$ in $\Re_+^{nm}$ through the point $(e_1, \ldots, e_m)$.

As coordinates for any point $x \in Y$ we may choose

$$x = (x_{11}, \ldots, x_{1n}, x_{21}, \ldots, x_{2n}, \ldots, x_{(m-1)1}, \ldots, x_{(m-1)n})$$

where it is implicit that the bundle of commodities available to agent $m$ is

$$x_m = (x_{m1}, \ldots, x_{mn})$$

where $x_{mj} = e_{\cdot j} - \sum_{i=1}^{m-1} x_{ij}$.

Now define $u_i^* : Y \to \Re$, the extended utility function of $i$ on $Y$ by

$$u_i^*(x) = u_i(x_{i1}, \ldots, x_{in}).$$

For $i \in M$, it is clear that the direction gradient of $i$ on $Y$ is

$$Du_i^*(x) = \left(0, \ldots, 0, \frac{\partial u_i}{\partial x_i}, \ldots, \frac{\partial u_i}{\partial x_{in}}, 0, \ldots\right) = (\ldots, 0, \ldots, Du_i(x), \ldots, 0).$$

For agent $m$, $\frac{\partial u_m^*}{\partial x_{ij}} = -\frac{\partial u_m}{\partial x_{mj}}$ for $i = 1, \ldots, m-1$; thus

$$Du_m^*(x) = -\left(\frac{\partial u_m}{\partial x_{m1}}, \ldots, \frac{\partial u_m}{\partial x_{mn}}, \ldots\right)$$
$$= -(Du_m(x), \ldots, \ldots, Du_m(x)).$$

If $p^*$ is a market-clearing price equilibrium, then by definition

$$\sum_{i=1}^m x_i^*(p^*) = \sum_{i=1}^m e_i.$$

Thus $x^*(p^*) = (x_1^*(p^*), \ldots, x_{m-1}^*(p^*))$ belongs to $Y$. But each $x_i^*$ is a critical point of $u_i : \Re_+^n \to \Re$ on the budget set $B_i(p^*)$ and $Du_i(x_i^*(p^*)) = \lambda_i p^*$.

Thus the Jacobian for $u^* = (u_1^*, \ldots, u_m^*) : Y \to \Re^m$ at $x^*(p^*)$ is

$$Du^*(x^*) = \begin{bmatrix} \lambda_1 p^* & 0 & \cdots & \cdots & 0 \\ 0 & \lambda_2 p^* & & & \\ \vdots & \vdots & & & \lambda_{m-1} p^* \\ -\lambda_m p^* & -\lambda_m p^* & & & -\lambda_m p^* \end{bmatrix}$$

Hence $\frac{1}{\lambda_1}Du_1^*(x^*) + \frac{1}{\lambda_2}Du_2^*(x^*) \cdots + \frac{1}{\lambda_m}Du_m^*(x^*) = 0$. But each $\lambda_i > 0$ for $i = 1, \ldots, m$. Then $dL(\mu, u^*)(x^*(p^*)) = 0$ where $L(\mu, u^*)(x) = \sum_{i=1}^m \mu_i u_i^*(x)$ and $\mu_i = \frac{1}{\lambda_i}$ and $\mu \in \overline{\Re_+^m}$.

By Theorem 4.21, $x^*(p^*)$ belongs to the critical Pareto set.

Clearly, if for each $i$, $u_i : \Re_+^n \to \Re$ is concave $C^1$- or strictly pseudo-concave then $u_1^* : Y \to \Re$ will be also. By previous results, the critical and global Pareto set will coincide, and so $x^*(p^*)$ will be Pareto optimal. □

One can also show that the *competitive allocation*, $x^*(p^*) \in \Re_+^{nm}$, constructed in this theorem is Pareto optimal in a very simple way. By definition $x^*(p^*)$ is characterised by the two properties:
1. $\sum_{i=1}^m x_i^*(p^*) = \sum_{i=1}^m e_i$ in $\Re^n$ (feasibility)
2. If $u_i(x_i) > u_i(x_i^*(p^*))$ then $\langle p^*, x_i \rangle > \langle p^*, e_i \rangle$ (by the optimality condition for agent $i$).

But if $x_i^*(p^*)$ is not Pareto optimal, then there exists a vector $x = (x_1, \ldots, x_m) \in \Re_+^{nm}$ such that $u_i(x_i) > u_i(x_i^*(p^*))$ for $i = 1, \ldots, m$. By (2), $\langle p^*, x_i \rangle > \langle p^*, e_i \rangle$ for each $i$ and so

$$\sum_{i=1}^m \langle p^*, x_i \rangle = \left\langle p^*, \sum_{i=1}^m x_i \right\rangle > \left\langle p^*, \sum_{i=1}^m e_i \right\rangle.$$

But if $x \in \Re_+^{nm}$ is feasible then $\sum_{i=1}^m x_i \leq \sum_{i=1}^m e_i$ which implies $\langle p^*, \sum_{i=1}^m x_i \rangle \leq \langle p^*, \sum_{i=1}^m e_i \rangle$.

By contradiction $x^*(p^*)$ must belong to the Pareto optimal set.

## 4.4 The Pareto Set and Price Equilibria

The observation has an immediate extension to a result on existence of a *core* of an economy.

**Definition 4.3** Let $e = (e_1, \ldots, e_m) \in \Re_+^{nm}$ be an initial endowment vector for a society $M$. Let $\mathcal{D}$ be any family of subsets of $M$, and let $P = (P_1, \ldots, P_m)$ be a profile of preferences for society $M$, where each $P_i : \Re_+^n \to \Re_+^n$ is a preference correspondence for $i$ on the $i$th consumption space $X_i \subset \Re_+^n$.

An *allocation* $x \in \Re_+^{nm}$ is $S$-feasible (for $S \subset M$) iff $x = (x_{ij}) \in \Re_+^{nm}$ and $\sum_{i \in S} x_{ij} = \sum_{i \in S} e_{ij}$ for each $j = 1, \ldots, n$.

Given $e$ and $P$, an allocation $x \in \Re_+^{nm}$ belongs to the $\mathcal{D}$-core of $(e, P)$ iff $x = (x_1, \ldots, x_m)$ is $M$-feasible and there exists no coalition $S \in \mathcal{D}$ and an allocation $y \in \Re_+^{nm}$ such that $y$ is $S$-feasible and of the form $y = (y_1, \ldots, y_m)$ with $y_i \in P_i(x_i), \forall i \in S$.

To clarify this definition somewhat, consider the set $Y$ from the proof of the welfare theorem. $Y = Y_M$ is a hyperplane of dimension $n(m-1)$ through the endowment point $e \in \Re_+^{nm}$. For any coalition $S \in \mathcal{D}$ of cardinality $s$, there is a hyperplane $Y_S$, say, of dimension $n(s-1)$ through the endowment point $e$, consisting of $S$-feasible trades among the members of $S$. Clearly $Y_S \subset Y_M$. If $x \in Y_M$ but there is some $y \in Y_S$ such that every member of $S$ prefers $y$ to $x$, then the members of $S$ could refuse to accept the allocation $x$. If there is no such point $x$, then $x$ is "unbeaten", and belongs to the $\mathcal{D}$-core of the economy described by $(e, P)$.

**Core Theorem** *Let $\mathcal{D}$ be any family of subsets of $M$. Suppose that $p^* \in \Delta$ is a market clearing price equilibrium for the economy $(e, P)$ and $x^*(B) \in \Re_+^{nm}$ is the demand vector, where*

$$x^*(p^*) = (x_1^*(p), \ldots, x_m^*(p)) \in Y_M, \quad \text{and} \quad P_i(x_i^*(p^*)) \cap B_i(p^*) = \Phi.$$

*Then $x^*(p^*)$ belongs to the $\mathcal{D}$-core, for the economy $(e, P)$.*

*Proof* Suppose that $x^*(p^*)$ is not in the core. Then there is some $y \in Y_S$ such that $y = (y_i : i \in S)$ and $y \in P_i(x_i)$ for each $i \in S$. Now $x_i^*(p^*)$ is a most preferred point for $i$ on $B_i(p^*)$ so $\langle p^*, y_i \rangle > \langle p^*, e_i \rangle$ for all $i \in S$.
Hence

$$\left\langle p^*, \sum_{i \in S} y_i \right\rangle = \sum_{i \in S} \langle p^*, y_i \rangle > \sum_{i \in S} \langle p^*, e_i \rangle.$$

However if $y \in Y_S$, then $\sum_{i \in S} y_i = \sum_{i \in S} e_i \in \Re_+^n$, which implies $\langle p^*, \sum_{i \in S}(y_i - e_i) \rangle = 0$. By contradiction, $x^*(p^*)$ must be in the core. □

The Core Theorem shows, even if a price mechanism is not used, that if a market clearing price equilibrium, $p^*$ does exist, then the core is non-empty. This means in essence that the competitive allocation, $x^*(p^*)$, is Pareto optimal for every coalition $S \in \mathcal{D}$.

By the results of Sect. 3.8, a market clearing price equilibrium $p^*$ will exist under certain conditions on preference. In particular suppose preference is representable by smooth utility functions that are concave or strictly pseudo-concave and monotonic in the individual consumption spaces. Then the conditions of the Welfare Theorem will be satisfied, and there will exist a market clearing price equilibrium, $p^*$, and a competitive allocation, $x^*(p^*)$, at $p^*$ which belongs to the Pareto set for the society $M$. Indeed, since the Core Theorem is valid when $\mathcal{D}$ consists of all subsets of $N$, the two results imply that $x^*(p^*)$ will then belong to the critical Pareto set $\Theta_S$, associated with each coalition $S \in M$. This in turn suggests for any $S$, there is a solution $x^* = x^*(p^*)$ to the Lagrangian problem $dL_S(\mu, u)(x^*) = 0$, where $L_S(\mu, u^*)(x) = \sum_{i \in S} \mu_i u_i^*(x)$ and $\mu_i \geq 0 \ \forall i \in S$.

Here $u_i^* : Y_S \to \Re$ is the extended utility function for $i$ on $Y_S$.

It is also possible to use the concept of a core in the more general context considered in Sect. 3.8, where preferences are defined on the full space $X = \Pi_i X_i \in \Re_+^{nm}$. In this case, however, a price equilibrium may not exist if the induced social preference violates convexity or continuity. It is then possible for the $\mathcal{D}$-core to be empty.

Note in particular that the model outlined in this section implicitly assumes that each economic agent chooses their demand so as to optimize a utility function on the budget set determined by the price vector. Thus prices are treated as exogenous variables. However, if agents treat prices as strategic variables then it may be rational for them to compute their effect on prices, and thus misrepresent their preferences. The economic game then becomes much more complicated than the one analyzed here.

A second consideration is whether the price equilibria are unique, or even locally unique. If there exists a continuum of pure equilibria, then prices may move around chaotically.

A third consideration concerns the attainment of the price equilibrium. In Sect. 3.8 we constructed an abstract preference correspondence for an "auctioneer" so as to adjust the price vector to increase the value of the excess supply of the commodities. We deal with these considerations in the next section and in Chap. 5.

### 4.4.2 Equilibria in an Exchange Economy

The Welfare Theorem gives an important insight into the nature of competitive allocations. The coefficients $\mu_i$ of the Lagrangean $L(\mu, u)$ of the social optimisation problem turn out to be inverse to the coefficients $\lambda_i$ in the individual optimisation problems, where $\lambda_i$ is equal to the marginal utility of income for the $i$th agent. This in turn suggests that it is possible for an agent to transform his utility function from $u_i$ to $u_i'$ in such a way as to decrease $\lambda_i$ and thus increase $\mu_i$, the "weight" of the $i$th agent in the social optimisation problem. This is called the *problem of preference manipulation* and is an interesting research problem with applications in trade theory.

Secondly the weights $\mu_i$ can be regarded as functionally dependent on the initial endowment vector $(e_1, \ldots, e_m) \in \Re_+^{nm}$. Thus the question of market equilibrium could be examined in terms of the functions $\mu_i : \Re_+^{nm} \to \Re, i = 1, \ldots, m$.

## 4.4 The Pareto Set and Price Equilibria

It is possible that one or a number of agents could destroy or exchange commodities so as to increase their weights. This is termed the problem of *resource manipulation* or the *transfer paradox* (see Gale 1974, and Balasko 1978).

*Example 4.13* To illustrate these observations consider a two person ($i = 1, 2$) exchange economy with two commodities ($j = 1, 2$).

As in Example 4.10, assume the preference of the $i$th agent is given by a utility function $f_i : \Re_+^2 \to \Re : f_i(x, y) = \beta_i \log x + (1 - \beta_i) \log y$ where $0 < \beta_i < 1$.

Let the initial endowment vector of $i$ be $e_i = (e_{i1}, e_{i2})$. At the price vector $p = (p_1, p_2)$, demand by agent $i$ is $d_i(p_1, p_2) = (\frac{I_i \beta_i}{p_1}, \frac{I_i(1-\beta_i)}{p_2})$ where $I = p_1 e_{i1} + p_2 e_{i2}$ is the value at $p$ of the endowment.

Thus agent $i$ "desires" to change his initial endowment from $e_i$ to $e_i'$:

$$(e_{i1}, e_{i2}) \to (e_{i1}', e_{i2}') = \left( \beta_i e_{i1} + \frac{\beta_i e_{i2} p_2}{p_1}, (1 - \beta_i) e_{i2} + (1 - \beta_i) \frac{e_{i1} p_1}{p_2} \right).$$

Another way of interpreting this is that $i$ optimally divides expenditure between the first and second commodities in the ratio $\beta : (1 - \beta)$. Thus agent $i$ offers to sell $(1 - \beta) e_{i1}$ units of commodity 1 for $(1 - \beta) e_{i1} p_1$ monetary units and buy $(1 - \beta_i) e_{i1} \frac{p_1}{p_2}$ units of the second commodity, and offers to sell $\beta_i e_{i2}$ units of the second commodity and buy $\beta_i e_{i2} \frac{p_2}{p_1}$ units of the first commodity.

At the price vector $(p_1, p_2)$ the amount of the first commodity on offer is $(1 - \beta_i) e_{11} + (1 - \beta_2) e_{21}$ and the amount on request is $\beta_1 e_{12} p_2^* + \beta_2 e_{22} p_2^*$; where $p_2^*$ is the ratio $p_2 : p_1$ of relative prices. For $(p_1, p_2)$ to be a market-clearing price equilibrium we require

$$e_{11}(1 - \beta_1) + e_{21}(1 - \beta_2) = p_2^*(e_{12}\beta_1 + e_{22}\beta_2).$$

Clearly if all endowments are increased by a multiple $\alpha > 0$, then the equilibrium relative price vector is unchanged. Thus $p_2^*$ is uniquely determined and so the final allocations $(e_{11}', e_{12}'), (e_{21}', e_{22}')$ can be determined.

As we showed in Example 4.10, the coefficients $\lambda_i$ for the individual optimisation problems satisfy $\lambda_i = \frac{1}{I_i}$, where $\lambda_i$ is the marginal utility of income for agent $i$.

By the previous analysis, the weights $\mu_i$ in the social optimisation problem satisfy $\mu_i = I_i$. After some manipulation of the price equilibrium equation we find

$$\frac{\mu_i}{\mu_k} = \frac{e_{i1}(e_{i2} + \beta_k e_{k2}) + e_{i2}(1 - \beta_k) e_{k1}}{e_{k1}(e_{k2} + \beta_i e_{i2}) + e_{k2}(1 - \beta_i) e_{i1}}.$$

Clearly if agent $i$ can increase the ratio $\mu_i : \mu_k$ then the relative utility of $i$ vis-à-vis $k$ is increased. However since the relative price equilibrium is uniquely determined in this example, it is not possible for agent $i$, say, to destroy some of the initial endowments $(e_{i1}, e_{i2})$ so as to bring about an advantageous final outcome. The interested reader is referred to Balasko (1978).

**Fig. 4.18** Multiple price equilibria in Example 4.13

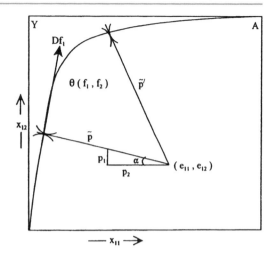

In this example the (relative) price equilibrium is unique, but this need not always occur. Consider the two person, two commodity case illustrated below in Fig. 4.18. As in the Welfare Theorem, the set of feasible outcomes $Y$ is the subset of $\Re_+^4 = (x_{11}, x_{12}, x_{21}, x_{22})$ such that $x_{11} + x_{21} = e_{\cdot 1}$; $x_{12} + x_{22} = e_{\cdot 2}$, and this is a two-dimensional hyperplane through the point $(e_{11}, e_{12}, e_{21}, e_{22})$. Thus $Y$ can be represented in the usual two-dimensional *Edgeworth* box where point $A$, the most preferred point for agent 1, satisfies $(x_{11}, x_{12}) = (e_{\cdot 1}, e_{\cdot 2})$.

The price ray $\tilde{p}$ is that ray through $(e_{11}, e_{12})$ where $\tan \alpha = \frac{p_1}{p_2}$. Clearly $(p_1, p_2)$ is an equilibrium price vector if the price ray intersects the critical Pareto set $\Theta(f_1, f_2)$ at a point $(x_{11}, x_{12})$ in $Y$ where $\tilde{p}$ is tangential to the indifference curves for $f_1$ and $f_2$ through $(x_{11}, x_{12})$. At such a point we then have $Df_1(x_{11}, x_{12}) + \mu Df_2(x_{11}, x_{12}) = 0$.

As Fig. 4.18 indicates there may well be a second price ray $\tilde{p}'$ which satisfies the tangency property. Indeed it is possible that there exists a family of such rays, or even an open set $V$ in the price simplex such that each $p$ in $V$ is a market clearing equilibrium. We now explore the question of local uniqueness of price equilibria.

To be more formal let $X = \Re_+^n$ be the commodity or consumption space. An *initial endowment* is a vector $e = (e_1, \ldots, e_m) \in X^m$. A $C^r$ utility function is a $C^r$-function $u = (u_1, \ldots, u_m) : X \to \Re^m$.

Let $C_r(X, \Re^m)$ be the set of $C^r$-profiles, and endow $C_r(X, \Re^m)$ with a topology in the following way. (See the next chapter for a more complete discussion of the *Whitney* topology.)

A neighbourhood of $f \in C_r(X, \Re^m)$ is a set

$$\left\{ g \in C_r(X, \Re^m) : \left\| d^k g(x) - d^k f(x) \right\| < \epsilon(x) \right\} \quad \text{for } k = 0, \ldots, r$$

where $\epsilon(x) > 0$ for all $x \in X$. (We use the notation that $d^0 g = g$ and $d^1 g = dg$.)

## 4.4 The Pareto Set and Price Equilibria

Write $C^r(X, \Re^m)$ for $C_r(X, \Re^m)$ with this topology. A property $K$ is called *generic* iff it is true for all profiles which belong to a *residual* set in $C^r(X, \Re^m)$. Here *residual* means that the set is the countable intersection of open dense sets in $C^r(X, \Re^m)$.

If a property is generic then we may say that almost all profiles in $C^r(X, \Re^m)$ have that property.

A smooth *exchange economy* is a pair $(e, u) \in X^m \times C^r(X, \Re^m)$. As before the feasible outcome set is

$$Y = \left\{ (x_1, \ldots, x_m) \in X^m : \sum_{i=1}^m x_i = \sum_{i=1}^m e_i \right\}$$

and a price vector $p$ belongs to the simplex $\Delta = \{p \in X : \|p\| = 1\}$, where $\Delta$ is an object of dimension $n - 1$.

As in the welfare theorem, the demand by agent $i$ at $p \in \Delta$ satisfies : $x_i^*(p)$ maximises $u_i$ on

$$B(p) = \left\{ x_i \in X : \langle p, x_i \rangle \leq \langle p, e_i \rangle \right\}.$$

As we have observed, under appropriate boundary conditions, we may assume $\forall i \in M, x_i^*(p)$ satisfies
1. $\langle p, x_i^*(p) \rangle = \langle p, e_i \rangle$
2. $D^* u_i(x_i^*(p)) = p \in \Delta$. Say $(x^*(p^*), p^*) = (x_1^*(p^*), \ldots, x_m^*(p^*), p^*) \in X^m \times \Delta$ is a market or *Walrasian equilibrium* iff $x^*(p^*)$ is the competitive allocation at $p^*$ and satisfies

$$\sum_{i=1}^m x_i^*(p^*) = \sum_{i=1}^m e_i \in \Re_+^n.$$

The economy $(e, u)$ is *regular* iff $(e, u)$ is such that the set of Walrasian equilibria is finite.

**Debreu-Smale Theorem on Generic Existence of Regular Economies** *There is a residual set $U$ in $C^r(X, \Re^m)$ such that for every profile $u \in U$, there is a dense set $V \in X^m$ with the property that $(e, u)$ is a regular economy whenever $(e, u) \in V \times U$.*

The proof of this theorem is discussed in the next chapter (the interested reader might also consult Smale 1976). However we can give a flavor of the proof here.

Consider a point $(e, x, p) \in X^m \times X^m \times \Delta$. This space is of dimension $2nm + (n - 1)$.

Now there are $n$ resource restrictions

$$\sum_{i=1}^m x_{ij} = \sum_{i=1}^m e_{ij}$$

for $j = 1, \ldots, n$, together with $(m - 1)$ budget restrictions $\langle p, x_i \rangle = \langle p, e_i \rangle$ for $i = 1, \ldots, m - 1$.

**Fig. 4.19** Finite Walrasian equilibria

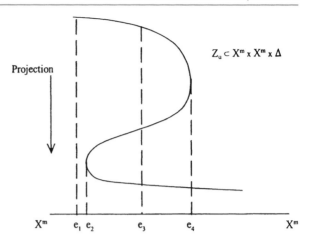

Note that the budget restriction for the $m$th agent is redundant.

Let $\Gamma = \{(e, x, p) \in X^m \times X^m \times \Delta\}$ be the set of points satisfying these various restrictions. Then $\Gamma$ will be of dimension $2nm + (n-1) - [n+m-1] = 2nm - m$.

However, we also have $m$ distinct vector equations $D^*u_i(x) = p$, $i = 1, \ldots, m$.

Since these vectors are normalised, each one consists of $(n-1)$ separate equations. Chapter 5 shows that singularity theory implies that for every profile $u$ in a residual set, each of these constraints is independent. Together these $m(n-1)$ constraints reduce the dimension of $\Gamma$ by $m(n-1)$. Hence the set of points in $\Gamma$ satisfying the first order optimality conditions is a smooth object $Z_u$ of dimension

$$2nm - m - m(n-1) = nm.$$

Now consider the projection

$$Z_u \subset X^m \times (X^m \times \Delta) \to X^m : (e, x, p) \to e,$$

and note that both $Z_u$ and $X^m$ have dimension $nm$.

A regular economy $(e, u)$ is one such that the projection map $proj : Z_u \to X^m$ : $(e, x, p) \to e$ has differential with maximal rank $nm$. Call $e$ a *regular value* in this case. From singularity theory it is known that for all $u$ in a residual set $U$, the set of regular values of the projection map is dense in $X^m$. Thus when $u \in U$, and $e$ is regular, the set of Walrasian equilibria for $(e, u)$ will be finite. Figure 4.19 illustrates this. At $e_1$ there is only one Walrasian equilibrium, while at $e_3$ there are three. Moreover in a neighbourhood of $e_3$ the Walrasian equilibria move continuously with $e$. At $e_4$ the Walrasian equilibrium set is one-dimensional. As $e$ moves from the right past $e_2$ the number of Walrasian equilibria drops suddenly from 3 to 1, and displays a discontinuity. Note that the points $(x, p)$ satisfying $(e, x, p) \in (proj)^{-1}(e)$ need not be Walrasian equilibria in the classical sense, since we have considered only the first order conditions. It is clearly the case that if there is non-convex preference, then the first order conditions are not sufficient for equilibrium. However, Smale's

## 4.4 The Pareto Set and Price Equilibria

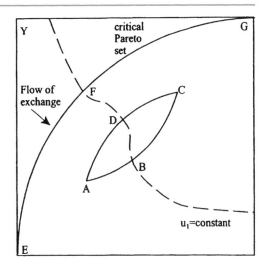

**Fig. 4.20** Two agent, two commodity exchange economy

theorem shows the existence of extended Walrasian equilibria. The same difficulty occurs in the proof that a Walrasian equilibrium gives a Pareto optimal point in $Y$.

Let $\overset{0}{\Theta}(u) = \overset{0}{\Theta}(u_1, \ldots, u_m)$ be the set of points satisfying the first order condition $dL(\lambda, u) = 0$ where $\lambda \in \overline{\mathfrak{R}_+^m}$. Suppose that we solve this with $\lambda_1 \neq 0$.

Then we may write $Du_1(x) + \sum_{i=2}^{m} \frac{\lambda_i}{\lambda_1} Du_i(x) = 0$.

Clearly there are $(m-1)$ degrees of freedom in this solution and indeed $\overset{0}{\Theta}(u_1, \ldots, u_m)$ can be shown to be a geometric object of dimension $(m-1)$ "almost always" (see Chap. 5). However $\overset{0}{\Theta}(u)$ will contain points that are the "social" equivalents of the minima of a real-valued function.

Note, by Theorem 4.21, that $\overset{0}{\Theta}(u)$ and the critical Pareto set $\Theta(u)$ coincide, except for boundary points. If the boundary of the space is smooth, then it is possible to define a Lagrangian which characterises the boundary points in $\Theta(u)$.

For example consider Fig. 4.20, of a two agent two commodity exchange economy.

Agent 1 has non-convex preference, and the critical Pareto set consists of three components $ABC$, $ADC$ and $EFG$.

On $ADC$ although the utilities satisfy the first order condition, there exist nearby points that both agents prefer. For example, both agents prefer a nearby point $y$ to $x$. See Fig. 4.21.

In Fig. 4.22 from an initial endowment such as $e = (e_{11}, e_{12})$, there exists three Walrasian extended equilibria, but at least one can be excluded. Note that if $e$ is the initial endowment vector, then the Walrasian equilibrium $B$ which is accessible by exchange along the price vector may be Pareto inferior to a Walrasian equilibrium, $F$, which is not readily accessible.

**Fig. 4.21** Unstable critical Paret point

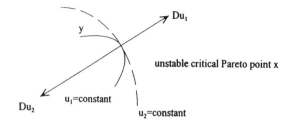

**Fig. 4.22** Unstable Pareto component

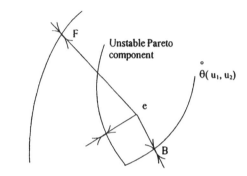

**Fig. 4.23** Example 4.14

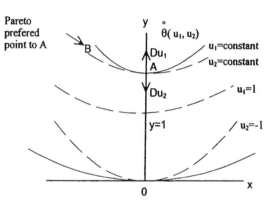

*Example 4.14* Consider the example due to Smale (as in Fig. 4.23). Let $Y = \Re^2$ and suppose

$$u_1(x, y) = y - x^2$$
$$u_2(x, y) = \frac{-y}{x^2 + 1}.$$

Then

$$Du_1(x, y) = (-2x, 1)$$
$$Du_2(x, y) = \left(\frac{2xy}{(x^2+1)^2}, \frac{-1}{x^2+1}\right).$$

Let $DL_\lambda(x, y) = \lambda_1(-2x, 1) + \lambda_2(\frac{2xy}{(x^2+1)^2}, \frac{-1}{x^2+1})$.

Clearly one solution will be $x = 0$, in which case $\lambda_1(0, 1) + \lambda_2(0, -1) = 0$ or $\lambda_1 = \lambda_2 = 1$.

The Hessian for $L$ at $x = 0$ is then

$$HL(0, y) = D^2 u_1(0, y) + D^2 u_2(0, y)$$
$$= \begin{pmatrix} -2 & 0 \\ 0 & 0 \end{pmatrix} + \begin{pmatrix} 2y & 0 \\ 0 & 0 \end{pmatrix}$$

which is negative semi-definite for $2(y - 1) < 0$ or $y < 1$.

## Further Reading

For a lucid account of economic equilibrium theory see:

Arrow, K. J., & Hahn, F. H. (1971). *General competitive analysis*. Edinburgh: Oliver and Boyd.
Hildenbrand, W., & Kirman, A. P. (1976). *Introduction to equilibrium analysis*. Amsterdam: North Holland.

For the ideas of preference or resource manipulation, see:

Balasko, Y. (1978). The transfer problem and the theory of regular economies. *International Economic Review, 19*, 687–694.
Gale, D. (1974). Exchange equilibrium and coalitions. *Journal of Mathematical Economics, 1*, 63–66.
Guesnerie, R., & Laffont, J.-J. (1978). Advantageous reallocations of initial resources. *Econometrica, 46*, 687–694.
Safra, Z. (1983). Manipulation by reallocating initial endowments. *Journal of Mathematical Economics, 12*, 1–17.

For a general introduction to the application of differential topology to economics see:

Smale, S. (1976). Dynamics in general equilibrium theory. *American Economic Review, 66*, 288–294. Reprinted in S. Smale (1980) *The Mathematics of Time*. Springer: Berlin.

# Singularity Theory and General Equilibrium 5

In the last section of the previous chapter we introduced the critical Pareto set $\Theta(u)$ of a smooth profile for a society, and the notion of a regular economy. Both ideas relied on the concept of a singularity of a smooth function $f : X \to \Re^m$, where a singularity is a point analogous to the critical point of a real-valued function.

In this chapter we introduce the fundamental result in singularity theory, that the set of singularity points of a smooth profile almost always has a particular geometric structure. We then go on to use this result to discuss the Debreu-Smale Theorem on the generic existence of regular economies. No attempt is made to prove these results in full generality. Instead the aim is to provide a geometric understanding of the ideas. Section 5.4 uses an example of Scarf (1960) to illustrate the idea of an excess demand function for an exchange economy. The example provides a general way to analyse a smooth adjustment process leading to a Walrasian equilibrium. Sections 5.5 and 5.6 introduce the more abstract topological ideas of structural stability and chaos in dynamical systems.

## 5.1 Singularity Theory

In Chap. 4 we showed that when $f : X \to \Re$ was a $C^2$-function on a normed vector space, then knowledge of the first and second differential of $f$ at a critical point, $x$, gave information about the local behavior (near $x$) of the function. In this section we discuss the case of a differentiable function $f : X \to Y$ between general normed vector spaces, and consider regular points (where the differential has maximal rank) and singularity points (where the differential has non-maximal rank). For both kinds of points we can locally characterise the behavior of the function.

### 5.1.1 Regular Points: The Inverse and Implicit Function Theorem

Suppose that $f : X \to Y$ is a function between normed vector spaces and that for all $x'$ in a neighbourhood $U$ of the point $x$ the differential $df(x')$ is defined. By the results of Sect. 3.2, if $df(x')$ is bounded, or if $X$ is finite-dimensional, then

$df(x')$ will be a *continuous* linear function. Suppose now that $X$ and $Y$ have the same finite dimension ($n$) and $df(x')$ has rank $n$ at all $x' \in U$. Then we know that $df(x')^{-1} : Y \to X$ is a linear map and thus continuous. We shall call $f$ a $C^2$-*diffeomorphism* on $U$ in this case.

In general, when $Y$ is an infinite-dimensional vector space, then even if $df(x')^{-1}$ exists it need not be continuous. However when $X$ and $Y$ are *complete* normed vector spaces then the existence of $df(x')^{-1}$ is sufficient to guarantee that $df(x')^{-1}$ is continuous.

Essentially a normed vector space $X$ is *complete* iff any "convergent" sequence $(x_k)$ does indeed converge to a point in $X$. More formally, a *Cauchy sequence* is a sequence $(x_k)$ such that for any $\epsilon > 0$ there exists some number $k(\epsilon)$ such that $r, s > k(\epsilon)$ implies that $\|x_r - x_s\| < \epsilon$. If $(x_k)$ is a sequence with a limit $x_0$ in $X$ then clearly $(x_k)$ must be a Cauchy sequence. On the other hand a Cauchy sequence need not in general converge to a point in the space $X$. If *every* Cauchy sequence has a limit in $X$, then $X$ is called *complete*. A complete normed vector space is called a *Banach* space. Clearly $\Re^n$ is a Banach space. Suppose now that $X, Y$ are normed vector spaces of the same dimension, and $f : U \subset X \to Y$ is a $C^r$-differentiable function on $U$, such that $df(x)$ has a continuous inverse $[df(x)]^{-1}$ at $x$. We call $f$ a $C^r$-diffeomorphism at $x$. Then we can show that $f$ has an inverse $f^{-1} : f(U) \to U$ with differential $df^{-1}(f(x)) = [df(x)]^{-1}$. Moreover there exists a neighbourhood $V$ of $x$ in $U$ such that $f$ is a $C^r$-diffeomorphism on $V$. In particular this means that $f$ has an inverse $f^{-1} : f(V) \to V$ with continuous differential $df^{-1}(f(x')) = [df(x')]^{-1}$ for all $x' \in V$, *and* that $f^{-1}$ is $C^r$-differentiable on $V$. To prove the theorem we need to ensure that $[df(x)]^{-1}$ is not only linear but continuous, and it is sufficient to assume $X$ and $Y$ are Banach spaces.

**Inverse Function Theorem** *Suppose $f : U \subset X \to Y$ is $C^r$-differentiable, where $X$ and $Y$ are Banach spaces of dimension $n$. Suppose that the linear map $df(x) : X \to Y$, for $x \in U$, is an isomorphism with inverse $[df(x)]^{-1} : Y \to X$. Then there exist open neighbourhoods $V$ of $x$ in $U$ and $V'$ of $f(x)$ such that $f : V \to V'$ is a bijection with inverse $f^{-1} : V' \to V$. Moreover $f^{-1}$ is itself $C^r$-differentiable on $V'$, and for all $x' \in V$, $df^{-1}(f(x')) = [df(x')]^{-1}$. $f$ is called a local $C^r$-diffeomorphism, at $x$.*

*Outline of Proof* Let $t = df(x) : X \to Y$. Since $[df(x)]^{-1}$ exists and is continuous, $t^{-1} : Y \to X$ is linear and continuous.

It is possible to choose a neighbourhood $V'$ of $f(x)$ in $f(U)$ and a closed ball $V_x$ in $U$ centered at $x$, such that, for each $y \in V'$, the function $g_y : V_x \subset U \subset X \to V_x \subset X$ defined by $g_y(x') = x' - t^{-1}[f(x') - y]$ is continuous. By Brouwer's Theorem, each $g_y$ has a fixed point.

That is to say for each $y \in V'$, there exists $x' \in V_x$ such that $g_y(x') = x'$. But then $t^{-1}[f(x') - y] = 0$. Since, by hypothesis, $t^{-1}$ is an isomorphism, its kernel $= \{0\}$, and so $f(x') = y$. Thus for each $y \in V'$ we establish $g_y(x') = x'$ is equivalent to $f(x') = y$. Define $f^{-1}(y) = g_y(x') = x'$, which gives the inverse function on $V'$. To show $f^{-1}$ is differentiable, proceed as follows.

## 5.1 Singularity Theory

Note that $dg_y(x') = \text{Id} - t^{-1} \circ df(x')$ is independent of $y$. Now $dg_y(x')$ is a linear and continuous function from $X$ to $X$ and is thus bounded. Since $X$ is Banach, it is possible to show that $\mathcal{L}(X, X)$, the topological space of linear and continuous maps from $X$ to $X$, is also Banach. Thus if $u \in \mathcal{L}(X, X)$, so is $(\text{Id} - u)^{-1}$. This follows since $(\text{Id} - u)^{-1}$ converges to an element of $\mathcal{L}(X, X)$.

Now $dg_y(x') \in \mathcal{L}(X, X)$ and so $(\text{Id} - dg_y(x'))^{-1} \in \mathcal{L}(X, X)$. But then $t^{-1} \circ df(x')$ has a continuous linear inverse. Now $t^{-1} \circ df(x') : X \to Y \to X$ and $t^{-1}$ has a continuous linear inverse. Thus $df(x')$ has a continuous linear inverse, for all $x' \in V$. Let $V$ be the interior of $V_x$. By the construction the inverse of $df(x')$, for $x' \in V$, has the required property. □

This is the fundamental theorem of differential calculus. Notice that the theorem asserts that if $f : \mathfrak{R}^n \to \mathfrak{R}^n$ and $df(x)$ has rank $n$ at $x$, then $df(x')$ has rank $n$ for all $x'$ in a neighbourhood of $x$.

*Example 5.1* (i) For a simple example, consider the function $\exp : \mathfrak{R} \to \mathfrak{R}_+ : x \to e^x$. Clearly for any finite $x \in \mathfrak{R}$, $d(\exp)(x) = e^x \neq 0$, and so the rank of the differential is 1. The inverse $\phi : \mathfrak{R}_+ \to \mathfrak{R}$ must satisfy

$$d\phi(y) = [d(\exp)(x)]^{-1} = \frac{1}{e^x}$$

where $y = \exp(x) = e^x$. Thus $d\phi(y) = \frac{1}{y}$.

Clearly $\phi$ must be the function $\log_e : y \to \log_e y$.

(ii) Consider $\sin : (0, 2\pi) \to [-1, +1]$.

Now $d(\sin)(x) \neq 0$. Hence there exist neighbourhoods $V$ of $x$ and $V'$ of $\sin x$ and an inverse $\phi : V' \to V$ such that

$$d\phi(y) = \frac{1}{\cos x} = \frac{1}{\sqrt{1-y^2}}.$$

This inverse $\phi$ is only locally a function. As Fig. 5.1 makes clear, even when $\sin x = y$, there exist distinct values $x_1, x_2$ such that $\sin(x_1) = \sin(x_2) = y$. However $d(\sin)(x_1) \neq d(\sin)(x_2)$.

The figure also shows that there is a neighbourhood $V'$ of $y$ such that $\phi : V' \to V$ is single-valued and differentiable on $V'$. Suppose now $x = \frac{\pi}{2}$. Then $d(\sin)(\frac{\pi}{2}) = 0$. Moreover there is no neighbourhood $V$ of $\frac{\pi}{2}$ such that $\sin : (\frac{\pi}{2} - \epsilon, \frac{\pi}{2} + \epsilon) = V \to V'$ has an inverse function.

Note one further consequence of the theorem. For $h$ small, we may write

$$f(x+h) = f(x) + df(x) \circ [df(x)]^{-1}(f(x+h) - f(x))$$
$$= f(x) + df(x)\psi(h),$$

where $\psi(h) = d[f(x)]^{-1}(f(x+h) - f(x))$. Now by a linear change of coordinates we can diagonalise $df(x)$. So that in the case $f = (f_1, \ldots, f_n) : \mathfrak{R}^n \to \mathfrak{R}^n$ we can ensure $\frac{\partial f_i}{\partial x_j}|_x = \partial_{ij}$ where $\partial_{ij} = 1$ if $i = j$ and 0 if $i \neq j$. Hence $f(x+h) = f(x) +$

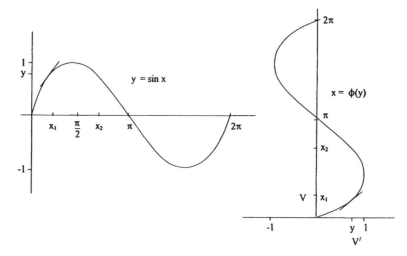

**Fig. 5.1** Example 5.1

$(\psi_1(h), \ldots, \psi_n(h))$. There is therefore a $C^r$-diffeomorphic change of coordinates $\phi$ near $x$ such that $\phi(\underline{0}) = x$ and

$$f\big(\phi(h_1, \ldots, h_n)\big) = f(x) + (h_1, \ldots, h_n).$$

In other words by choosing coordinates appropriately $f$ may be represented by its linear differential.

Suppose now that $f : U \subset \Re^n \to \Re^m$ is a $C^1$-function. The maximal rank of $df$ at a point $x$ in $U$ is $\min(n, m)$. If indeed $df(x)$ has maximal rank then $x$ is called a *regular point* of $f$, and $f(x)$ a *regular value*. In this case we write $x \in S_0(f)$. The inverse function theorem showed that when $n = m$ and $x \in S_0(f)$ then $f$ could be regarded as an identity function near $x$.

In particular this means that there is a neighbourhood $U$ of $x$ such that $f^{-1}[f(x)] \cap U = \{x\}$ is an isolated point.

In the case that $n \neq m$ we use the inverse function theorem to characterise $f$ at regular points.

**Implicit Function Theorem for Vector Spaces**
1. (Surjective Version). *Suppose that $f : U \subset \Re^n \to \Re^m$, $n \geq m$, and rank $(df(x)) = m$, with $f(x) = \underline{0}$ for convenience. If $f$ is $C^r$-differentiable at $x$, then there exists a $C^r$-diffeomorphism $\phi : \Re^n \to \Re^n$ on a neighbourhood of the origin such that $\phi(\underline{0}) = x$, and $f \circ \phi(h_1, \ldots, h_n) = (h_1, \ldots, h_m)$.*
2. (Injective Version). *If $f : U \subset \Re^n \to \Re^m$, $n \leq m$, rank $(df(x)) = n$, with $f(\underline{0}) = y$, and $f$ is $C^r$-differentiable at $x$ then there exists a $C^r$-diffeomorphism $\psi : \Re^m \to \Re^m$ such that $\psi(y) = \underline{0}$ and*

$$\psi f(h_1, \ldots, h_n) = (h_1, \ldots, h_n, 0, \ldots, 0).$$

## 5.1 Singularity Theory

*Proof*

1. Now $df(x) = [B\ C]$, with respect to some coordinate system, where $B$ is an $(m \times m)$ non singular matrix and $C$ is an $(n-m) \times m$ matrix. Define $F : \Re^n \to \Re^n$ by

$$F(x_1, \ldots, x_n) = \big(f(x_1, \ldots, x_n), x_{m+1}, \ldots, x_n\big).$$

Clearly $DF(x)$ has rank $n$, and by the inverse function theorem there exists an inverse $\phi$ to $F$ near $x$. Hence $F \circ \phi(h_1, \ldots, h_n) = (h_1, \ldots, h_n)$. But then $f \circ \phi(h_1, \ldots, h_n) = (h_1, \ldots, h_m)$.

2. Follows similarly. □

As an application of the theorem, suppose $f : \Re^m \times \Re^{n-m} \to \Re^m$. Write $x$ for a vector in $\Re^m$, and $y$ for a vector in $\Re^{n-m}$, and let $df(x, y) = [B\ C]$ where $B$ is an $m \times m$ matrix and $C$ is an $(n-m) \times m$ matrix. Suppose that $B$ is invertible, at $(\bar{x}, \bar{y})$, and that $f(\bar{x}, \bar{y}) = 0$. Then the implicit function theorem implies that there exists an open neighbourhood $U$ of $\bar{y}$ in $\Re^{n-m}$ and a differentiable function $g : U \to \Re^m$ such that $g(y') = x'$ and $f(g(y'), y') = 0$ for all $y' \in V$.

To see this define

$$F : \Re^m \times \Re^{n-m} \to \Re^m \times \Re^{n-m}$$

by $F(x, y) = (f(x, y), y)$.

Clearly $dF(\bar{x}, \bar{y}) = \begin{bmatrix} B & C \\ O & I \end{bmatrix}$ and so $dF(\bar{x}, \bar{y})$ is invertible. Thus there exists a neighbourhood $V$ of $(\bar{x}, \bar{y})$ in $\Re^n$ on which $F$ has a diffeomorphic inverse $G$. Now $F(\bar{x}, \bar{y}) = (0, \bar{y})$. So there is a neighbourhood $V'$ of $(0, \bar{y})$ and a neighbourhood $V$ of $(\bar{x}, \bar{y})$ s.t. $G : V \subset \Re^n \to V' \subset \Re^n$ is a diffeomorphism.

Let $g(y')$ be the $x$ coordinate of $G(0, y')$ for all $y'$ such that $(0, y') \in V'$.

Clearly $g(y')$ satisfies $G(0, y') = (g(y'), y')$ and so $F \circ G(0, y') = F(g(y'), y) = (f(g(y'), y')), y') = (0, y')$.

Now if $(x', y') \in V'$ then $y' \in U$ where $U$ is open in $\Re^{n-m}$. Hence for all $y' \in U$, $g(y')$ satisfies $f(g(y'), y') = 0$. Since $G$ is differentiable, so must be $g : U \subset \Re^{n-m} \to \Re^m$. Hence $x' = g(y')$ *solves* $f(x', y') = 0$.

*Example 5.2*

1. Let $f : \Re^3 \to \Re^2$ where

$$f_1(x, y, z) = x^2 + y^2 + z^2 - 3$$
$$f_2(x, y, z) = x^3 y^3 z^3 - x + y - z.$$

At $(x, y, z) = (1, 1, 1)$, $f_1 = f_2 = 0$.

We seek a function $g : \Re \to \Re^2$ such that $f(g_1(z), g_2(z), z) = 0$ for all $z$ in a neighbourhood of 1.

Now

$$df(x, y, z) = \begin{pmatrix} 2x & 2y & 2z \\ 3x^2 y^3 z^3 - 1 & 3x^3 y^2 z^3 + 1 & 3x^3 y^3 z^2 - 1 \end{pmatrix}$$

and so
$$df(1,1,1) = \begin{pmatrix} 2 & 2 & 2 \\ 2 & 4 & 2 \end{pmatrix}.$$

The matrix
$$\begin{pmatrix} 2 & 2 \\ 2 & 4 \end{pmatrix}$$

is non-singular. Hence there exists a diffeomorphism $G : \Re^3 \to \Re^3$ such that $G(0, 0, 1) = (1, 1, 1)$, and $G(0, 0, z') = (g_1(z'), g_2(z'), z')$ for $z'$ near 1.

2. In a simpler example consider $f : \Re^2 \to \Re : (x, y) \to (x-a)^2 + (y-b)^2 - 25 = 0$, with $df(x, y) = (2(x-a), 2(y-b))$.

Now let $F(x, y) = (x, f(x, y))$, where $F : \Re^2 \to \Re^2$, and suppose $y \neq b$. Then

$$dF(x, y) = \begin{pmatrix} 1 & 0 \\ \frac{\partial f}{\partial x} & \frac{\partial f}{\partial y} \end{pmatrix}$$
$$= \begin{pmatrix} 1 & 0 \\ 2(x-a) & 2(y-b) \end{pmatrix}$$

with inverse
$$dG(x, y) = \frac{1}{2(y-b)} \begin{pmatrix} 2(y-b) & 0 \\ -2(x-a) & 1 \end{pmatrix}.$$

Define $g(x')$ to be the $y$-coordinate of $G(x', 0)$. Then $F \circ G(x', 0) = F(x', g(x')) = F(x', f(x', y')) = (x', 0)$, and so $y' = g(x')$ for $f(x', y') = 0$ and $y'$ sufficiently close to $y$. Note also that

$$\left. \frac{dg}{dx} \right|_{x'} = \left. \frac{dG_2}{\partial x} \right|_{(x', g')} = -\frac{(x'-a)}{(y'-b)}.$$

In Example 5.2(1), the "solution" $g(z) = (x, y)$ to the smooth constraint $f(x, y, z) = 0$ is, in a topological sense, a one-dimensional object (since it is given by a single constraint in $\Re^2$).

In the same way in Example 5.2(2) the solution $y = g(x)$ is a one-dimensional object (since it is given by a single constraint in $\Re^2$).

More specifically say that an open set $V$ in $\Re^n$ is an *r-dimensional smooth manifold* iff there exists a diffeomorphism

$$\phi : V \subset \Re^n \to U \subset \Re^r.$$

When $f : \Re^n \to \Re^m$ and rank $(df(x)) = m \leq n$ then say that $f$ is a *submersion* at $x$. If rank $(df(x)) = n \leq m$ then say $f$ is an *immersion* at $x$.

One way to interpret the implicit function theorem is as follows:

## 5.1 Singularity Theory

(1) When $f$ is a submersion at $x$, then the inverse $f^{-1}(f(x))$ of a point $f(x)$ "looks like" an object of the form $\{x, h_{m+1}, \ldots, h_n\}$ and so is a smooth manifold in $\Re^n$ of dimension $(n-m)$.

(2) When $f$ is an immersion at $x$, then the image of an ($n$-dimensional) neighborhood $U$ of $x$ "looks like" an $n$-dimensional manifold, $f(u)$, in $\Re^m$. These observations can be generalized to the case when $f : X^n \to Y^m$ is "smooth", and $X, Y$ are themselves smooth manifolds of dimension $n, m$ respectively. Without going into the formal details, $X$ is a smooth manifold of dimension $n$ if it is a paracompact topological space and for any $x \in X$ there is a neighborhood $V$ of $x$ and a smooth "chart", $\phi : V \subset X \to U \subset \Re^n$. In particular if $x \in V_i \cap V_j$ for two open neighborhoods, $V_i, V_j$ of $x$ then

$$\phi_i \circ \phi_j^{-1} : \phi_j(V_i \cap V_j) \subset \Re^n \to \phi_i(V_i \cap V_j) \subset \Re^n$$

is a diffeomorphism. A smooth structure on $X$ is an *atlas*, namely a family $\{(\phi_i, V_i)\}$ of charts such that $\{V_i\}$ is an open cover of $X$. The purpose of this definition is that if $f : X^n \to Y^m$ then there is an induced function near a point $x$ given by

$$f_{ij} = \psi_i \circ f \circ \phi_j^{-1} : \Re^n \to \phi_j^{-1}(V_j) \to Y \to \Re^m.$$

Here $(\phi_j, V_j)$ is a chart at $x$, and $(\psi_i, V_i)$ is a chart at $f(x)$. If the induced functions $\{f_{ij}\}$ at every point $x$ are differentiable then $f$ is said to be differentiable, and the "induced" differential of $f$ is denoted by $df$. The charts thus provide a convenient way of representing the differential $df$ of $f$ at the point $x$. In particular once $(\phi_j, V_j)$ and $(\psi_i, V_i)$ are chosen for $x$ and $f(x)$, then $df(x)$ can be represented by the Jacobian matrix $Df(x) = (\partial f_{ij})$. As before $Df(x)$ will consist of $n$ columns and $m$ rows. Characteristics of the Jacobian, such as rank, will be independent of the choices for the charts (and thus coordinates) at $x$ and $f(x)$. (See Chillingsworth (1976), for example, for the details.)

If the differential $df$ of a function $f : X^n \to Y^m$ is defined and continuous then $f$ is called $C^1$. Let $C_1(X, Y)$ be the collection of such $C^1$-maps. Analogous to the case of functions between real vector spaces, we may also write $C_r(X, Y)$ for the class of $C^r$-differentiable functions between $X$ and $Y$.

The implicit function theorem also holds for members of $C_1(X, Y)$.

**Implicit Function Theorem for Manifolds** *Suppose that $f : X^n \to Y^m$ is a $C^1$-function between smooth manifolds of dimension $n, m$ respectively.*

1. *If $n \geq m$ and $f$ is a submersion at $x$ (i.e., rank $df(x) = m$) then $f^{-1}(f(x))$ is (locally) a smooth manifold in $X$ of dimension $(n - m)$. Moreover, if $Z$ is a manifold in $Y^n$ of dimension $r$, and $f$ is a submersion at each point in $f^{-1}(Z)$, then $f^{-1}(Z)$ is a submanifold of $X$ of dimension $n - m + r$.*
2. *If $n \leq m$ and $f$ is an immersion at $x$ (i.e., rank $df(x) = n$) then there is a neighbourhood $U$ of $x$ in $X$ such that $f(U)$ is an $n$-dimensional manifold in $Y$ and in particular $f(U)$ is open in $Y$.*

**Fig. 5.2** Example 5.3

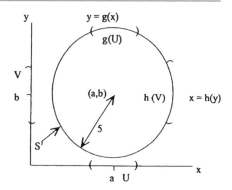

The proof of this theorem is considerably beyond the scope of this book, but the interested reader should look at Golubitsky and Guillemin (1973, p. 9) or Hirsch (1976, p. 22). This theorem is a smooth analogue of the isomorphism theorem for linear functions given in Sect. 2.2. For a linear function $T : \Re^n \to \Re^m$ when $n \geq m$ and $T$ is surjective, then $T^{-1}(y)$ has the form $x_0 + K$ where $K$ is the $(n-m)$-dimensional kernel. Conversely if $T : \Re^n \to \Re^m$ and $n \leq m$ when $T$ is injective, then image $(T)$ is an $n$-dimensional subspace of $\Re^m$. More particularly if $U$ is an $n$-dimensional open set in $\Re^n$ then $T(U)$ is also an $n$-dimensional open set in $\Re^m$.

*Example 5.3* To illustrate, consider Example 5.2(2) again. When $y \neq b$, $df$ has rank 1 and so there exists a "local" solution $y' = g(x')$ such that $f(x', g(x')) = 0$. In other words

$$f^{-1}(0) = \{(x', g(x')) \in \Re^2 : x' \in U\},$$

which essentially is a copy of $U$ but deformed in $\Re^2$. Thus $f^{-1}(0)$ is "locally" a one-dimensional manifold. Indeed the set $S^1 = \{(x, y) : f(x, y) = 0\}$ itself is a 1-dimensional manifold in $\Re^2$.

If $y \neq b$, and $(x, y) \in S^1$ then there is a neighbourhood $U$ of $x$ and a diffeomorphism $g : S^1 \to \Re : (x', y') \to g(y')$ and this parametrises $S^1$ near $(x, y)$.

If $y = b$, then we can do the same thing through a local solution $x' = h(y')$ satisfying $f(h(y'), y') = 0$.

### 5.1.2 Singular Points and Morse Functions

When $f : X^n \to Y^m$ is a $C^1$-function between smooth manifolds, and rank $df(x)$ is maximal $(= \min(n, m))$ then as before write $x \in S_0(f)$.

The set of *singular points* of $f$ is $S(f) = X \backslash S_0(f)$. Let $z = \min(n, m)$ and say that $x$ is a *corank r singularity*, or $x \in S_r(f)$, if rank $(df(x)) = z - r$.

Clearly $S(f) = \bigcup_{r \geq 1} S_r(f)$.

In the next section we shall examine the corank $r$ singularity sets of a $C^1$-function and show that they have a nice geometric structure. In this section we consider the case $m = 1$.

## 5.1 Singularity Theory

In the case of a $C^2$-function $f : X^n \to \Re$, either $x$ will be regular (in $S_0(f)$) or a critical point (in $S_1(f)$) where $df(x) = 0$. We call a critical point of $f$ *non-degenerate* iff $d^2 f(x)$ is non-singular. A $C^2$-function all of whose critical points are non degenerate is called a *Morse* function. A Morse function, $f$, near a critical point has a very simple representation.

A local system of coordinates at a point $x$ in $X$ is a smooth assignment

$$y \xrightarrow{\phi} (h_1, \ldots, h_n)$$

for every $y$ in some neighbourhood $U$ of $x$ in $X$.

**Lemma 5.1** (Morse) *If $f : X^n \to \Re$ is $C^2$ and $x$ is a non-degenerate critical point of index $k$, then there exists a local system of coordinates (or chart $(\phi, V)$) at $x$ such that $f$ is given by*

$$y \xrightarrow{\phi} (h_1, \ldots, h_n) \xrightarrow{g} f(x) - \sum_{i=1}^{k} h_i^2 + \sum_{i=k+1}^{n} h_i^2.$$

As before the *index* of the critical point is the number of negative eigenvalues of the Hessian $Hf$ at $x$. The $C^2$-function $g$ has Hessian

$$Hg(0) = \begin{pmatrix} -2 & & & & & \\ & \ddots & & & & \\ & & -2 & & & \\ & & & 2 & & \\ & & & & \ddots & \end{pmatrix} \begin{matrix} \uparrow \\ k \\ \downarrow \end{matrix}$$

with $k$ negative eigenvalues. Essentially the Morse lemma implies that when $x$ is a non-degenerate critical point of $f$, then $f$ is topologically equivalent to the function $g$ with a similar Hessian at the point.

By definition, if $f$ is a Morse function then all its critical points are non-degenerate. Moreover if $x \in S_1(f)$ then there exists a neighbourhood $V$ of $x$ such that $x$ is the only critical point of $f$ in $V$.

To see this note that for $y \in V$,

$$df(y) = dg(h_1, \ldots, h_n) = (-2h_1, \ldots, 2h_n) = 0$$

iff $h_1 = \cdots = h_n = 0$, or $y = x$. Thus each critical point of $f$ is isolated, and so $S_1(f)$ is a set of isolated points and thus a zero-dimensional object.

As we shall see almost any smooth function can be approximated arbitrarily closely by a Morse function.

To examine the regular points of a differentiable function $f : X \to \Re$, we can use the Sard Lemma.

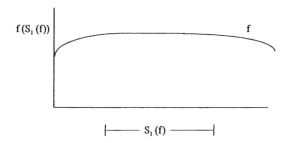

**Fig. 5.3** Illustration of Sard's Lemma

First of all a set $V$ in a topological space $X$ is called *nowhere dense* if its closure, $clos(V)$, contains no non-empty open set. Alternatively $X \backslash clos(V)$ is dense.

If $X$ is a complete metric space then the union of a countable collection of closed nowhere dense sets is nowhere dense. This also means that a residual set (the intersection of a countable collection of open dense sets) is dense. (See Sect. 3.1.2). A set $V$ is of *measure zero* in $X$ iff for any $\epsilon > 0$ there exists a family of cubes, with volume less than $\epsilon$, covering $V$. If $V$ is closed, of measure zero, then it is nowhere dense.

**Lemma 5.2** (Sard) *If $f : X^n \to \Re$ is a $C^r$-map where $r \geq n$, then the set $f(S_1(f))$ of critical values of $f$ has measure zero in $\Re$. Thus $f(S_0(f))$, the set of regular values of $f$, is the countable intersection of open dense sets and thus is dense.*

To illustrate this consider Fig. 5.3. $f$ is a quasi-concave $C^1$-function $f : \Re \to \Re$. The set of critical points of $f$, namely $S_1(f)$, clearly does not have measure zero, since $S_1(f)$ has a non-empty interior. Thus $f$ is not a Morse function. However $f(S_1(f))$ is an isolated point in the image.

*Example 5.4* To illustrate the Morse lemma let $Z = S^1 \times S^1$ be the *torus* (the skin of a doughnut) and let $f : Z \to \Re$ be the height function.

Point $s$, at the bottom of the torus, is a minimum of the function, and so the index of $s = 0$. Let $f(s) = 0$.

Then near $s$, $f$ can be represented by

$$(h_1, h_2) \to 0 + h_1^2 + h_2^2.$$

Note that the Hessian of $f$ at $s$ is $\begin{bmatrix} 2 & 0 \\ 0 & 2 \end{bmatrix}$, and so is positive definite.

The next critical point, $t$, is obviously a saddle, with index 1, and so we can write $(h_1, h_2) \to f(t) + h_1^2 - h_2^2$. Clearly $Hf(t) = \begin{bmatrix} 2 & 0 \\ 0 & -2 \end{bmatrix}$.

Suppose now that $a \in (f(s), f(t))$. Clearly $a$ is a regular value, and so any point $x \in Z$ satisfying $f(x) = a$ is a regular point, and $f$ is a submersion at $x$. By the implicit function theorem $f^{-1}(a)$ is a one-dimensional manifold. Indeed it is a single copy of the circle, $S^1$. See Fig. 5.4.

**Fig. 5.4** Critical points on Z

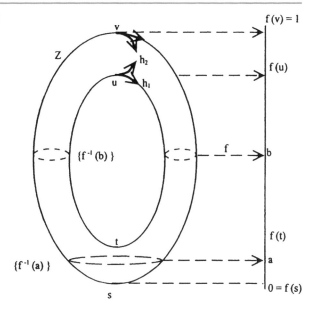

The next critical point is the saddle, $u$, near which $f$ is represented as

$$(h_1, h_2) \to f(u) - h_1^2 + h_2^2.$$

Now for $b \in (f(t), f(u))$, $f^{-1}(b)$ is a one-dimensional manifold, but this time it is *two* copies of $S^1$. Finally $v$ is a local maximum and $f$ is represented near $v$ by $(h_1, h_2) \to f(u) - h_1^2 - h_2^2$. Thus the index of $v$ is 2.

We can also use this example to introduce the idea of the Euler characteristic $\chi(X)$ of a manifold $X$. If $X$ has dimension, $n$, let $c_i(X, f)$ be the number of critical points of index $i$, of the function $f : X \to \Re$ and let

$$\chi(X, f) = \sum_{i=0}^{n} (-1)^i c_i(X, f).$$

For example the height function, $f$, on the torus $Z$ has

(i) $c_0(Z, f) = 1$, since $s$ has index 0
(ii) $c_1(Z, f) = 2$, since both $t$ and $u$ have index 1
(iii) $c_2(Z, f) = 1$, since $v$ has index 2.

Thus $\chi(Z, f) = 1 - 2 + 1 = 0$. In fact, it can be shown that $\chi(X, f)$ is independent of $f$, when $X$ is a compact manifold. It is an *invariant* of the smooth manifold $X$, labelled $(\chi(X))$. Example 5.4 illustrates the fact that $\chi(Z) = 0$.

*Example 5.5*
(1) The sphere $S^1$. It is clear that the height function $f : S^1 \to \Re$ has an index 0 critical point at the bottom and an index 1 critical point at the top, so $\chi(S^1) = 1 - 1 = 0$.

**Fig. 5.5** Critical points in $S^2$

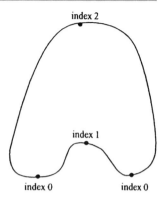

(2) The sphere $S^2$ has an index 0 critical point at the bottom and an index 2 critical point at the top, so $\chi(S^2) = c_0 + c_2 = 1 + 1 = 2$. See Fig. 5.5.

It is possible to deform the sphere, $S^2$, so as to induce a saddle, but this creates an index 0 critical point. In this case $c_0 = 2$, $c_1 = 1$, and $c_2 = 1$ as in Fig. 5.5. Thus $\chi(S^2) = 2 - 1 + 1 = 2$ again.

(3) More generally, $\chi(S^n) = 0$ if $n$ is odd and $= 2$ if $n$ is even.

(4) To compute $\chi(B^n)$ for the closed $n$-ball, take the sphere $S^n$ and delete the top hemisphere. The remaining bottom hemisphere is diffeomorphic to $B^n$. By this method we have removed the index $n$ critical point at the top of $S^n$.

For $n = 2k + 1$ odd, we know

$$\chi\left(S^{2k+1}\right) = \sum_{i=0}^{2k}(-1)^i c_i\left(S^n\right) - c_n\left(S^n\right) = 0,$$

so $\chi(B^{2k+1}) = \chi(S^{2k+1}) + 1 = 1$. For $n = 2k$, even we have

$$\sum_{i=0}^{2k-1}(-1)c_i\left(S^n\right) + c_n\left(S^n\right) = 2, \quad \text{so } \chi\left(B^{2k}\right) = \chi\left(S^{2k}\right) - 1 = 1.$$

## 5.2 Transversality

To examine the singularity set $S(f)$ of a smooth function $f : X \to Y$ we introduce the idea of *transversality*.

A linear manifold $V$ in $\Re^n$ of dimension $v$ is of the form $x_0 + K$, where $K$ is a vector subspace of $\Re^n$ of dimension $v$. Intuitively if $V$ and $W$ are linear manifolds in $\Re^n$ of dimension $v$, $w$ then typically they will not intersect if $v + w < n$.

On the other hand if $v + w \geq n$ then $V \cap W$ will typically be of dimension $v + w - n$.

For example two lines in $\Re^2$ will intersect in a point of dimension $1 + 1 - 2$.

## 5.2 Transversality

Another way of expressing this is to define the codimension of $V$ in $\Re^n$ to be $n - v$. Then the codimension of $V \cap W$ in $W$ will typically be $w - (v + w - n) = n - v$, the same codimension.

Suppose now that $f : X^n \to Y^m$ where $X, Y$ are vector spaces of dimension $n, m$ respectively. Let $Z$ be a $z$-dimensional linear manifold in $Y$. Say that $f$ is *transversal* to $Z$ iff for all $x \in X$, either (i) $f(x) \notin Z$ or (ii) the image of $df(x)$, regarded as a vector subspace of $Y^m$, together with $Z$ span $Y$. In this case write $f \stackrel{T}{\cap} Z$. The same idea can be extended to the case when $X, Y, Z$, are all manifolds. Whenever $f \stackrel{T}{\cap} Z$, then if $x \in f^{-1}(Z)$, $f$ will be a submersion at $x$, and so $f^{-1}(Z)$ will be a smooth manifold in $X$ of codimension equal to the codimension of $Z$ in $Y$. Another way of interpreting this is that the number of constraints which determine $Z$ in $Y$ will be equal to the number of constraints which determine $f^{-1}(Z)$ in $X$. Thus $\dim(f^{-1}(Z)) = n - (m - z)$.

In the previous chapter we put the Whitney $C^s$-topology on the set of $C^s$-differentiable maps $X^n \to Y^m$, and called this $C^s(X, Y)$. In this topological space a residual set is dense. The fundamental theorem of singularity theory is that transversal intersection is generic.

**Thom Transversality Theorem** *Let $X^n, Y^m$ be manifolds and $Z^z$ a submanifold of $Y$. Then the set*

$$\{f \in C^s(X, Y) : f \stackrel{T}{\cap} Z\} = \stackrel{T}{\cap}(X, Y; Z)$$

*is a residual (and thus dense) set in the topological space $C^s(X, Y)$.*

Note that if $f \in \stackrel{T}{\cap}(X, Y; Z)$ then $f^{-1}(Z)$ will be a manifold in $X$ of dimension $n - m + z$.

Moreover if $g \in C^s(X, Y)$ but $g$ is not transversal to $Z$, then there exists some $C^s$-map, as near to $g$ as we please in the $C^s$ topology, which is transversal to $Z$. Thus transversal intersection is typical or *generic*.

Suppose now that $f : X^n \to Y^m$, and corank $df(x) = r$, so rank $df(x) = \min(n, m) - r$. In this case we said that $x \in S_r(f)$, the corank $r$ singularity set of $f$. We seek to show that $S_r(f)$ is a manifold in $X$, and compute its dimension.

Suppose $X^n, Y^m$ are vector spaces, with dimension $n, m$ respectively. As before $\mathcal{L}(X, Y)$ is the normed vector space of linear maps from $X$ to $Y$. Let $\mathcal{L}_r(X, Y)$ be the subset of $\mathcal{L}(X, Y)$ consisting of linear maps with corank $r$.

**Lemma 5.3** *$\mathcal{L}_r(X, Y)$ is a submanifold of $\mathcal{L}(X, Y)$ of codimension $(n - z + r)(m - z + r)$ where $z = \min(n, m)$.*

*Proof* Choose bases such that a corank $r$ linear map $S$ is represented by a matrix $\begin{pmatrix} A & B \\ C & D \end{pmatrix}$ where rank $(A) = k = z - r$.

Let $U$ be a neighbourhood of $S$ in $\mathcal{L}(X,Y)$, such that for all $S' \in U$, corank $(S') = r$.

Define $F : U \to \mathcal{L}(\Re^{n-k}, \Re^{m-k})$ by $F(S') = D' - C(A')^{-1}B'$.

Now $S' \in F^{-1}(0)$ iff rank $(S') = k$. The codimension of 0 in $\mathcal{L}(\Re^{n-k}, \Re^{m-k})$ is $(n-k)(m-k)$. Since $F$ is a submersion $F^{-1}(0) = \mathcal{L}_r(X, Y)$ is of codimension $(n-z+r)(m-z+r)$. □

Now if $f \in C^s(X, Y)$, then $df \in C^{s-1}(X, \mathcal{L}(X,Y))$. If $df(x) \in \mathcal{L}_r(X,Y)$ then $x \in S_r(f)$. By the Thom Transversality Theorem, there is a residual set in $C^{s-1}(X, \mathcal{L}(X,Y))$ such that $df(x)$ is transversal to $\mathcal{L}_r(X,Y)$. But then $S_r(f) = df^{-1}(\mathcal{L}_r(X,Y))$ is a submanifold of $X$ of codimension $(n-z+r)(m-z+r)$. Thus we have:

**The Singularity Theorem** *There is the residual set $V$ in $C^s(X, Y)$ such that for all $f \in V$, the corank $r$ singularity set of $f$ is a submanifold in $X$ of codimension $(n-z+r)(m-z+r)$.*

If $f : X^n \to \Re$ then codim $S_1(f) = (n-1+1)(1-1+1) = n$. Hence generically the set of critical points will be zero-dimensional, and so critical points will be isolated. Now a Morse function has isolated, non-degenerate critical points. Let $M^s(X)$ be the set of Morse functions on $X$ with the topology induced from $C^s(X, \Re)$.

**Morse Theorem** *The set $M^s(X)$ of $C^s$-differentiable Morse functions (with non-degenerate critical points), is an open dense set in $C^s(X, \Re)$. Moreover if $f \in M^s(X)$, then the critical points of $f$ are isolated.*

More generally if $f : X^n \to Y^m$ with $m \leq n$ then in the generic case, $S_1(f)$ is of codimension $(n-m+1)(n-m+1) = n-m+1$ in $X$ and so $S_1(f)$ will be of dimension $(m-1)$.

Suppose now that $n > 2m - 4$, and $n \geq m$. Then $2n - 2m + 4 > n$ and so $r(n-m+r) > n$ for $r \geq 2$. But codimension $(S_r(f)) = r(n-m+r)$, and since codimension $(S_r(f)) >$ dimension $X$, $S_r(f) = \Phi$ for $r \geq 2$.

**Submanifold Theorem** *If $Z^z$ is a submanifold of $Y^m$ and $z < m$ then $Z$ is nowhere dense in $Y$ (here $z = \dim(Z)$ and $m = \dim(Y)$).*

In the case $n \geq m$, the singularity set $S(f)$ will generically consist of a union of the various co-rank $r$ singularity submanifolds, for $r \geq 1$. The highest dimension of these is $m - 1$. We shall call an $S(f)$ a *stratified manifold* of dimension $(m-1)$. Note also that $S(f)$ will then be nowhere dense in $X$. We also require the following theorem.

**Morse Sard Theorem** *If $f : X^n \to Y^m$ is a $C^s$-map, where $s > n - m$, then $f(S(f))$ has measure zero in $Y$ and $f(S_0(f))$ is residual and therefore dense in $Y$.*

We are now in a position to apply these results to the critical Pareto set. Suppose then that $u = (u_1, \ldots, u_m) : X^n \to \Re^m$ is a smooth profile on the manifold of feasible states $X$.

Say $x \in \overset{0}{\Theta}(u_1, \ldots, u_m) = \overset{0}{\Theta}(u)$ iff $dL(\lambda, u)(x) = 0$, where $L(\lambda, u) = \sum_{i=1}^{m} \lambda_i u_i$ and $\lambda \in \overline{\Re_+^m}$.

By Theorem 4.21, the critical Pareto set, $\Theta(u)$, contains $\overset{0}{\Theta}(u)$ but possibly also contains further points on the boundary of $X$. However we shall regard $\overset{0}{\Theta}$ as the differential analogue of the Pareto set. By Lemma 4.19, $\overset{0}{\Theta}(u)$ must be closed in $X$. Moreover when $n \geq m$ and $x \in \overset{0}{\Theta}(u)$ then the differentials $\{du_i(x) : i \in M\}$ must be linearly dependent. Hence $\overset{0}{\Theta}(u)$ must belong to $S(u)$. But also $S(u)$ will be nowhere dense in $X$. Thus we obtain the following theorem (Smale 1973 and Debreu 1970).

**Pareto Theorem** *There exists a residual set $U$ in $C^1(X, \Re^m)$, for $\dim(X) \geq m$, such that for any profile $u \in U$, the closed critical Pareto set $\overset{0}{\Theta}(u)$ belongs to the nowhere dense stratified singularity manifold $S(u)$ of dimension $(m-1)$. Moreover if $\dim(X) > 2m - 4$, then $\Theta(u)$ is itself a manifold of dimension $(m-1)$, for all $u \in C^1(X, \Re^m)$.*

As we have already observed this result implies that $\Theta(u)$ can generally be regarded as an $(m-1)$ dimensional object parametrised by $(m-1)$ coefficients $(\frac{\lambda_2}{\lambda_1}, \ldots, \frac{\lambda_m}{\lambda_1})$ say. Since points in $\overset{0}{\Theta}(u)$ are characterised by first order conditions alone, it is necessary to examine the Hessian of $L$ to find the Pareto optimal points.

## 5.3 Generic Existence of Regular Economies

In this section we outline a proof of the Debreu-Smale Theorem on the Generic Existence of Regular Economies (see Debreu 1970, and Smale 1974a).

As in Sect. 4.4, let $u = (u_1, \ldots, u_m) : X^m \to \Re^m$ be a smooth profile, where $X = \Re_+^n$, the commodity space facing each individual. Let $e = (e_1, \ldots, e_m) \in X^m$ be an initial endowment vector. Given $u$, define the *Walras manifold* to be the set

$$Z_u = \{(e, x, p) \in X^m \times X^m \times \Delta\}$$

(where $\Delta$ is the price simplex) such that $(x, p)$ is a *Walrasian equilibrium* for the economy $(e, u)$. That is, $(x, p) = (x_1, \ldots, x_m, p) \in X^m \times \Delta$ satisfies:
1. individual optimality: $D^* u_i(x_i) = p$ for $i \in M$,
2. individual budget constraints: $\langle p, x_i \rangle = \langle p, e_i \rangle$ for $i \in M$,
3. social resource constraints: $\sum_{i=1}^{m} x_{ij} = \sum_{i=1}^{m} e_{ij}$ for each commodity $j = 1, \ldots, n$.

Note that we implicitly assume that each individual's utility function, $u_i$, is defined on a domain $X_i \equiv X \subset \Re_+^n$, so that the differential $du_i(x_i)$ at $x_i \in X_i$ can be represented by a vector $Du_i(x_i) \in \Re^n$. As we saw in Chap. 4, we may normalize

$Du_i$ and $p$ so the optimality condition for $i$ becomes $D^*u_i(x) = p$ for $p \in \Delta$. For the space of normalized price vectors, we may identify $\Delta$ with $\{p \in \Re_+^n : \|p\| = 1\}$. Observe that $\dim(\Delta) = n - 1$.

We seek to show that there is a residual set $U$ in $C^s(X, \Re^m)$ such that the Walras manifold is a smooth manifold of dimension $mn$.

Now define the Debreu projection

$$\pi : Z_u \subset X^m \times X^m \times \Delta \to X^m : (e, x, p) \to e.$$

Note that both $Z_u$ and $X^m$ will then have dimension $mn$.

By the Morse Sard Theorem the set

$$V = \{e \in X^m : d\pi \text{ has rank } nm \text{ at}(e, x, p)\}$$

is dense in $X^m$.

Say the economy $(e, u)$ is *regular* if $\pi(e, x, p) = e$ is a regular value of $\pi$ (or rank $d\pi = mn$) for all $(x, p) \in X^m \times \Delta$ such that $((e, x, p) \in Z_u)$.

When $e$ is a regular value of $\pi$, then by the inverse function theorem,

$$\pi^{-1}(e) = \{(e, x, p)^1, (e, x, p)^2, \ldots, (e, x, p)^k\}$$

is a zero-dimensional object, and thus will consist of a finite number of isolated points. Thus for each $e \in V$, the Walrasian equilibria associated with $e$ will be finite in number. Moreover there will exist a neighbourhood $N$ of $e$ in $V$ such that the Walrasian equilibria move continuously with $e$ in $N$.

*Proof of the Generic Regularity of the Debreu Projection* Define $\psi_u : X^m \times \Delta \to \Delta^{m+1}$ where $u \in C^r(X, \Re^m)$ by $\psi_u(x, p) = (D^*u_1(x_1), \ldots, D^*u_m(x_m), p)$ where $x = (x_1, \ldots, x_m)$ and $u = (u_1, \ldots, u_m)$. Let $I$ be the diagonal $\{(p, \ldots, p)\}$ in $\Delta^{m+1}$. If $(x, p) \in \psi_u^{-1}(I)$ then for each $i$, $D^*u_i(x_i) = p$ and so the first order individual optimality conditions are satisfied. By the Thom Transversality Theorem there is a residual set (in fact an open dense set) of profiles $U$ such that $\psi_u$ is transversal to $I$ for each $u \in U$. But then the codimension of $\psi_u^{-1}(I)$ in $X^m \times \Delta$ equals the codimension of $I$ in $\Delta^{m+1}$.

Now $\Delta$ and $I$ are both of dimension $(n-1)$ and so codimension $(I)$ in $\Delta^{m+1}$ is $(m+1)(n-1) - (n-1) = m(n-1)$. Thus $\dim(X^m \times \Delta) - \dim(\psi_u^{-1}(I)) = m(n-1)$ and $\dim(\psi_u^{-1}(I)) = mn + (n-1) - m(n-1) = n + m - 1$, for all $u \in U$.

Now let $e \in X^m$ be the initial endowment vector and

$$Y(e) = \left\{(x_1, \ldots, x_m) \in X^m : \sum_{i=1}^m x_i = \sum_{i=1}^m e_i\right\}$$

be the set of feasible outcomes, a hyperplane in $\Re_+^{nm}$ of dimension $n(m-1)$. For each $i$, let $B_i(p) = \{x_i \in X : \langle p, x_i \rangle = \langle p, e_i \rangle\}$, be the hyperplane through the boundary of the $i$th budget set at the price vector $p$.

## 5.3 Generic Existence of Regular Economies

Define

$$\Sigma(e) = \{(x, p) \in X^m \times \Delta : x \in Y(e), x_i \in B_i(p), \forall i \in M\}, \quad \text{and}$$
$$\Gamma = \{(e, x, p) : e \in X^m, (x, p) \in \Sigma(e)\}.$$

As discussed in Chap. 4, $Y(e)$ is characterized by $n$ linear equations, while the budget restrictions induce a further $(m-1)$ linear restraints (the $m$th budget restraint is redundant). Thus the dimension of $\Gamma$ is $2mn + (n-1) - n - (m-1) = 2mn - m$. (In fact, if $\Delta$ is taken to be the $(n-1)$ dimensional simplex, then $\Gamma$ will be a linear manifold of dimension $2mn - m$. More generally, $\Gamma$ will be a submanifold of $X^m \times X^m \times \Delta$ of dimension $2mn - m$. At each point the projection is a regular map (i.e., the rank of the differential of $(e, x, p) \to (x, p)$ is maximal).

To see this define $\phi : X^m \times X^m \times \Delta \to \Re^n \times \Re^{m-1}$ by

$$\phi(e, x, p) = \left(\sum_{i=1}^{m} x_i - \sum_{i=1}^{m} e_i, \langle p, x_1 \rangle - \langle p, e_1 \rangle, \ldots, \langle p, x_{m-1} \rangle - \langle p, e_{m-1} \rangle \right).$$

Clearly if $\phi(e, x, p) = \underline{0}$ then $x \in Y(e)$ and $x_i \in B_i(p)$ for each $i$. But $\underline{0}$ is of codimension $n + m - 1$ in $\Re^n \times \Re^{m-1}$; thus $\phi^{-1}(\underline{0})$ is of the same codimension in $X^{2m} \times \Delta$. Thus $\dim(X^{2m} \times \Delta) - \dim \phi^{-1}(\underline{0}) = n + m - 1$ and $\dim \phi^{-1}(\underline{0}) = 2nm + (n-1) - (n + m - 1) = 2mn - m$ (giving the dimension of $\Gamma$). In a similar fashion, for $(x, p) \in \Sigma(e), \phi(e, x, p) = \underline{0}$, and so

$$\dim(X^m \times \Delta) - \dim \phi^{-1}(\underline{0}) = n + m - 1.$$

Thus $\Sigma(e)$ is a submanifold of $X^m \times \Delta$ of dimension

$$nm + (n-1) - (n + m - 1) = mn - m.$$

Finally define $Z_u = \{(e, x, p) \in \Gamma : \psi_u(x, p) \in I\}$. For each $u \in U$, $Z_u$ is a submanifold of $X^{2m} \times \Delta$ of dimension $mn$.

To see this, let $f_u(e, x, p) = \psi_u(x, p)$. Then

$$f_u : \Gamma \to X^m \times X^m \times \Delta \to X^m \times \Delta \xrightarrow{\psi_u} \Delta^{m+1}.$$

As we observed for all $u \in U$, $\psi_u$ is transversal to $I$ in $\Delta^{m+1}$. But the codimension of $I$ in $\Delta^{m+1}$ is $m(n-1)$. Since $f_u$ will be transversal to $I$,

$$\dim(\Gamma) - \dim(f_u^{-1}(I)) = m(n-1).$$

Hence $\dim(f_u^{-1}(I)) = mn$. Clearly $Z_u = f_u^{-1}(I)$.

Thus for all $u \in U$, the Debreu projection $\pi : Z_u \to X^m$ will be a $C^1$-map between manifolds of dimension $mn$. The Morse Sard Theorem gives the result. $\square$

**Fig. 5.6** The Debreu map

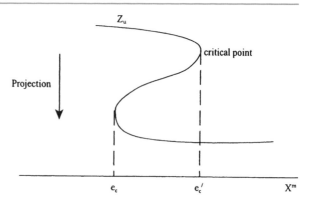

Thus we have shown that for each smooth profile $u$ in an open dense set $U$, there exists an open dense set $V$ of initial endowments such that $(e, u)$ is a regular economy for all $e \in V$.

The result is also related to the existence of a demand function for an economy. A demand function for $i$ (with utility $u_i$) is the function

$$f_i : \Delta \times \Re_+ \to X$$

where $f_i(p, I)$ is that $x_i \in X$ which maximizes $u_i$ on

$$B_i(p, I) = \{x \in X : \langle p, x \rangle = I\}.$$

Now define $\phi_i : X \to \Delta \times \Re_+$ by $\phi_i(x) = (D^*u_i(x), \langle D^*u_i(x), x \rangle)$.

But the optimality condition is precisely that $D^*u_i(x) = p$ and $\langle D^*u_i(x), x \rangle = \langle p, x \rangle = I$. Thus when $\phi_i$ has maximal rank, it is locally invertible (by the inverse function theorem) and so locally defines a demand function.

On the other hand if $f_i$ is a $C^1$-function then $\phi_i$ must be locally invertible (by $f_i$). If this is true for all the agents, then $\psi_u : X^m \times \Delta \to \Delta^{m+1}$ must be transversal to $I$. Consequently if $u = (u_1, \ldots, u_m)$ is such that each $u_i$ defines a $C^1$-demand function $f_i : \Delta \times \Re_+ \to X$ then $u \in U$, the open dense set of the regular economy theorem.

As a final note suppose that $u \in U$ and $e$ is a critical value of the Debreu projection. Then it is possible that $\pi^{-1}(e) = e \times W$ where $W$ is a continuum of Walrasian equilibria. Another possibility is that there is a continuum of singular or *catastrophic* endowments $C$, so that as the endowment vector crosses $C$ the number of Walrasian equilibria changes suddenly. As we discussed in Sect. 4.4, at a "catastrophic" endowment, stable and unstable Walrasian equilibria may merge (see Balasko 1975).

Another question is whether for every smooth profile, $u$, and every endowment vector, $e$, there exists a Walrasian equilibrium $(x, p)$. This is equivalent to the requirement that for every $u$ the projection $Z_u \to X^m$ is *onto* $X^m$.

In this case for each $e \in X^m$ there will exist some $(e, x, p) \in Z_u$. This is clearly necessary if there is to exist a market clearing price equilibrium $p$ for the economy $(e, u)$.

The usual general equilibrium arguments to prove existence of a market clearing price equilibrium typically rely on convexity properties of preference (see Chap. 3). However weaker assumptions on preference permit the use of topological arguments to show the existence of an extended Walrasian equilibrium (where only the first order conditions are satisfied).

When the market-clearing price equilibrium does exist, it is useful to consider a price adjustment process (or "auctioneer") to bring about equilibrium.

## 5.4 Economic Adjustment and Excess Demand

To further develop the idea of a demand function and price adjustment process, we consider the following famous example of Scarf (1960).

*Example 5.6* There are three individuals $i \in M = \{1, 2, 3\}$ and three commodities. Agent $i$ has utility $u_i(x_1^i, x_2^i, x_3^i) = \min(x_i^i, x_j^i)$ where $x^i = (x_1^i, x_2^i, x_3^i) \in \mathfrak{R}_+^3$ is the $i$th commodity space.

At income $I$, and price vector $p = (p_1, p_2, p_3)$, $i$ demands equal amounts of $x_i^i, x_j^j$ and zero of $x_k^i$: thus $(p_i + p_j)x = I$, so $x_i^i = I(p_i + p_j)^{-1} = x_j^i$.

Suppose the initial endowment $e_i$ of agent $i$ is 1 unit of the $i$th good, and nothing of the $j$th and $k$th. Then $I = p_i$ and so $i$'s demand function $f_i$ has the form $f_i(p) = (f_{ii}(p), f_{ij}(p), f_{ik}(p)) = (\frac{p_i}{p_i + p_j}, \frac{p_i}{p_i + p_j}, 0) \in \mathfrak{R}^3$.

The *excess demand* function by $i$ is $\xi_i(p) = f_i(p) - e_i$.

Since $e_i = (e_{ii}, e_{ij}, e_{ik}) = (1, 0, 0)$ this gives

$$\xi_i(p) = \left(\frac{-p_j}{p_i + p_j}, \frac{p_i}{p_i + p_j}, 0\right) = (\xi_{ii}, \xi_{ij}, \xi_{ik}).$$

Suppose now the other two consumers are described by cyclic permutation of subscripts, e.g., $j$ has 1 unit of the $j$th good and utility $u_j(x_j^j, x_k^j, x_i^j) = \min(x_j^j, x_k^j)$, etc., then the total excess demand at $p$ is

$$\xi(p) = \sum_{i=1}^{3} \xi_i(p) \in \mathfrak{R}^3.$$

For example, the excess demand in commodity $j$ is:

$$\xi_j = \xi_{1j} + \xi_{2j} + \xi_{3j} = \frac{p_i}{p_i + p_j} - \frac{p_k}{p_j + p_k}.$$

Since each $i$ chooses $f_i(p)$ to maximize utility subject to $\langle p_1, f_i(p) \rangle = I = \langle p, e_i \rangle$ we expect $\sum_{i=1}^{3} \langle p, f_i(p) - e_i \rangle = 0$.

To see this note that

$$\langle p, \xi(p) \rangle = \left\langle p, \left(\frac{p_3}{p_3 + p_1} - \frac{p_2}{p_1 + p_2}, \frac{p_1}{p_1 + p_2} - \frac{p_3}{p_2 + p_3}, \frac{p_2}{p_2 + p_3} - \frac{p_1}{p_1 + p_3}\right)\right\rangle$$
$$= 0.$$

The equation $\langle p, \xi(p)\rangle = 0$ is known as Walras' Law. To interpret it, suppose we let $\Delta$ be the simplex in $\Re^3$ of price vectors such that $\|p\| = 1$, and $p_i \geq 0$. Walras' Law says that the excess demand vector $\xi(p)$ is orthogonal to the vector $p$. In other words $\xi(p)$ may be thought of as a tangent vector in $\Delta$. (This is easier to see if we identify $\Delta$ with a quadrant of the sphere, $S^2$.)

We may therefore consider a *price adjustment process*, which changes the price vector $p(t)$, at time $t$ by the differential equation

$$\frac{dp(t)}{dt} = \xi(p) \qquad (*)$$

This adjustment process is a vector field on $\Delta$: that is at every $p$ there exists a rule that changes $p(t)$ by the equation $\frac{dp(t)}{dt} = \xi(p)$.

If at a vector $p^*$, the excess demand $\xi(p^*) = 0$ then $\frac{dp(t)}{dt}|_{p^*} = 0$, and the price adjustment process has a stationary point. The flow on $\Delta$ can be obtained by integrating the differential equation. It is easy to see that if $p^*$ satisfies $p_1^* = p_2^* = p_3^*$ then $\xi(p^*) = 0$, so there clearly is a price equilibrium where excess demand is zero.

The price adjustment equation $(*)$ does not result in a flow into the price equilibrium.

To see this, compute the scalar product

$$\langle (p_2 p_3, p_1 p_3, p_1 p_2), \xi(p) \rangle$$
$$= -p_3 \frac{(p_1{}^2 - p_2{}^2)}{p_1 + p_2} + p_2 \frac{(p_3{}^2 - p_1{}^2)}{p_1 + p_3} + p_1 \frac{(p_1{}^2 - p_3{}^2)}{p_2 + p_3}$$
$$= p_3(p_1 - p_2) + p_2(p_3 - p_1) + p_1(p_2 - p_1)$$
$$= 0.$$

But if $\xi(p) = \frac{dp}{dt}$ then we obtain $p_2 p_3 \frac{dp}{dt} + p_1 p_3 \frac{dp}{dt} + p_1 p_2 \frac{dp}{dt} = 0$.

The obvious solution to this equation is that $p_1(t) p_2(t) p_3(t) = \text{constant}$.

In other words when the adjustment process satisfies $(*)$ then the price vector $p$ (regarded as a function of time, $t$,) satisfies the equation $p_1(t) p_2(t) p_3(t) = \text{constant}$. The flow through any point $p = (p_1, p_2, p_3)$, other than the equilibrium price vector $p^*$, is then homeomorphic to a circle $S^1$, inside $\Delta$.

Just to illustrate, consider a vector $p$ with $p_3 = 0$.

Then $\xi(p) = (\frac{-p_2}{p_1+p_2}, \frac{p_1}{p_1+p_2}, 0)$.

Because we have drawn the flow on the simplex $\Delta = \{p \in \Re_+^3 : \sum p_i = 1\}$ the flow $\frac{dp}{dt}(t) = \xi(p)$ is discontinuous in $p$ at the three vertices of $\Delta$.

However in the interior of $\Delta$ the flow is essentially circular (and anti-clockwise). See Fig. 5.7.

To examine the nature of the flow given by the differential equation $\frac{dp}{dt}(t) = \xi(p)$, define a Lyapunov function at $p(t)$ to be $L(p(t)) = \frac{1}{2} \sum_{i=1}^{3} (p_i(t) - p_i^*)^2$ where $p^* = (p_1^*, p_2^*, p_3^*)$ is the equilibrium price vector satisfying $\xi(p^*) = 0$. Since

## 5.4 Economic Adjustment and Excess Demand

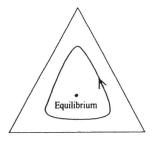

**Fig. 5.7** Flow on the price simplex

$p^* \in \Delta$, we may choose $p^* = (\frac{1}{3}, \frac{1}{3}, \frac{1}{3})$. Then

$$\frac{dL}{dt} = \sum_{i=1}^{3} -(p_i(t) - p_i^*)\frac{dp_i}{dt}$$

$$= \sum_{i=1}^{3} \xi_i(p(t))p_i(t) - \sum_{i=1}^{3} p_i^* \xi_i(p(t))$$

$$= -\frac{1}{3}\sum_{i=1}^{3} \xi_i(p(t)).$$

(This follows since $\langle p, \xi(p)\rangle = 0$.)

If $\xi_i(p(t)) > 0$ for $i = 1, 2, 3$ then $\frac{dL}{dt} < 0$ and so the Lyapunov distance $L(p(t))$ of $p(t)$ from $p^*$ decreases as $t \to \infty$. In other words if $\delta p(t) = p(t) - p^*$ then the distance $\|\delta p(t)\| \to 0$ as $t \to \infty$. The equilibrium $p^*$ is then said to be *stable*.

If on the contrary $\xi_i(p(t)) < 0$ $\forall i$, then $\frac{dL}{dt} > 0$ and $\|\delta p(t)\|$ increases as $t \to \infty$. In this case $p^*$ is called *unstable*.

However it is easy to show that the equilibrium point $p^*$ is neither stable nor unstable. To see this consider the price vector $p(t) = (\frac{2}{3}, \frac{1}{6}, \frac{1}{6})$. It is then easy to show that $\xi = (0, \frac{3}{10}, \frac{-3}{10})$ so the flow through $p$ (locally) keeps the distance $L(p(t))$ constant. To see how $L(p(t))$ behaves near $p(t)$, consider the points $p(t - \delta t) = (\frac{2}{3}, \frac{1}{6} - \frac{1}{20}, \frac{1}{6} + \frac{1}{10})$ and $p(t + \delta t) = (\frac{2}{3}, \frac{1}{6} + \frac{1}{10}, \frac{1}{6} - \frac{1}{10})$. After some easy arithmetic we find that $\xi(t - \delta t) = (0.195, 0.409, -0.814)$ so that $\|\delta p(t - \delta t)\|$ is increasing at $p(t - \delta t)$. On the other hand $\xi(t + \delta t) = (-0.195, 0.814, -0.409)$ so $\|\delta p(t + \delta t)\|$ is decreasing at $p(t + \delta t)$. In other words the total excess demand $\sum_{i=1}^{3} \xi_i(p(t))$ oscillates about zero as we transcribe one of the closed orbits, so the distance $\|\delta p(t)\|$ increases then decreases.

The Scarf Example gives a way to think more abstractly about the process of price adjustment in an economy.

As we have observed, the differential equation $\frac{dp}{dt} = \xi(p)$ on $\Delta$ defines a *flow* in $\Delta$. That is if we consider any point $p^0 \in \Delta$ and then solve the equation for $p$, we obtain an "orbit"

$$\{p(t) \in \Delta, t \in (-\infty, \infty); \; p(0) = p^0 \text{ and } dp = \xi(p(t))\}$$

**Fig. 5.8** Smoothing the Scarf profile

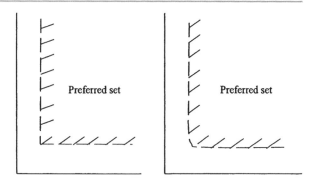

that commences at the point $p(0) = p^0$, and gives the past and future trajectory. Because the differential equation has a unique solution, any point $p_0$ can belong to only one orbit. As we saw, each orbit in the example satisfies the equation $p_1(t) \cdot p_2(t) \cdot p_3(t) = $ constant. The *phase portrait* of the differential equation is its collection of orbits.

The differential equation $\frac{dp}{dt} = \xi(p)$ assigns to each point $p \in \Delta$ a vector $\xi(p) \in \Re^n$, and so $\xi$ may be regarded as a function $\xi : \Delta \to \Re^n$. In fact $\xi$ is a continuous map except at the boundary of $\Delta$. This discontinuity only occurs because $\Delta$ itself is not smooth at its vertices. If we ignore this boundary feature, then we may write $\xi \in C_0(\Delta, \Re^n)$, where $C_0$ as before stands for the set of continuous maps. In fact if we examine $\xi$ as a function of $p$ then it can be seen to be differentiable, so $\xi \in C_1(\Delta, \Re^n)$. Obviously $C_1(\Delta, \Re^n)$ has a natural metric and therefore the set $C_1(\Delta, \Re^n)$ can be given the $C^1$-topology. A differential equation $\frac{dp}{dt} = \xi(p)$ of this kind can thus be treated as an element of $C_1(\Delta, \Re^n)$ in which case it is called a *vector field*. $C_1(\Delta, \Re^n)$ with the $C^1$-topology is written $C^1(\Delta, \Re^n)$ or $\mathcal{V}^1(\Delta)$. We shall also write $\mathcal{P}(\Delta)$ for the collection of phase portraits on $\Delta$. Obviously, once the vector field, $\xi$, is specified, then this defines the *phase portrait*, $\tau(\xi)$, of $\xi$.

In the example, $\xi$ was determined by the utility profile $u$ and endowment vector $e \in \Re^{3 \times 3}$. As Fig. 5.8 illustrates the profile $u$ can be smoothed by rounding each $u_i$ without changing the essence of the example.

## 5.5 Structural Stability of a Vector Field

More abstractly then we can view the excess demand function $\xi$ as a map from $C^s(X, \Re^m) \times X^m$ to the metric space of vector fields on $\Delta$: that is

$$\xi : C^s(X, \Re^m) \times X^m \longrightarrow \mathcal{V}^1(\Delta).$$

The genericity theorem given above implies that, in fact, there is an open dense set $U$ in $C^s(X, \Re^m)$ such that $\xi$ is indeed a $C^1$ vector field on $\Delta$. Moreover $\xi$ is an excess demand function obtained from the individual demand functions $\{f_i\}$ as described above.

## 5.5 Structural Stability of a Vector Field

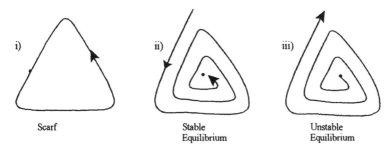

**Fig. 5.9** Dissimilar phase portraits

An obvious question to ask is how $\xi$ "changes" as the parameters $u \in C^s(X, \Re^m)$ and $e \in X^m$ change. One way to do this is to consider small perturbations in a vector field $\xi$ and determine how the phase portrait of $\xi$ changes.

It should be clear from the Scarf example that small perturbations in the utility profile or in $e$ may be sufficient to change $\xi$ so that the orbits change in a qualitative way. If two vector fields, $\xi_1$, and $\xi_2$ have phase portraits that are homeomorphic, then $\tau(\xi_1)$ and $\tau(\xi_2)$ are qualitatively identical (or similar). Thus we say $\xi_1$ and $\xi_2$ are *similar* vector fields if there is a homeomorphism $h : \Delta \to \Delta$ such that each orbit in the phase portrait $\tau(\xi_1)$ of $\xi_1$ is mapped by $h$ to an orbit in $\tau(\xi_2)$.

As we saw in the Scarf example, each of the orbits of the excess demand function, $\xi_1$, say, comprises a closed orbit (homeomorphic to $S^1$). Now consider the vector field $\xi_2$ whose orbits approach an equilibrium price vector $p^*$. The phase portraits of $\xi_1$ and $\xi_2$ are given in Fig. 5.9.

The price equilibrium in Fig. 5.9(b) is *stable* since $\lim_{t \to \infty} p(t) \to p^*$. Obviously each of the orbits of $\xi_2$ are homeomorphic to the half open interval $(-\infty, 0]$. Moreover $(-\infty, 0]$ and $S^1$ are not homeomorphic, so $\xi_1$ and $\xi_2$ are not similar.

It is intuitively obvious that the vector field, $\xi_2$ can be obtained from $\xi_1$ by a "small perturbation", in the sense that $\|\xi_1 - \xi_2\| < \delta$, for some small $\delta > 0$. When there exists a small perturbation $\xi_2$ of $\xi_1$, such that $\xi_1$ and $\xi_2$ are dissimilar, then $\xi_1$ is called *structurally unstable*. On the other hand, it should be plausible that, for any small perturbation $\xi_3$ of $\xi_2$ then $\xi_3$ will have a phase portrait $\tau(\xi_3)$ homeomorphic to $\tau(\xi_2)$, so $\xi_2$ and $\xi_3$ will be similar. Then $\xi_2$ is called *structurally stable*. Notice that structural stability of $\xi_2$ is a much more general property than stability of the equilibrium point $p^*$ (where $\xi_2(p^*) = 0$).

All that we have said on $\Delta$ can be generalised to the case of a smooth manifold $Y$. So let $\mathcal{V}^1(Y)$ be the topological space of $C^1$-vector fields on $Y$ and $\mathcal{P}(Y)$ the collection of phase portraits on $Y$.

**Definition 5.1**
(1) Let $\xi_1, \xi_2 \in \mathcal{V}^1(Y)$. Then $\xi_1$ and $\xi_2$ are said to be *similar* (written $\xi_1 \sim \xi_2$) iff there is a homeomorphism $h : Y \to Y$ such that an orbit $\sigma$ is the phase portrait $\tau(\xi_1)$ of $\xi_1$ iff $h(\sigma)$ is in the phase portrait of $\tau(\xi_2)$.
(2) The vector field $\xi$ is *structurally stable* iff there exists an open neighborhood $V$ of $\xi$ in $\mathcal{V}^1(Y)$ such that $\xi' \sim \xi$ for all $\xi' \in V$.

**Fig. 5.10** A source

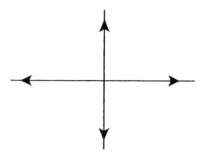

(3) A property $K$ of vector fields in $\mathcal{V}^1(2)$ is *generic* iff the set $\{\xi \in \mathcal{V}^1(Y) : \xi \text{ satisfies } K\}$ is residual in $\mathcal{V}^1(Y)$.

As before, a residual set, $V$, is the countable intersection of open dense sets, and, when $\mathcal{V}^1(Y)$ is a "Baire" space, $V$ will itself be dense.

It was conjectured that structural stability is a generic property. This is true if the dimension of $Y$ is 2, but is false otherwise (Smale 1966; Peixoto 1962).

Before discussing the Peixoto-Smale Theorems, it will be useful to explore further how we can qualitatively "classify" the set of phase portraits on a manifold $Y$. The essential feature of this classification concerns the nature of the critical or *singularity* points of the vector field on $Y$ and how these are constrained by the topological nature of $Y$.

*Example 5.7* Let us return to the example of the torus $Z = S^1 \times S^1$ examined in Example 5.4. We defined a height function $f : Z \to \Re$ and considered the four critical points $\{s, t, u, v\}$ of $f$. To remind the reader $v$ was an index 2 critical point (a local maximum of $f$). Near $v$, $f$ could be represented as

$$f(h_1, h_2) = f(v) - h_1{}^2 - h_2{}^2.$$

Now $f$ defines a gradient vector field $\xi$ where

$$\xi(h_1, h_2) = -df(h_1, h_2)$$

Looking down on $v$ we see the flow near $v$ induced by $\xi$ resembles Fig. 5.10.

The field $\xi$ may be interpreted as the law of motion under a potential energy field, $f$, so that the system flows from the "source", $v$, towards the "sink", $s$, at the bottom of $Z$.

Another way of characterizing the source, $v$, is by what we can call the "degree" of $v$. Imagine a small ball $B^2$ around $v$ and consider how the map $g : S^1 \to S^1 : (h_1, h_2) \to \frac{\xi(h_1,h_2)}{\|\xi(h_1,h_2)\|}$ behaves as we consider points $(h_1, h_2)$ on the boundary $S^1$ of $B^2$.

At point 1, $\xi(h'_1, h'_2)$ points "north" so $1 \to 1'$. Similarly at 2, $(\xi(h_1{}^2, h_2{}^2))$ points east, so $2 \to 2'$. As we traverse the circle once, so does $g$. The degree of $g$ is $+1$, and the degree of $v$ is also $+1$. However the saddle, $u$, is of degree $-1$.

## 5.5 Structural Stability of a Vector Field

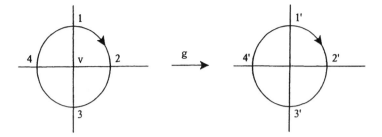

**Fig. 5.11** Degree of the map $g$

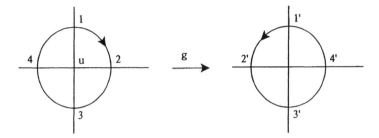

**Fig. 5.12** A saddle of degree $-1$

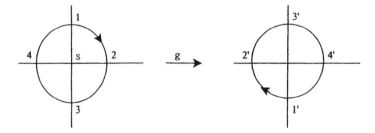

**Fig. 5.13** A sink of degree $+1$

At 1, the field points north, but at 2 the field points west, so as we traverse the circle on the left in a clockwise direction, we traverse the circle on the right in an anti-clockwise direction. Because of this change of orientation, the degree of $u$ is $-1$.

It can also easily be shown that the sink $s$ has degree $+1$.

The rotation at $s$ induced by $g$ is clockwise. It can be shown that, in general, the Euler characteristic can be interpreted as the sum of the degrees of the critical points. Thus $\chi(Z) = 1 - 1 - 1 + 1$, since the degree at each of the two saddle points is $-1$, and the degree at the source and sink is $+1$.

**Fig. 5.14** Computing the degree

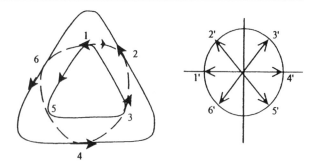

*Example 5.8* It is obvious that the flow for the Scarf example is not induced by a gradient vector field. If there were a function $f : \Delta \to \Re$ satisfying $\xi(p) = -df(p)$, then the orbits of $\xi$ would correspond to decreasing values of $f$. As we saw however, there are circular orbits for $\xi$. It is clearly impossible to have a circular flow such that $f$ decreases round the circle. However the Euler characteristic still determines the nature of the zeros (or *singularities*) of $\xi$.

First we compute the degree of the singularity $p^*$ where $\xi(p^*) = 0$.

As we saw in Example 5.5, the orbits look like smoothed triangles homeomorphic to $S^1$. Let $S^1$ be a copy of the circle (see Fig. 5.14). At 1, $g$ points west, while at 2, $g$ points northwest. At 3, $g$ points northeast, at 4, $g$ points east. Clearly the degree is $+1$ again.

As we showed, the Euler characteristic $\chi(\Delta)$ of the simplex is $+1$, and the degree of the only critical point $p^*$ of the vector field $\xi$ is $+1$. This suggests that again there is a relationship between the Euler characteristic $\chi(Y)$ of a manifold $Y$ and the sum of the degrees of the critical points of any vector field $\xi$ on $Y$. One technical point should be mentioned, concerning the nature of the flow on the boundary of $\Delta$. In the Scarf example the flow of $\xi$ was "along" the boundary, $\partial \Delta$, of $\Delta$. In a real economy one would expect that as the price vector approaches the boundary $\partial \Delta$ (so that $p_i \to 0$ for some price $p_i$), then excess demand $\xi_i$ for that commodity would rapidly increase as ($\xi_i \to \infty$). This essentially implies that the vector field $\xi$ would point towards the interior of $\Delta$. So now consider perturbations of the Scarf example as in Fig. 5.15.

In Fig. 5.15(a) is a perturbation where one of the circular orbits (called $S$) are retained; only flow commencing near to the boundary approaches $S$, as does any flow starting near to the zero $p^*$ where $\xi(p^*) = 0$. In this case the boundary $\partial \Delta$ is a *repellor*; the closed orbit is an *attractor*, and the singularity point or zero, $p^*$, is a source (or point repellor). In Fig. 5.15(b) the flow is reversed. The closed orbit $S$ is a repellor; $p^*$ is a sink (or attractor), while $\partial \Delta$ is an attractor. Now consider a copy $\Delta'$ of $\Delta$ inside $\Delta$ (given by the dotted line in Fig. 5.15(b)). On the boundary of $\Delta'$ the flow points outwards. Then $\chi(\Delta')$ is still 1 and the degree of $p^*$ is still 1. This illustrates the following theorem.

**Poincaré-Hopf Theorem** *Let $Y$ be a compact smooth manifold with boundary $\partial Y$. Suppose $\xi \in \mathcal{V}^1(Y)$ has only a finite number of singularities, and points outwards on $\partial Y$. Then $\chi(Y)$ is equal to the sum of the degrees of the singularities of $\xi$.*

**Fig. 5.15** Illustration of the Poincare Hopf Theorem

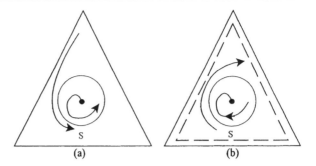

To apply this theorem, suppose $\xi$ is the vector field given by the excess demand function. Suppose that $\xi$ points towards the interior of $\Delta$. Then the vector field $(-\xi)$ points *outward*. By the Debreu-Smale Theorem, we can generically assume that $\xi$ has (at most) a finite number of singularities. Since $\chi(\Delta) = 1$, there must be at least one singularity of $(-\xi)$ and thus of $\xi$. Unfortunately this theorem does not allow us to infer whether or not there exists a singularity $p^*$ which is stable (*i.e.*, an attractor).

The Poincaré-Hopf Theorem can also be used to understand singularities of vector fields on manifolds without boundary. As we have suggested, the Euler characteristic of a sphere is 2 for the even dimensional case and 0 for the odd dimensional case. This gives the following result.

**The "Hairy Ball" Theorem** *Any vector field $\xi$ on $S^{2n}$ (even dimension) must have a singularity. However there exists a vector field $\xi$ on $S^{2n+1}$ (odd dimension) such that $\xi(p) = 0$ for no $p \in S^{2n+1}$.*

To illustrate Fig. 5.16 shows a vector field on $S^2$ where the flow is circular on each of the circles of latitude, but both north and south poles are singularities. The flow is evidently non-gradient, since no potential function, $f$, can increase around a circular orbit.

*Example 5.9* As an application, we may consider a more general type of flow, by defining at each point $x \in Y$ a *set* $h(x)$ of vectors in the tangent space at $x$. As discussed in Chap. 4, $h$ could be induced by a family of utility functions $\{u_i : Y \to \Re, i \in M\}$, such that

$$h(x) = \{v \in \Re^n : \langle du_i(x), v \rangle > 0, \ \forall i \in M\}.$$

That is to say $v \in h(x)$ iff each utility function increases in the direction $v$. We can interpret Theorem 4.21 to assert that $h(x) = \Phi$ whenever $x \in \overset{0}{\Theta}(u_1, \ldots, u_m)$, the critical Pareto set.

Suppose that $\overset{0}{\Theta} = \Phi$. Then in general we can use a selection theorem to select from $h(x)$ a non-zero vector $\xi(x)$, at every $x \in Y$, such that $\xi$ is a continuous vector field. That is, $\xi \in \mathcal{V}^1(Y)$, but $\xi$ has no singularities. However if $\chi(Y) \neq 0$ then any

**Fig. 5.16** Zeros of a field on $S^2$

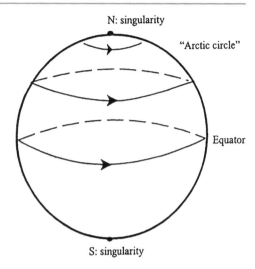

vector field $\xi$ on $Y$ has critical points whose degrees sum to $\chi(Y)$, subject to the boundary condition of the Poincaré-Hopf Theorem.

**Pareto Theorem** *If $\chi(Y) \neq 0$ then $\overset{0}{\Theta}(u) \neq \Phi$ for any smooth profile $u$ on $Y$.*

The Euler characteristic can be interpreted as an *obstruction* to the non-existence of equilibria, or of fixed points. For example suppose that $\chi(Y) = 0$. Then it is possible to construct a vector field $\xi$ on $Y$ without zeros. It is then possible to construct a function $f : Y \to Y$ which follows the trajectories of $\xi$ a small distance, $\epsilon$, say. But then the function $f$ is homotopic to the identity.

That is to say, from each point $x$ construct a path $c_x : [0, 1] \to Y$ with $c_x(0) = x$ and $c_x(1) = f(x)$ whose gradient $\frac{dc}{dt}(t)$ at time $t$ is given by the vector field $\xi$ at the point $x' = c_x(t)$. Say that $f$ is *induced* by the vector field, $\xi$. The homotopy $F : [0, 1] \times Y \to Y$ is then given by $F(0, x) = x$ and $F(t, x) = c_x(t)$. Since $c_x$ is continuous, so is $F$. Thus $F$ is a homotopy between $f$ and the identity on $Y$. A function $f : Y \to Y$ which is homotopic to the identity is called a *deformation* of $Y$.

If $\chi(Y) = 0$ then it is possible to find a vector field $\xi$ on $Y$ without singularities and then construct a deformation $f$ of $Y$ induced by $\xi$. Since $\xi(x) = 0$ for no $x$, $f$ will not have a fixed point. Conversely if $f$ is a deformation on $Y$ and $\chi(Y) \neq 0$, then the homotopy between $f$ and the identity generates a vector field $\xi$. Were $f$ to have no fixed point, then $\xi$ would have no singularity. If $f$ and thus $\xi$ have the right behavior on the boundary, then $\xi$ must have at least one singularity. This contradicts the fixed point free property of $f$.

**Lefschetz Fixed Point Theorem** *If $Y$ is a manifold with $\chi(Y) = 0$ then there exists a fixed point free deformation of $Y$. If $\chi(Y) \neq 0$ then any deformation of $Y$ has a fixed point.*

## 5.5 Structural Stability of a Vector Field

Note that the Lefschetz Fixed Point Theorem does not imply that any function $f : S^2 \to S^2$ has a fixed point. For example the "antipodal" map $f(x) = -x$, for $\|x\| = 1$, is fixed point free. However $f$ cannot be induced by a continuous vector field, and is therefore not a deformation. The Lefschetz fixed point theorem just stated is in fact a corollary of a deeper result: For any continuous map $f : Y \to Y$, with $Y$ compact, there is an "obstruction" called the Lefschetz number $\lambda(f)$. If $\lambda(f) \neq 0$, then $f$ must have a fixed point. If $Y$ is "homotopy equivalent" to the compact ball then $\lambda(f) \neq 0$ for every continuous function on the ball, so the ball has the fixed point property. On the other hand if $f$ is homotopic to the identity, $Id$, on $Y$ then $\lambda(f) = \lambda(Id)$, and it can be shown that $\lambda(Id) = \chi(Y)$ the Euler characteristic of $Y$. It therefore follows that $\chi(Y)$ is an obstruction to the existence of a fixed point free deformation of the compact manifold $Y$.

*Example 5.10* To illustrate an application of this theorem in social choice, suppose that $Y$ is a compact manifold of dimension at most $k(\sigma) - 2$, where $k(\sigma)$ is the Nakamura number of the social choice rule, $\sigma$ (see Sect. 3.8). Suppose $(u_1, \ldots, u_m)$ is a smooth profile on $Y$. Then it can be shown (Schofield 1984) that if the choice $C_{\sigma(\pi)}(Y)$ is empty, then there exists a fixed point free deformation on $Y$. Consequently if $\chi(X) \neq 0$, then $C_{\sigma(\pi)}(Y)$ must be non-empty.

*Example 5.11* It would be useful to be able to use the notion of the Lefschetz obstruction to obtain conditions under which the singularities of the excess demand function, $\xi$, of an economy were stable. However, as Scarf's example showed, it is entirely possible for there to be a single attractor, or a single repellor (as in Fig. 5.15) or even a situation with an infinite number of closed orbits. However, consider a more general price adjustment process as follows. At each $p$ in the interior, Int $\Delta$, of $\Delta$ let

$$\xi^*(p) = \{v \in \Re^n : \langle v, \xi(p) \rangle > 0\}.$$

A vector field $v \in \mathcal{V}^1(\Delta)$ is *dual* to $\xi$ iff $v(p) \in \xi^*(p)$ for all $p \in$ Int $\Delta$, and $v(p) = 0$ iff $\xi(p) = 0$ for $p \in$ Int $\Delta$. It may be possible to find a vector field, $v$, dual to $\xi$ which has attractors. Suppose that $v$ is dual to $\xi$, and that $f : \Delta \to \Delta$ is induced by $v$. As we have seen, the Lefschetz number of $f$ gives information about the singularities of $v$.

Dierker (1972) essentially utilized the following hypothesis: there exists a dual vector field, $v$, and a function $f : \Delta \to \Delta$ induced by $v$ such that $f = f_0$ is homotopic to the constant map $f_1 : \Delta \to \{(\frac{1}{n}, \ldots, \frac{1}{n})\}$ such that $\{p \in \Delta : f_t(p) = p \text{ for } t \in [0, 1]\}$ is compact. Under the assumption that the economy is regular (so the number of singularities of $\xi$ is finite), then he showed that the number of such singularities must be odd. Moreover, if it is known that $\xi$ only has stable singularities, then there is only one. The proof of the first assertion follows by observing that $\lambda(f_0) = \lambda(f_1)$. But $f_1$ is the constant map on $\Delta$ so $\lambda(f_1) = 1$. Moreover $\lambda(f_0)$ is equal to the sum of the degrees of the singularities of $v$, and Dierker shows that at each singularity of $v$, the degree is $\pm 1$. Consequently the number of singularities must be odd. Finally if there are only stable singularities, each has degree $+1$, so it must be unique.

**Fig. 5.17** A flow on the Torus

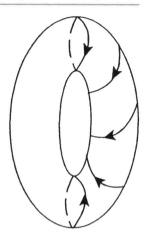

*Example 5.12* As a further application of the Lefschetz obstruction, suppose, contrary to the usual assumption that negative prices are forbidden, that $p \in S^{n-1}$ rather than $\Delta$. It is natural to suppose that $\xi(p) = \xi(-p)$ for any $p \in S^{n-1}$. Suppose now that $\xi(p) = 0$ for no $p \in S^{n-1}$. This defines an (even) spherical map $g : S^{n-1} \to S^{n-1}$ by $g(p) = \xi(p)/\|\xi(p)\|$. Thus $g(p) = g(-p)$. The degree (deg $(g)$) of such a $g$ can readily be seen to be an even integer, and it follows that the Lefschetz obstruction of $g$ is $\lambda(g) = 1 + (-1)^{n-1} \deg(g)$.

Clearly $\lambda(g) \neq 0$ and so $g$ has a fixed point $\overline{p}$ such that $g(\overline{p}) = \overline{p}$. But then $\xi(p) = \alpha p$ for some $\alpha > 0$. This violates Walras' Law, since $\langle p, \xi(p) \rangle = \alpha \|p\|^2 \neq 0$, so $\xi(p) = 0$ for some $p \in S^{n-1}$. Keenan (1992) goes on to develop some of the earlier work by Rader (1972) to show, in this extended context, that for generic, regular economies there must be an odd number of singularities.

The above examples have all considered flows on the simplex or the sphere. To return to the idea of structural stability, let us consider once again examples of a vector field on the torus.

*Example 5.13* (1) For a more interesting deformation of the torus $Z = S^1 \times S^1$, consider Fig. 5.17.

The closed orbit at the top of the torus is a repellor, $R$, say. Any flow starting near to $R$ winds towards the bottom closed orbit, $A$, an attractor. There are no singularities, and the induced deformation is fixed point free.

(2) Not all flows on the torus $Z$ need have closed orbits. Consider the flow on $Z$ given in Fig. 5.18. If the tangent of the angle, $\theta$, is rational, then the orbit through $x$ is closed, and will consist of a specific number of turns round $Z$. However suppose this flow is perturbed. There will be, in any neighborhood of $\theta$, an irrational angle. The orbits of an irrational flow will not close up. To relate this to the Peixoto Theorem which follows, with rational flow there will be an infinite number of closed orbits. However the phase portrait for rational flow cannot be homeomorphic to the portrait for irrational flow. Thus any perturbations of rational flow gives a non-

## 5.5 Structural Stability of a Vector Field

**Fig. 5.18** Structurally unstable rational flow on the Torus

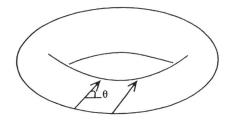

homeomorphic irrational flow. Clearly any vector field on the torus which gives rational flow is structurally unstable.

**Structural Stability Theorem**
1. *If* $\dim Y = 2$ *and* $Y$ *is compact, then structural stability of vector fields on* $Y$ *is generic.*
2. *If* $\dim Y \geq 3$, *then structural stability is non-generic.*

Peixoto (1962) proved part (1) by showing that structurally stable vector fields on compact $Y$ (of dimension 2) must satisfy the following properties:
(1) there are a finite number of non-degenerate isolated singularities (that is, critical points which can be sources, sinks, or saddles)
(2) there are a finite number of attracting or repelling closed orbits
(3) every orbit (other than closed orbits) starts at a source or saddle, or winds away from a repellor and finishes at a saddle or sink, or winds towards an attractor
(4) no orbit connects saddle points.

Peixoto showed that for any vector field $\xi$ on $Y$ and any neighborhood $V$ of $\xi$ in $\mathcal{V}^1(Y)$ there was a vector field $\xi'$ in $V$ that satisfied the above four conditions and thus was structurally stable.

Although we have not carefully defined the terms used above, they should be intuitively clear. To illustrate, Fig. 5.19(a) shows an orbit connecting saddles, while Fig. 5.19(b) shows that after perturbation a qualitatively different phase portrait is obtained.

In Fig. 5.19(a), $A$ and $B$ are connected saddles, C is a repellor (orbits starting near to C leave it) and $D$ is a closed orbit. A small perturbation disconnects $A$ and $B$ as shown in Fig. 5.19(b), and orbits starting near to $D$ (either inside or outside) approach $D$, so it is an attractor.

The excess demand function, $\xi$, of the Scarf example clearly has an infinite number of closed orbits (all homeomorphic to $S^1$). Thus $\xi$ cannot be structurally stable. From Peixoto's Theorem, small perturbations of $\xi$ will destroy this feature. As we suggested, a small perturbation may change $\xi$ so that $p^*$ becomes a stable equilibrium (an attractor) or an unstable equilibrium (a repellor).

Smale's (1966) proof that structural stability was non-generic in three or more dimensions was obtained by constructing a diffeomorphism $f : Y^3 \to Y^3$ (with $Y = S^1 \times S^1 \times S^1$). This induced a vector field $\xi \in \mathcal{V}^1(Y)$ that had the property that for a neighborhood $V$ of $\xi$ in $\mathcal{V}^1(Y)$, no $\xi'$ in $V$ was structurally stable. In other

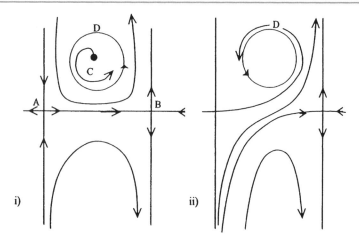

**Fig. 5.19** (i) shows an orbit connecting saddle points, while (ii) shows that this property is destroyed by a small perturbation

words every $\xi'$ when perturbed led to a qualitatively different phase portrait. We could say that $\xi$ was *chaotic*. Any attempt to model $\xi$ by an approximation $\xi'$, say, results in an essentially different vector field. The possibility of chaotic flow and its ramifications will be discussed in general terms in the next section. The consequence for economic theory is immediate, however. Since it can be shown that *any* excess demand function $\xi$ and thus any vector field can result from an economy, $(u, e)$, it is possible that the price adjustment process is chaotic.

As we observed after Example 5.8, an economically realistic excess demand function $\xi$ on $\Delta$ should point into the price simplex at any price vector on $\partial \Delta$. This follows because if $p_i \to 0$ then $\xi_i$ would be expected to approach $\infty$. Let $\mathcal{V}_0^1(\Delta)$ be the topological space of vector fields on $\Delta$, of the form $\frac{dp}{dt}|_p = \xi(p)$, such that $\frac{dp}{dt}|_p$ points into the interior of $\Delta$ for $p$ near $\partial \Delta$.

**The Sonnenschein-Mantel-Debreu Theorem** *The map*

$$\xi : C^s(X, \mathfrak{R}^m) \times X^m \to \mathcal{V}_0^1(\Delta)$$

*is onto if $m \geq n$.*

Suppose that there are at least as many economic agents ($m$) as commodities. Then it is possible to construct a well-behaved economy $(u, e)$ with monotonic, strictly convex preferences induced from smooth utilities, and an endowment vector $e \in X^m$, such that *any* vector field in $\mathcal{V}_0^1(\Delta)$ is generated by the excess demand function for the economy $(u, e)$.

Versions of the theorem were presented in Sonnenschein (1972), Mantel (1974), and Debreu (1974). A more recent version can be found in Mas-Colell (1985). As we have discussed in this section, because the simplex $\Delta$ has $\chi(\Delta) = 1$, then the "excess demand" vector field $\xi$ will always have at least one singularity. In fact, from the Debreu-Smale theorem, we expect $\xi$ to generically exhibit only a finite number of singularities. Aside from these restrictions, $\xi$, is essentially unconstrained. If there are at least four commodities (and four agents) then it is always possible to construct $(u, e)$ such that the vector field induced by excess demand is "chaotic".

As we saw in Sect. 5.4, the vector field $\xi$ of the Scarf example was structurally unstable, but any perturbation of $\xi$ led to a structurally stable field $\xi'$, say, either with an attracting or repelling singularity. The situation with four commodities is potentially much more difficult to analyze. It is possible to find $(u, e)$ such that the induced vector field $\xi$ on $\Delta$ is chaotic—in some neighborhood $V$ of $\xi$ there is no structurally stable field. Any attempt to model $\xi$ by $\xi'$, say, must necessarily incorporate some errors, and these errors will multiply in some fashion as we attempt to map the phase portrait. In particular the flow generated by $\xi$ through some point $x \in \Delta$ can be very different from the flow generated by $\xi'$ through $x$. This phenomenon has been called "sensitive dependence on initial conditions."

## 5.6 Speculations on Chaos

It is only in the last twenty years or so that the implications of "chaos" (or failure of structural stability in a profound way) have begun to be realized. In a recent book Kauffman commented on the failure of structural stability in the following way.

"One implication of the occurrence or non-occurrence of structural stability is that, in structurally stable systems, smooth walks in parameter space must (result in) smooth changes in dynamical behavior. By contrast, chaotic systems, which are not structurally stable, adapt on uncorrelated landscapes. Very small changes in the parameters pass through many interlaced bifurcation surfaces and so change the behavior of the system dramatically."[1]

The whole point of the Debreu-Smale Theorem is that generically the Debreu map is regular. Thus there are open sets in the parameter space (of utility profiles and endowments) where the number, $E(\xi)$, of singularities of $\xi$ is finite and constant. As the Scarf example showed, however, even though $E(\xi)$ may be constant in a neighborhood, the vector field $\xi$ can be structurally unstable. The structurally unstable circular vector field of the Scarf example is not particularly surprising. After all, similar structurally unstable systems are common (the oscillator or pendulum is one example). These have the feature that, when perturbed, they become structurally unstable. Thus the dynamical system of a pendulum with friction is structurally stable. Its phase portrait shows an attractor, which still persists as the friction is increased or decreased. Smale's Structural Instability Theorem together with the Sonnenschien-Mantel-Debreu Theorem suggests that the vector field generated by excess demand can indeed be chaotic when there are at least four commodities and agents. This does not necessarily mean that chaotic price adjustment processes are pervasive. As in Dierker's example, even though the vector field $\xi(p) = \frac{dp}{dt}$ can be

---

[1] S. Kauffman, *The Origins of Order* (1993) Oxford University Press: Oxford.

chaotic, it may be possible to find a structurally stable vector field, $v$, dual to $\xi$, which is structurally stable.

This brief final section will attempt to discuss in an informal fashion, whether or not it is plausible for economies to exhibit chaotic behavior.

It is worth mentioning that the idea of structural stability is not a new one, though the original discussion was not formalized in quite the way it is today. Newton's great work *Philosophiae Naturalis Principia Mathematica* (published in 1687) grew out of his work on gravitation and planetary motion. The laws of motion could be solved precisely giving a vector field and the orbits (or phase portrait) in the case of a planet (a point mass) orbiting the sun. The solution accorded closely with Kepler's (1571–1630) empirical observations on planetary motion. However, the attempt to compute the planetary orbits for the solar system had to face the problem of perturbations. Would the perturbations induced in each orbit by the other planets cause the orbital computations to converge or diverge? With convergence, computing the orbit of Mars, say, can be done, by approximating the effects of Jupiter, Saturn perhaps, on the Mars orbit. The calculations would give a prediction very close to the actual orbit. Using the approximations, the planetary orbits could be computed far into the future, giving predictions as precise as calculating ability permitted. Without convergence, it would be impossible to make predictions with any degree of certainty. Laplace in his work "Mécanique Céleste" (published between 1799 and 1825) had argued that the solar system (viewed as a formal dynamical system) is structurally stable (in our terms). Consistent with his view was the use of successive approximations to predict the perihelion (a point nearest the sun) of Haley's comet, in 1759, and to infer the existence and location of Neptune in 1846.

Structural stability in the three-body problem (of two planets and a sun) was the obvious first step in attempting to prove Laplace's assertion. In 1885 a prize was announced to celebrate the King of Sweden's birthday. Henri-Poincaré submitted his entry "Sur le problème des trois corps et les Equations de la Dynamique." This attempted to prove structural stability in a restricted three body problem. The prize was won by Poincaré's entry, although it was later found to contain an error. Poincaré had obtained his doctorate in mathematics in Paris in 1878, had briefly taught at Caen and later became professor at Paris. His work on differential equations in the 1880s and his later work on Celestial Mechanics in the 1890s developed new qualitative techniques (in what we now call differential topology) to study dynamical equations.

In passing it is worth mentioning that since there is a natural periodicity to any rotating celestial system, the state space in some sense can be viewed as products of circles (that is tori). Many of the examples mentioned in the previous section, such as periodic (rational) or a-periodic (non-rational) flow on the torus came up naturally in celestial mechanics.

One of the notions implicitly emphasized in the previous sections of this chapter is that of *bifurcation*: namely a dynamical system on the boundary separating qualitatively different systems. At such a bifurcation, features of the system separate out in pairs. For example, in the Debreu map, a bifurcation occurs when two of the price equilibria coalesce. This is clearly linked to the situation studied by Dierker, where the number of price equilibria (in $\Delta$) is odd. At a bifurcation, two equilibria with opposite degrees coalesce. In a somewhat similar fashion Poincaré showed that, for the three-body problem, if there is some value $\mu_0$ (of total mass, say) such that periodic solutions exist for $\mu \leq \mu_0$ but not for $\mu > \mu_0$, then two periodic solutions must have coalesced at $\mu_0$. However Poincaré also discovered that the bifurcation could be associated with the appearance of a new solution with period double that of the original. This phenomenon is central to the existence of a period-doubling cascade as one of the characteristics of chaos. Near the end of his *Celestial Mechanics*, Poincaré writes of this phenomenon:

"Neither of the two curves must ever cut across itself, but it must bend back upon itself in a very complex manner an infinite number of times.... Nothing is more suitable for providing us with an idea of the complex nature of the three body problem."[2]

Although Poincaré was led to the possibility of chaos in his investigations into the solar system, it appears that the system is in fact structurally stable. Arnol'd showed in 1963 that for a system with small planets, there is an open set of initial conditions leading to bounded orbits for all time. Computer simulations of the system far into time also suggests it is structurally stable.[3] Even so, there are events in the system that affect us and appear to be chaotic (perhaps catastrophic would be a more appropriate term). The impact of large asteroids may have a dramatic effect on the biosphere of the earth, and these have been suggested as a possible cause of mass extinction. The onset and behavior of the ice ages over the last 100,000 years is very possibly chaotic, and it is likely that there is a relationship between these violent climatic variations and the recent rapid evolution of human intelligence.[4]

More generally, evolution itself is often perceived as a gradient dynamical process, leading to increasing complexity. However Stephen Jay Gould has argued over a number of years that evolution is far from gradient-like: increasing complexity coexists with simple forms of life, and past life has exhibited an astonishing variety.[5] Evolution itself appears to proceed at a very uneven rate.[6]

---

[2] My observations and quotations are taken from D. Goroff's introduction and the text of a recent edition of Poincaré's *New Methods of Celestial Mechanics,* (1993) American Institute of Physics: New York.

[3] See I. Peterson, *Newton's Clock: Chaos in the Solar System* (1993) Freeman: New York.

[4] See W. H. Calvin, *The Ascent of Mind.* Bantam: New York.

[5] S. J. Gould, *Full House* (1996) Harmony Books: New York; S. J. Gould, *Wonderful Life* (1989) Norton: New York.

[6] N. Eldredge and S. J. Gould, "Punctuated Equilibria: An Alternative to Phyletic Gradualism," in *Models in Paleobiology* (1972), T. J. M. Schopf, ed. Norton: New York.

**Fig. 5.20** The butterfly

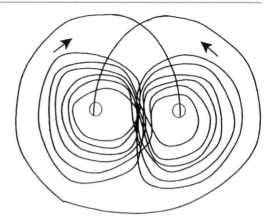

"Empirical" chaos was probably first discovered by Edward Lorenz in his efforts to numerically solve a system of equations representative of the behavior of weather.[7] A very simple version is the non-linear vector equation

$$\frac{dx}{dt} = \begin{pmatrix} dx_1 \\ dx_2 \\ dx_3 \end{pmatrix} = \begin{pmatrix} -a(x_1 - x_2) \\ -x_1 x_3 + a_2 x_1 - x_2 \\ x_1 x_2 - a_3 x_3 \end{pmatrix}$$

which is chaotic for certain ranges of the three constants, $a_1, a_2, a_3$.

The resulting "butterfly" portrait winds a number of times about the left hole (A in Fig. 5.20), then about the right hole (B), then the left, *etc.* Thus the phase portrait can be described by a sequence of winding numbers ($w_l^1, w_k^1, w_l^2, w_k^2$, etc.). Changing the constants $a_1, a_2, a_3$ slightly changes the winding numbers.

Given that chaos can be found in such a simple meteorological system, it is worthwhile engaging in a thought experiment to see whether "climatic" chaos is a plausible phenomenon. Weather occurs on the surface of the earth, so the spatial context is $S^2 \times I$, where $I$ is an interval corresponding to the depth of the atmosphere. As we know, $\chi(S^2) = \chi(S^2 \times I) = 2$ so we would expect singularities. Secondly there are temporal periodicities, induced by the distance from the sun and earth's rotation. Thirdly there are spatial periodicities or closed orbits. Chief among these must be the jet stream and the oceanic orbit of water from the southern hemisphere to the North Atlantic (the Gulf Stream) and back. The most interesting singularities are the hurricanes generated each year off the coast of Africa and channeled across the Atlantic to the Caribbean and the coast of the U.S.A. Hurricanes

---

[7]E.N. Lorenz, "The Statistical Prediction of Solutions of Dynamical Equations," *Proceedings Int. Symp. Num. Weather Pred* (1962) Tokyo; E. N. Lorenz "Deterministic Non Periodic Flow," *J. Atmos. Sci.* (1963): 130–141.

are self-sustaining heat machines that eventually dissipate if they cross land or cool water. It is fairly clear that their origin and trajectory is chaotic.

Perhaps we can use this thought experiment to consider the global economy. First of all there must be local periodicities due to climatic variation. Since hurricanes and monsoons, *etc.* effect the economy, one would expect small chaotic occurrences. More importantly, however, some of the behavior of economic agents will be based on their future expectations about the nature of economic growth, *etc.* Thus one would expect long term expectations to affect large scale "decisions" on matters such as fertility. The post-war "baby boom" is one such example. Large scale periodicities of this kind might very well generate smaller chaotic effects (such as, for example, the oil crisis of the 1970s), which in turn may trigger responses of various kinds.

It is evident enough that the general equilibrium (GE) emphasis on the existence of price equilibria, while important, is probably an incomplete way to understand economic development. In particular, GE theory tends to downplay the formation of expectations by agents, and the possibility that this can lead to unsustainable "bubbles".

Remember, it is a key assumption of GE that agents' preferences are defined on the commodity space alone. If, on the contrary, these are defined on commodities *and* prices, then it is not obvious that the assumptions of the Ky Fan Theorem (*cf.*, Chap. 3) can be employed to show existence of a price equilibrium. Indeed manipulation of the kind described in Chap. 4 may be possible. More generally one can imagine energy engines (very like hurricanes) being generated in asset markets, and sustained by self-reinforcing beliefs about the trajectory of prices. It is true that modern decentralised economies are truly astonishing knowledge or data-processing mechanisms. From the perspective of today, the argument that a central planning authority can be as effective as the market in making "rational" investment decisions appears to have been lost. Hayek's case, the so-called "calculation" argument, with von Mises and against Lange and Schumpeter, was based on the observation that information is dispersed throughout the economy and is, in any case, predominantly subjective. He argued essentially that only a market, based on individual choices, can possibly "aggregate" this information.[8]

Recently, however, theorists have begun to probe the degree of consistency or convergence of beliefs in a market when it is viewed as a game. It would seem that when the agents "know enough about each other", then convergence in beliefs is a consequence.[9]

---

[8] See F. A. Hayek, "The Use of Knowledge in Society," *American Economic Review* (1945) 55: 519–530, and the discussion in A. Gamble, *Hayek: The Iron Cage of Liberty* (1996) Westview: Boulder, Colorado.

[9] See R. J. Aumann, "Agreeing to Disagree," *Annals of Statistics* (1976) 1236–1239 and K. J. Arrow "Rationality of Self and Others in an Economic System," *Journal of Business* (1986) 59: S385–S399.

In fact the issue about the "truth-seeking" capability of human institutions is very old and dates back to the work of Condorcet.[10] Nonetheless it is possible for belief cascades or bubbles to occur under some circumstances.[11] It is obvious enough that economists writing after the Great Crash of the 1930s might be more willing than those writing today to consider the possibility of belief cascades and collapse. John Maynard Keynes' work on *The General Theory of Employment, Interest and Money* (1936) was very probably the most influential economic book of the century. What is interesting about this work is that it does appear to have grown out of work that Keynes did in the period 1906 to 1914 on the foundation of probability, and that eventually was published as the *Treatise on Probability* (1921). In the *Treatise*, Keynes viewed probability as a degree of belief. He also wrote: "The old assumptions, that all quantity is numerical and that all quantitative characteristics are additive, can no longer be sustained. Mathematical reasoning now appears as an aid in its symbolic rather than its numerical character. I, at any rate, have not the same lively hope as Condorcet, or even as Edgeworth, 'Eclairer le Science morales et politiques par le flambeau de l'Algèbre.'"[12]

Macro-economics as it is practiced today tends to put a heavy emphasis on the empirical relationships between economic aggregates. Keynes' views, as I infer from the *Treatise*, suggest that he was impressed neither by econometric relationships nor by algebraic manipulation. Moreover, his ideas on "speculative euphoria" and crashes[13] would seem to be based on an understanding of the economy grounded not in econometrics or algebra but in the qualitative aspects of its dynamics.

Obviously I have in mind a dynamical representation of the economy somewhere in between macro-economics and general equilibrium theory. The laws of motion of such an economy would be derived from modeling individuals' "rational" behavior as they process information, update beliefs and locally optimise. At present it is not possible to construct such a micro-based macro-economy because the laws of motion are unknown. Nonetheless, just as simulation of global weather systems can be based on local physical laws, so may economic dynamics be built up from local "rationality" of individual agents. In my view, the qualitative theory of dynamical systems will have a major rôle in this enterprise. The applications of this theory, as outlined in the chapter, are intended only to give the reader a taste of how this theory might be developed.

---

[10] See his work on the so-called Jury Theorem in his *Essai* of 1785. A discussion of Condorcet's work can be found in I. McLean and F. Hewitt, *Condorcet: Foundations of Social Choice and Political Theory* (1994) Edward Elgar: Aldershot, England.

[11] See S. Bikhchandani, D. Hirschleifer and I. Welsh, "A Theory of Fads, Fashion, Custom, and Cultural Change as Information Cascades," *Journal of Political Economy* (1992) 100: 992–1026.

[12] John Maynard Keynes, *Treatise on Probability* (1921) Macmillan: London pp. 349. The two volumes by Robert Skidelsky on *John Maynard Keynes* (1986, 1992) are very useful in helping to understand Keynes' thinking in the *Treatise* and the *General Theory*.

[13] See, for example, the work of Hyman Minsky *John Maynard Keynes* (1975) Columbia University Press: New York, and *Stabilizing an Unstable Economy* (1986) Yale University Press: New Haven.

# Further Reading

A very nice though brief survey of the applications of global analysis (or differential topology) to economics is:

Debreu, G. (1976). The application to economics of differential topology and global analysis: regular differentiable economies. *The American Economic Review, 66*, 280–287.


An advanced and detailed text on the use of differential topology in economics is:

Mas-Colell, A. (1985). *The theory of general economic equilibrium*. Cambridge: Cambridge University Press.


Background reading on differential topology and the ideas of transversality can be found in:

Chillingsworth, D. R. J. (1976). *Differential topology with a view to applications*. Pitman: London.
Golubitsky, M., & Guillemin, V. (1973). *Stable mappings and their singularities*. Berlin: Springer.
Hirsch, M. (1976). *Differential topology*. Berlin: Springer.


For the Debreu-Smale Theorem see:

Balasko, Y. (1975). Some results on uniqueness and on stability of equilibrium in general equilibrium theory. *Journal of Mathematical Economics, 2*, 95–118.
Debreu, G. (1970). Economies with a finite number of equilibria. *Econometrica, 38*, 387–392.
Smale, S. (1974a). Global analysis of economics IV: finiteness and stability of equilibria with general consumption sets and production. *Journal of Mathematical Economics, 1*, 119–127.
Smale, S. (1974b). Global analysis and economics IIA: extension of a theorem of Debreu. *Journal of Mathematical Economics, 1*, 1–14.


The Smale-Pareto theorem on the generic structure of the Pareto set is in:

Smale, S. (1973). Global analysis and economics I: Pareto optimum and a generalization of Morse theory. In M. Peixoto (Ed.), *Dynamical systems*. New York: Academic Press.


Scarf's example and the use of the Euler characteristic are discussed in Dierker's notes on topological methods.

Dierker, E. (1972). Two remarks on the number of equilibria of an economy. *Econometrica*, 951–953.
Dierker, E. (1974). *Lecture notes in economics and mathematical systems: Vol. 92. Topological methods in Walrasian economics*. Berlin: Springer.
Scarf, H. (1960). Some examples of global instability of the competitive equilibrium. *International Economic Review, 1*, 157–172.

The Lefschetz fixed point theorem as an obstruction theory for the existence of fixed point free functions and continuous vector fields can be found in:

Brown, R. (1971). *The Lefschetz fixed point theorem*. Glenview: Scott and Foresman.

Some applications of these techniques in economics are in:

Keenan, D. (1992). Regular exchange economies with negative prices. In W. Nenuefeind & R. Riezman (Eds.), *Economic theory and international trade*. Berlin: Springer.
Rader, T. (1972). *Theory of general economic equilibrium*. New York: Academic Press.

An application of this theorem to existence of a voting equilibrium is in:

Schofield, N. (1984). Existence of equilibrium on a manifold. *Journal of Operations Research, 9*, 545–557.

The results on structural stability are given in:

Peixoto, M. (1962). Structural stability on two-dimensional manifolds. *Topology, 1*, 101–120.
Smale, S. (1966). Structurally stable systems are not dense. *American Journal of Mathematics, 88*, 491–496.

A very nice and fairly elementary demonstration of Peixoto's Theorem in two dimensions, together with the much earlier classification results of Andronov and Pontrjagin, is given in:

Hubbard, J. H., & West, B. H. (1995). *Differential equations: a dynamical systems approach*. Berlin: Springer.

René Thom's early book applied these "topological" ideas to development:

Thom, R. (1975). *Structural stability and morphogenesis*. Reading: Benjamin.

The result that any excess demand function is possible can be found in:

Debreu, G. (1974). Excess demand functions. *Journal of Mathematical Economics, 1*, 15–21.
Mantel, R. (1974). On the characterization of aggregate excess demand. *Journal of Economic Theory, 12*, 197–201.
Sonnenschein, H. (1972). Market excess demand functions. *Econometrica, 40*, 549–563.

The idea of chaos has become rather fashionable recently. For a discussion see:

Gleick, J. (1987). *Chaos: making a new science*. New York: Viking.

An excellent background to the work of Poincaré, Birkhoff and Smale, and applications in meteorology is:

Lorenz, E. N. (1993). *The essence of chaos*. Seattle: University of Washington Press.


For applications of the idea of chaos in various economic and social choice contexts see:

Saari, D. (1985). Price dynamics, social choice, voting methods, probability and chaos. In D. Aliprantis, O. Burkenshaw, & N. J. Rothman (Eds.), *Lecture notes in economics and mathematical systems* (Vol. 244). Berlin: Springer.

# Topology and Social Choice 6

In Chap. 3 we showed the Nakamura Theorem that a social choice could be guaranteed as long as the dimension of the space did not exceed $k(\sigma) - 2$. We now consider what can happen in dimension above $k(\sigma) - 1$. We then go on to consider "probabilistic" social choice, where there is some uncertainty over voters' preferences.

## 6.1 Existence of a Choice

First we repeat some of the results presented in Chap. 3.

As we showed in Chap. 3, arguments for the existence of an equilibrium or choice are based on some version of Brouwer's fixed point theorem, which we can regard as a variant of the Fan Choice Theorem. Brouwer's theorem asserts that any continuous function $f : B \to B$ between the finite dimensional ball, $B$ or indeed any compact convex set in $\Re^w$, has the fixed point property.

This section will consider the use of variants of the Brouwer theorem, to prove existence of an equilibrium of a general social choice mechanism. We shall argue that the condition for existence of an equilibrium will be violated if there are cycles in the underlying mechanism.

Let $W \subset \Re^w$ be the set of alternatives and, and let $2^W$ be the set of all subsets of $W$. A *preference correspondence*, $P$, on $W$ assigns to each point $x \in W$, its *preferred set* $P(x)$. Write $P: W \twoheadrightarrow W$ to denote that the image of $x$ under $P$ is a set (possibly empty) in $W$. For any subset $V$ of $W$, the restriction of $P$ to $V$ gives a correspondence $P_V : V \twoheadrightarrow V$. Define $P_V^{-1} : V \twoheadrightarrow V$ such that for each $x \in V$,

$$P_V^{-1}(x) = \{y : x \in P(y)\} \cap V.$$

The sets $P_V(x), P_V^{-1}(x)$ are sometimes called the *upper* and *lower* preference sets of $P$ on $V$. When there is no ambiguity we delete the suffix $V$. The *choice* of $P$ from $W$ is the set

$$C(W, P) = \{x \in W : P(x) = \varnothing\}.$$

Here $\varnothing$ is the empty set. The choice of $P$ from a subset, $V$, of $W$ is the set

$$C(V, P) = \{x \in V : P_V(x) = \varnothing\}.$$

Call $C_P$ a *choice function* on $W$ if $C_P(V) = C(V, P) \neq \varnothing$ for every subset $V$ of $W$. We now seek general conditions on $W$ and $P$ which are sufficient for $C_P$ to be a choice function on $W$. Continuity properties of the preference correspondence are important and so we require the set of alternatives, $W$, to be a topological space.

**Definition 6.1** Let $W, Y$ be two topological spaces. A correspondence $P : W \twoheadrightarrow Y$ is

(i) *Lower hemi-continuous* (*lhc*) iff, for all $x \in W$, and any open set $U \subset Y$ such that $P(x) \cap U \neq \varnothing$ there exists an open neighborhood $V$ of $x$ in $W$, such that $P(x') \cap U \neq \varnothing$ for all $x' \in V$.

(ii) *Upper hemi-continuous* (*uhc*) iff, for all $x \in W$ and any open set $U \subset Y$ such that $P(x) \subset U$, there exists an open neighborhood $V$ of $x$ in $W$ such that $P(x') \subset U$ for all $x' \in V$.

(iii) *Lower demi-continuous* (*ldc*) iff, for all $x \in Y$, the set

$$P^{-1}(x) = \{y \in W : x \in P(y)\}$$

is open (or empty) in $W$.

(iv) *Upper demi-continuous* (*udc*) iff, for all $x \in W$, the set $P(x)$ is open (or empty) in $Y$.

(v) *Continuous* iff $P$ is both *ldc* and *udc*.

(vi) *Acyclic* if it is impossible to find a cycle $x_t \in P(x_{t-1}), x_{t-1} \in P(x_{t-2}), \ldots, x_1 \in P(x_t)$.

We shall use lower demi-continuity of a preference correspondence to prove existence of a choice. In some cases, however, it is possible to make use of lower hemi-continuity. Note that if $P$ is *ldc* then it is *lhc*.

We shall now show that if $W$ is compact, and $P$ is an acyclic and *ldc* preference correspondence $P : W \twoheadrightarrow W$, then $C(W, P) \neq \varnothing$. First of all, say a preference correspondence $P : W \twoheadrightarrow W$ satisfies the *finite maximality property* (FMP) on $W$ iff for every finite set $V$ in $W$, there exists $x \in V$ such that $P(x) \cap V = \varnothing$.

**Lemma 6.1** (Walker 1977) *If $W$ is a compact, topological space and $P$ is an ldc preference correspondence that satisfies FMP on $W$, then $C(W, P) \neq \varnothing$. This follows readily, using compactness to find a finite subcover, and then using FMP.*

**Corollary 6.1** *If $W$ is a compact topological space and $P$ is an acyclic, ldc preference correspondence on $W$, then $C(W, P) \neq \varnothing$.*

As Walker (1977) noted, when $W$ is compact and $P$ is *ldc*, then $P$ is acyclic iff $P$ satisfies FMP on $W$, and so either property can be used to show existence of a choice. A alternative method of proof to show that $C_P$ is a choice function is to substitute a convexity property for $P$ rather than acyclicity.

**Definition 6.2**
 (i) If $W$ is a subset of a vector space, then the *convex hull* of $W$ is the set, Con$[W]$, defined by taking all convex combinations of points in $W$.
 (ii) $W$ is *convex* iff $W = \text{Con}[W]$. (The empty set is also convex.)
 (iii) $W$ is *admissible* iff $W$ is a compact, convex subset of a topological vector space.
 (iv) A preference correspondence $P: W \twoheadrightarrow W$ on a convex set $W$ is *convex* iff, for all $x \in W$, $P(x)$ is convex.
 (v) A preference correspondence $P: W \twoheadrightarrow W$ is *semi-convex* iff, for all $x \in W$, it is the case that $x \notin \text{Con}(P(x))$.

As we showed in Chap. 3, Fan (1961) has demonstrated that if $W$ is admissible and $P$ is ldc and semi-convex, then $C(W, P)$ is non-empty.

**Choice Theorem** (Fan 1961; Bergstrom 1975) *If $W$ is an admissible subset of a Hausdorff topological vector space, and $P: W \twoheadrightarrow W$ a preference correspondence on $W$ which is ldc and semi-convex then $C(W, P) \neq \varnothing$.*

As demonstrated in Chap. 3, the proof uses the KKM lemma due to Knaster et al. (1929). There is a useful corollary to the Fan Choice theorem. Say a preference correspondence on an admissible space $W$ satisfies the *convex maximality property* (CMP) iff for any finite set $V$ in $W$, there exists $x \in \text{Con}(V)$ such that $P(x) \cap \text{Con}(V) = \varnothing$.

**Corollary 6.2** *Let $W$ be admissible and $P: W \twoheadrightarrow W$ be ldc and semi-convex. Then $P$ satisfies the convex maximality property.*

Numerous applications of the procedure have been made to show existence of such an economic equilibrium. Note however, that these results all depend on semi-convexity of the preference correspondences.

## 6.2 Dynamical Choice Functions

We now consider a *generalized preference field* $H: W \twoheadrightarrow TW$, on a smooth manifold $W$.

We use this notation to mean that at any $x \in W$, $H(x)$ is a *cone* in the tangent space $T_x W$ above $x$. That is, if a vector $v \in H(x)$, then $\lambda v \in H(x)$ for any $\lambda > 0$. If there is a smooth curve, $c: [-1, 1] \to W$, such that the differential $\frac{dc(t)}{dt} \in H(x)$, whenever $c(t) = x$, then c is called an integral curve of $H$. An integral curve of $H$ from $x = c(o)$ to $y = \lim_{t \to 1} c(t)$ is called an *H-preference curve* from $x$ to $y$. The preference field $H$ is called *S-continuous* iff, for any $v \in H(x) \neq \varnothing$ then there is an integral curve, $c$, in a neighborhood of $x$ with $\frac{dc(0)}{dt} = v$. The choice $C(W, H)$ of $H$ on $W$ is defined by

$$C(W, H) = \{x \in W : H(x) = \varnothing\}.$$

Say $H$ is *half open* if at every $x \in W$, either $H(x) = \emptyset$ or there exists a vector $v' \in T_x W$ such that $(v' \cdot v) > 0$ for all $v \in H(x)$. We can say in this case that there is, at $x$, a *direction gradient* $d$ in the cotangent space $T_x^* W$ of linear maps from $T_x W$ to $\Re$ such that $d(v) > 0$ for all $v \in H(x)$. If $H$ is S-continuous and half-open, then there will exist such a continuous direction gradient $d\, V \to T^*V$ on a neighborhood $V$ of $x$.

**Choice Theorem** *If $H$ is an S-continuous half open preference field, on a finite dimensional compact manifold, $W$, then $C(W, H) \neq \emptyset$. If $H$ is not half open then there exists an $H$-preference cycle through $\{x_1, x_2, x_3, ..x_r.x_1\}$. For each arc $(x_s, x_{s+1})$ there is an $H$-preference curve from $x_s$ to $x_{s+1}$, with a final $H$-preference curve from $x_r$ to $x_1$.*

The Choice Theorem implies the existence of a *singularity* of the field, $H$.

**Existence of Nash Equilibrium** Let $\{W_1, \ldots, W_n\}$ be a family of compact, contractible, smooth, strategy spaces with each $W_i \subset \Re^w$. A smooth profile is a function $u : W^N = W_1 \times W_2 \times \cdots \times W_n \twoheadrightarrow \Re^n$. Let $H_i : W_i \twoheadrightarrow TW_i$ be the induced $i$-preference field in the tangent space over $W_i$. If each $H_i$ is S-continuous and half open in $TW_i$ then there exists a *critical Nash equilibrium*, $\mathbf{z} \in W^N$ such that $H^N(\mathbf{z}) = (H_1 \times \cdots \times H_n)(\mathbf{z}) = \emptyset$.

This follows from the previous choice theorem because the product preference field, $H^N$, will be half-open and S-continuous Below we consider existence of local Nash equilibrium. With smooth utility functions, a local Nash equilibrium can be found by checking the second order conditions on the Hessians. We now repeat Example 3.12. from Chap. 3.

*Example 6.1* To illustrate the Choice Theorem, consider the example due to Kramer (1973), with $N = \{1, 2, 3\}$. Let the preference relation $P_\mathbb{D}: W \twoheadrightarrow W$ be generated by a set of decisive coalitions, $\mathbb{D} = \{\{1, 2\}, \{1, 3\}, \{2, 3\}\}$, so that $y \in P_\mathbb{D}(x)$ whenever two voters prefer $y$ to $x$. Suppose further that the preferences of the voters are characterized by the direction gradients

$$\{du_i(x): i = 1, 2, 3\}$$

as in Fig. 6.1.

As the figure makes evident, it is possible to find three points $\{a, b, c\}$ in $W$ such that

$$u_1(a) > u_1(b) = u_1(x) > u_1(c)$$
$$u_2(b) > u_2(c) = u_2(x) > u_2(a)$$
$$u_3(c) > u_3(a) = u_3(x) > u_3(b).$$

That is to say, preferences on $\{a, b, c\}$ give rise to a *Condorcet cycle*. Note also that the set of points $P_\mathbb{D}(x)$, preferred to $x$ under the voting rule, are the shaded "win sets" in the figure. Clearly $x \in \text{Con}\, P_\mathbb{D}(x)$, so $P_\mathbb{D}(x)$ is not semi-convex. Indeed it

## 6.2 Dynamical Choice Functions

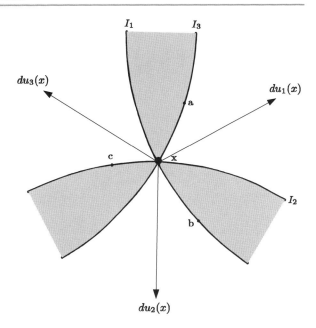

**Fig. 6.1** Cycles near a point $x$

should be clear that in *any* neighborhood $V$ of $x$ it is possible to find three points $\{a', b', c'\}$ such that there is *local* voting cycle, with $a' \in P_\mathbb{D}(b'), b' \in P_\mathbb{D}(c'), c' \in P_\mathbb{D}(a')$. We can write this as

$$a' \to c' \to b' \to a'.$$

Not only is there a voting cycle, but the Fan theorem fails, and we have no reason to believe that $C(W, P_\mathbb{D}) \neq \varnothing$.

We can translate this example into one on preference fields by writing

$$H_\mathbb{D}(u) = \cup H_M(u) : W \twoheadrightarrow TW$$

where each $M \in \mathbb{D}$ and

$$H_M(u)(x) = \{v \in T_x W : \big(du_i(x) \cdot v\big) > 0, \ \forall i \in M\}.$$

Figure 6.2 shows the three difference preference fields $\{H_i : i = 1, 2, 3\}$ on $W$, as well as the intersections $H_M$, for $M = \{1, 2\}$ etc.

Obviously the joint preference field $H_\mathbb{D} : W \twoheadrightarrow TW$ fails the half open property at $x$. Although $H_\mathbb{D}$ is S-continuous, we cannot infer that $C(W, H) \neq \varnothing$. If we define

$$Cycle(W, H) = \{x \in W : H(x) \text{ is not half open}\},$$

then at any point in $Cycle(W, H)$ it is possible to construct local cycles in the manner just described.

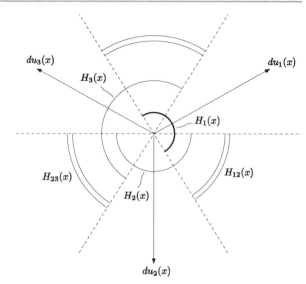

**Fig. 6.2** The failure of half-openness of a preference field

The choice theorem can then be interpreted to mean that for any S-continuous field on $W$, then

$$Cycle(W, H) \cup C(W, H) \neq \emptyset.$$

Chilchinisky (1995) has obtained similar results for markets, where the condition that the dual is non-empty was termed *market arbitrage*, and defined in terms of global market co-cones associated with each player. Such a dual co-cone, $[H_i(u)]^*$ is precisely the set of prices in the cotangent space that lie in the dual of the preferred cone, $[H_i(u)]$, of the agent. By analogy with the above, she identifies this condition on non-emptiness of the intersection of the family of co-cones as one which is necessary and sufficient to guarantee an equilibrium.

The following Theorem implies that Fig. 6.2 is "generic." As in Chap. 4, a property is *generic* if it is true of all profiles in a residual set in $C^r(W, \Re^n)$, where this is the Whitney topology on smooth profiles, for a society of size $n$, on the policy space $W$. Now consider a non-collegial voting game, $\mathbb{D}$ with Nakamura number $\kappa(\mathbb{D})$ Then we have the following Theorem by Saari (1997).

**Saari Theorem** *For any non-collegial $\mathbb{D}$, there exists an integer $w(\mathbb{D}) > \kappa(\mathbb{D})$ such that $\dim(W) > w(\mathbb{D})$ implies that $C(W, H_{\mathbb{D}}(u)) = \emptyset$ for all $u$ in a residual subspace of $C^r(W, \Re^n)$.*

This result was essentially proved by Saari (1997), building on earlier results by Banks (1995), McKelvey (1976, 1979), Kramer (1973), Plott (1967), Schofield (1978, 1983) and McKelvey and Schofield (1987). Although this result formally applies to voting rules, Schofield (2010) argues that it is applicable to any non-collegial social mechanism, and as a result can be interpreted to imply that chaos is

**Fig. 6.3** The heart with a uniform electorate on the triangle

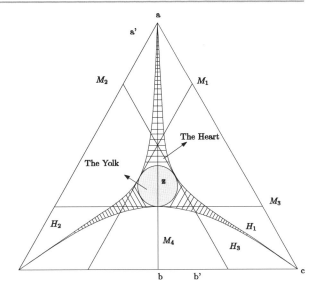

a generic phenomenon in coalitional systems. Since an equilibrium may not exist, we now introduce a more general social solution concept, called the *heart*.

**Definition 6.3** (The Heart)
(i) If $W$ is a topological space, then $x \in W$ is *locally covered* (under a preference correspondence $\mathbf{Q}$) iff for any neighborhood $Y$ of $x$ in $W$, there exists $y \in Y$ such that (a) $y \in \mathbf{Q}(x)$, and (b) there exists a neighborhood $Y'$ of $y$, with $Y' \subseteq Y$ such that $Y' \cap \mathbf{Q}(y) \subset \mathbf{Q}(x)$.
(ii) The *heart* of $\mathbf{Q}$, written $\mathcal{H}(\mathbf{Q})$, is the set of locally uncovered points in $W$.

This notion can be applied to a preference correspondence $P_\mathbb{D}$ or to the preference field, $H_\mathbb{D}(u)$, in which case we write $\mathcal{H}(P_\mathbb{D})$ or $\mathcal{H}(H_\mathbb{D}(u))$. Schofield (1999) shows that the heart will belong to the Pareto set, and is lower hemi-continuous when regarded as a correspondence.

*Example 6.2* To illustrate the heart, Fig. 6.3 gives a simple artificial example where the utility profile, $u$, is one where society has "Euclidean" preferences, based on distance, and the ideal points are uniformly distributed on the boundary of the equilateral triangle. Under majority rule, $\mathbb{D}$, the heart $\mathcal{H}(\mathbf{H}_\mathbb{D}(u))$, is the star-shaped figure inside the equilateral triangle (the Pareto set), and contains the "yolk" McKelvey (1986). The heart is generated by an infinite family of "median lines," such as $\{M_1, M_2, \ldots\}$. The shape of the heart reflects the asymmetry of the distribution. Inside the heart, voting trajectories can wander anywhere. Outside the heart the dual cones intersect, so any trajectory starting outside the heart converges to the heart. Thus the heart is an "attractor" of the voting process. Figure 6.4 gives a similar example, this time where preferences are defined by the pentagon, and the heart is the small centrally located ball.

**Fig. 6.4** The heart with a uniform electorate on the pentagon

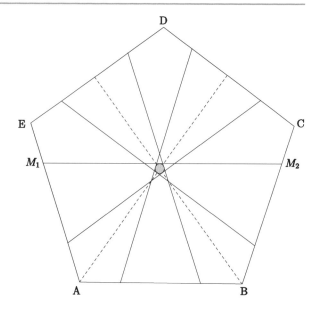

## 6.3 Stochastic Choice

To construct such a social preference field, we first consider political choice on a compact space, $W$, of political proposals. This model is an extension of the standard multiparty stochastic model, modified by inducing asymmetries in terms of the preferences of voters.

We define a stochastic electoral model, $\mathbb{M}(\lambda, \mu, \theta, \alpha, \beta)$, which utilizes sociodemographic variables and voter perceptions of character traits. For this model we assume that voter $i$ utility is given by the expression

$$u_{ij}(x_i, z_j) = \lambda_j + \mu_j(z_j) + (\theta_j \cdot \eta_i) + (\alpha_j \cdot \tau_i) - \beta \|x_i - z_j\|^2 + \varepsilon_j \quad (6.1)$$
$$= \left[u_{ij}^*(x_i, z_j)\right] + \varepsilon_j \quad (6.2)$$

The points $\{x_i \in W : i \in N\}$ are the preferred policies of the set, $N$, of voters in the political or policy space $W$, and $\mathbf{z} = \{z_j \in W : j \in Q\}$ are the positions of the agents/candidates. The term $\|x_i - z_j\|$ is simply the Euclidean distance between $x_i$ and $z_j$. The error vector $(\varepsilon_1, \ldots, \varepsilon_j, \ldots, \varepsilon_p)$ is distributed by the iid type I extreme value distribution, as assumed in empirical multinomial logit estimation (MNL). The symbol $\theta$ denotes a set of $k$-vectors $\{\theta_j : j \in Q\}$ representing the effect of the $k$ different sociodemographic parameters (class, domicile, education, income, religious orientation, etc.) on voting for agent $j$ while $\eta_i$ is a $k$-vector denoting the $i$th individual's relevant "sociodemographic" characteristics. The compositions $\{(\theta_j \cdot \eta_i)\}$ are scalar products, called the *sociodemographic valences* for $j$. These scalar terms characterize the various types of the voters. See Lin et al. (1999) for work with the stochastic model.

## 6.3 Stochastic Choice

The terms $\{(\alpha_j \cdot \tau_i)\}$ are scalars giving voter $i$'s perceptions and beliefs. These can include perceptions of the character traits of candidate or agent $j$, or beliefs about the state of the economy, etc. We let $\alpha = (\alpha_q, \ldots, \alpha_1)$. A *trait score* can be obtained by factor analysis from a set of survey questions asking respondents about the traits of the agent, including 'moral', 'caring', 'knowledgeable', 'strong', 'honest', 'intelligent', etc. The perception of traits can be augmented with voter perception of the state of the economy, etc. in order to examine how anticipated changes in the economy affect each agent's electoral support.

Finally the *exogenous* valence vector $\lambda = (\lambda_1, \lambda_2, \ldots, \lambda_q)$ gives the general perception of the quality of the various candidates, $\{1, \ldots, q\}$. This vector satisfies $\lambda_q \geq \lambda_{q-1} \geq \cdots \geq \lambda_2 \geq \lambda_1$, where $(1, \ldots, q)$ label the candidates, and $\lambda_j$ is the exogenous valence of agent or candidate $j$. In empirical multinomial logit models, the valence vector $\lambda$ is given by the intercept terms for each agent. Finally $\{\mu_j(z_j)\}$ represent the *endogenous* valences of the candidates. These valences depend on the positions $\{z_j \in W : j \in Q\}$ of the agents.

In the model, the probability that voter $i$ chooses candidate $j$, when party positions are given by $\mathbf{z}$ is:

$$\rho_{ij}(\mathbf{z}) = \Pr\big[[u_{ij}(x_i, z_j) > u_{il}(x_i, z_l)]\big], \text{ for all } l \neq j].$$

A local Nash equilibrium (LNE) is one that locally maximizes the expectation.

A *strict local Political Nash equilibrium* (SLNE) is a vector, $\mathbf{z}$, such that each candidate, $j$, has chosen $z_j$ to locally strictly maximize the expectation $\Sigma_i \rho_{ij}(\mathbf{z})$.

The type I extreme value distribution, $\Psi$, has a cumulative distribution the closed form

$$\Psi(h) = \exp\big[-\exp[-h]\big],$$

while its pdf has variance $\frac{1}{6}\pi^2$.

With this distribution it follows, for each voter $i$, and candidate, $j$, that

$$\rho_{ij}(\mathbf{z}) = \frac{\exp[u^*_{ij}(x_i, z_j)]}{\sum_{k=1}^{q} \exp u^*_{ik}(x_i, z_k)}. \tag{6.3}$$

This game is an example of what is known as a *Quantal response game* McKelvey and Palfrey (1995), Levine and Palfrey (2007). Note that the utility expressions $\{u^*_{ik}(x_i, z_k)\}$ can be estimated from surveys that include vote intentions. We can use the American National Election Survey (ANES for 2008) which gives gave individual perceptions of the important political policy questions. As indicated in Table 6.1, we are able to use factor analysis of these responses to construct a two dimensional policy space. ANES 2008 also gave voter perceptions of the character traits of the candidates, in terms of "moral", "caring", "knowledgeable", "strong" and "honest". We performed a factor analysis of these perceptions as shown in Table 6.2. Further details of the model can be found in Schofield et al. (2011)). Table 6.3 gives estimates of the average voter positions for the two parties . We also obtained data on those voters who declared they provided support for the candidates. these we designated as activists. Figure 6.5 gives an estimate of the voter estimated positions as well as the two presidential candidate positions in 2008.

**Table 6.1** Factor loadings for economic and social policy for the 2008 election

| Question | Economic policy | Social policy |
|---|---|---|
| Less Government services | 0.53 | 0.12 |
| Oppose Universal health care | 0.51 | 0.22 |
| Oppose Bigger Government | 0.50 | 0.14 |
| Prefer Market to Government | 0.56 | |
| Decrease Welfare spending | 0.24 | |
| Less government | 0.65 | |
| Worry more about Equality | 0.14 | 0.37 |
| Tax Companies Equally | 0.28 | 0.10 |
| Support Abortion | | 0.55 |
| Decrease Immigration | 0.12 | 0.25 |
| Civil right for gays | | 0.60 |
| Disagree Traditional values | | 0.53 |
| Gun access | 0.36 | |
| Support Afr. Amer | 0.14 | 0.45 |
| Conservative v Liberal | 0.30 | 0.60 |
| Eigenvalue | 1.93 | 1.83 |

**Table 6.2** Factor loadings for candidate traits scores 2008

| Question | Obama traits | McCain traits |
|---|---|---|
| Obama Moral | 0.72 | −0.01 |
| Obama Caring | 0.71 | −0.18 |
| Obama Knowledgeable | 0.61 | −0.07 |
| Obama Strong | 0.69 | −0.13 |
| Obama Honest | 0.68 | −0.09 |
| Obama Intelligent | 0.61 | 0.08 |
| Obama Optimistic | 0.55 | 0.00 |
| McCain Moral | −0.09 | 0.67 |
| McCain Cares | −0.17 | 0.63 |
| McCain Knowledgeable | −0.02 | 0.65 |
| McCain Strong | −0.10 | 0.70 |
| McCain Honest | −0.03 | 0.63 |
| McCain Intelligent | 0.11 | 0.68 |
| McCain Optimistic | −0.07 | 0.57 |
| Eigenvalue | 3.07 | 3.00 |

These survey data allow us to construct a spatial logit model of the election as in Table 6.4. This table has Obama as the baseline candidate.

## 6.3 Stochastic Choice

**Table 6.3** Estimated voter, activist and candidate positions in 2008

|  | Econ policy | | | Social policy | | | n |
|---|---|---|---|---|---|---|---|
|  | Mean | s.e. | 95 % C.I | Mean | s.e. | 95 % C.I |  |
| Activists |  |  |  |  |  |  |  |
| Democrats | −0.20 | 0.09 | [−0.38, −0.02] | 1.14 | 0.11 | [0.92, 1.37] | 80 |
| Republicans | 1.41 | 0.13 | [1.66, 1.16] | −0.82 | 0.09 | [−0.99, −0.65] | 40 |
| Non-activists |  |  |  |  |  |  |  |
| Democrats | −0.17 | 0.03 | [−0.24, −0.11] | 0.36 | 0.04 | [0.29, 0.44] | 449 |
| Republicans | 0.72 | 0.06 | [0.60, 0.84] | −0.56 | 0.05 | [−0.65, −0.46] | 219 |
|  |  |  |  |  |  |  | 788 |

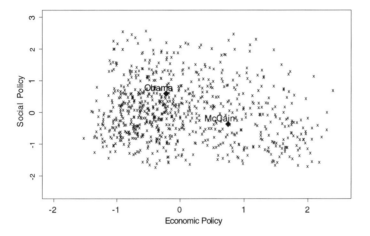

**Fig. 6.5** Distribution of voter ideal points and candidate positions in the 2008 presidential election

Once the voter probabilities over a given set $\mathbf{z} = \{z_j : j \in Q\}$ are computed, then estimation procedures allow these probabilities to be computed for each possible vector $\mathbf{z}$. This allows the determination and proof of existence of *local Nash equilibrium* (LNE), namely a vector, $\mathbf{z}^* = (z_1^*, \ldots, z_j^*, \ldots, z_q^*)$ such that each candidate, $j$, chooses $z_j^*$ to locally maximize its expected vote share, given by the expectation $V_j(\mathbf{z}^*) = \sum_i \rho_{ij}(\mathbf{z}^*)$.

Schofield (2006) shows that the first order condition for a LNE is that the *marginal electoral pull* at $\mathbf{z}^* = (z_1^*, \ldots, z_j^*, \ldots, z_q^*)$ is zero. For candidate $j$, this is defined to be

$$\frac{d\mathcal{E}_j^*}{dz_j}(z_j^*) = [z_j^{el} - z_j^*]$$

$$\text{where } z_j^{el} \equiv \sum_{i=1}^{n} \varpi_{ij} x_i$$

**Table 6.4** Spatial logit models for USA 2008[a]

| Variable | (1) Spatial | (2) Sp. & Traits | (3) Sp. & Dem. | (4) Full |
|---|---|---|---|---|
| McCain valence $\lambda$ | −0.84*** | −1.08*** | −2.60** | −3.58*** |
|  | (7.6) | (8.3) | (2.8) | (3.4) |
| Spatial $\beta$ | 0.85*** | 0.78*** | 0.86*** | 0.83*** |
|  | (14.1) | (10.1) | (12.3) | (10.3) |
| McCain traits |  | 1.30*** |  | 1.36*** |
|  |  | (7.6) |  | (7.15) |
| Obama traits |  | −1.02*** |  | −1.16*** |
|  |  | (6.8) |  | (6.44) |
| Age |  |  | −0.01 | −0.01 |
|  |  |  | (1.0) | (1.0) |
| Gender (F) |  |  | 0.29 | 0.44 |
|  |  |  | (1.26) | (0.26) |
| African American |  |  | −4.16*** | −3.79*** |
|  |  |  | (3.78) | (3.08) |
| Hispanic |  |  | −0.55 | −0.23 |
|  |  |  | (1.34) | (0.51) |
| Education |  |  | 0.15* | 0.22*** |
|  |  |  | (2.5) | (3.66) |
| Income |  |  | 0.03 | 0.01 |
|  |  |  | (1.5) | (0.50) |
| Working Class |  |  | −0.54* | −0.70** |
|  |  |  | (2.25) | (2.59) |
| South |  |  | 0.36 | −0.02 |
|  |  |  | (1.5) | (0.07) |
| Observations | 788 |  |  |  |
| log likelihood (LL) | −299 | −243 | −250 | −207 |
| AIC | 601 | 494 | 521 | 438 |
| BIC | 611 | 513 | 567 | 494 |

[a]Obama is the baseline for this model

is the *weighted electoral mean* of candidate $j$.

Here the weights $\{\varpi_{ij}\}$ are individual specific, and defined at the vector $\mathbf{z}^*$ by:

$$[\varpi_{ij}] = \left[ \frac{[\rho_{ij}(\mathbf{z}^*) - \rho_{ij}(\mathbf{z}^*)^2]}{\sum_{k \in N}[\rho_{kj}(\mathbf{z}^*) - \rho_{kj}(\mathbf{z}^*)^2]} \right] \quad (6.4)$$

Because the candidate utility functions $\{V_j : W \to \Re\}$ are differentiable, the second order condition on the Hessian of each $V_j$ at $\mathbf{z}^*$ can then be used to determine whether $\mathbf{z}^*$ is indeed an LNE. Proof of existence of such an LNE will

## 6.3 Stochastic Choice

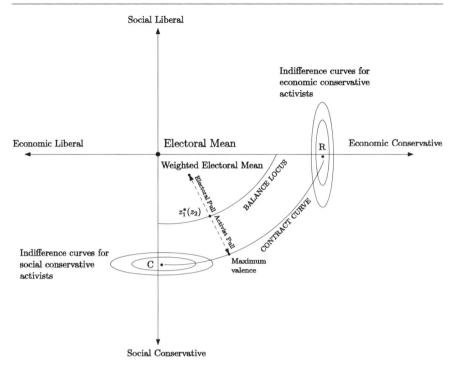

**Fig. 6.6** Optimal Republican position

then follow from some version of the Choice Theorem. For example, in the simpler model $\mathbb{M}(\lambda, \beta)$, without activists, all weights are equal to $\frac{1}{n}$, so the electoral mean $x_0 = \frac{1}{n} \sum x_i$ satisfies the first order condition, as suggested by Hinich (1977). Schofield (2007) gives the necessary and sufficient second order conditions for an LNE at the mean.

The underlying idea of this model is that each candidate, $j$, will be attracted to the respective weighted electoral mean, $z_j^{el}$, but will also be pulled away to a position preferred by the party activists. Figure 6.6 suggests the location of the weighted electoral mean for a Republican candidate. The contract curve in this figure is a heuristic estimated locus of preferred activist positions. See also Schofield et al. (2011).

Because the candidate utility functions $\{V_j : W \to \Re\}$ are differentiable, the second order condition on the Hessian of each $V_j$ at $\mathbf{z}^*$ can then be used to determine whether $\mathbf{z}^*$ is indeed an LNE. Proof of existence of such an LNE will then follow from some version of the Choice Theorem. For example, in the simpler model $\mathbb{M}(\lambda, \beta)$, without activists, all weights are equal to $\frac{1}{n}$, so the electoral mean $x_0 = \frac{1}{n} \sum x_i$ satisfies the first order condition, as suggested by Hinich (1977). Schofield (2007) gives the necessary and sufficient second order conditions for an LNE at the mean.

The underlying idea of this model is that each candidate, $j$, will be attracted to the respective weighted electoral mean, $z_j^{el}$, but will also be pulled away to a position preferred by the party activists. Figure 6.6 suggests the location of the weighted electoral mean for a Republican candidate. The contract curve in this figure is a heuristic estimated locus of preferred activist positions. See also Schofield et al. (2011).

We can compute the necessary and sufficient second order conditions for an LNE at the electoral mean. In the case that the activist valence functions and sociodemographic terms are identically zero, we call this the *pure spatial model*, denoted $\mathbb{M}(\lambda, \beta)$.

In this case, the first order condition is

$$\frac{dV_j(\mathbf{z})}{dz_j} = \frac{1}{n}\sum_{i\in N}\frac{d\rho_{ij}}{dz_j} \tag{6.5}$$

$$= \frac{1}{n}\sum_{i\in N}\{2\beta(x_i - z_j)\}[\rho_{ij} - \rho_{ij}^2] = 0. \tag{6.6}$$

Suppose that all $z_j$ are identical. Then all $\rho_{ij}$ are independent of $\{x_i\}$ and thus of $i$, and $\rho_{ij}$ may be written as $\rho_j$. Then for each fixed $j$, the first order condition is

$$\frac{dV_j(\mathbf{z})}{dz_j} = 2\beta[\rho_j - \rho_j^2]\sum_{i\in N}[(x_i - z_j)] = 0. \tag{6.7}$$

The second order condition for an LNE at z* depends on the negative definiteness of the Hessian of the activist valence function. If the eigenvalues of these Hessians are negative at a balance solution, and of sufficient magnitude, then this will guarantee that a vector **z**\* which satisfies the balance condition will be a SLNE. Indeed, this condition can ensure concavity of the vote share functions, and thus of existence of a PNE.

### 6.3.1 The Model Without Activist Valence Functions

We now apply the Theorem to the pure spatial model $\mathbb{M}(\lambda, \beta)$, by setting $\boldsymbol{\mu} = \boldsymbol{\theta} = \boldsymbol{\alpha} \equiv \mathbf{0}$.

As we have shown above, the *joint electoral mean* $\mathbf{z}_0$ satisfies the first order condition for a LNE. We now consider the second order condition.

**Definition 6.4** (The Convergence Coefficient of the Model $\mathbb{M}(\lambda, \beta)$) When the space $W$ has dimension $w$.
  (i) Define

$$\rho_1 = \left[1 + \sum_{k=2}^{p}\exp[\lambda_k - \lambda_1]\right]^{-1}. \tag{6.8}$$

(ii) Let $W$ be endowed with an orthogonal system of coordinate axes $(1, \ldots, s, \ldots, t, \ldots, w)$. For each coordinate axis let $\xi_t = (x_{1t}, x_{2t}, \ldots, x_{nt}) \in \mathbb{R}^n$ be the vector of the $t$th coordinates of the set of $n$ voter ideal points. Let $(\xi_s, \xi_t) \in \mathbb{R}$ denote scalar product. The covariance between the $s$th and $t$th axes is denoted $(\sigma_s, \sigma_t) = \frac{1}{n}(\xi_s, \xi_t)$ and $\sigma_s^2 = \frac{1}{n}(\xi_s, \xi_s)$ is the electoral variance on the $s$th axis. Note that these variances and covariances are taken about the electoral means on each axis.

(iii) The symmetric $w \times w$ electoral covariance matrix $\nabla_0$ is defined to be $\frac{1}{n}[(\xi_s, \xi_t)]_{t=1,\ldots,w}^{s=1,\ldots,w}$.

(iv) The *electoral variance* is

$$\sigma^2 = \sum_{s=1}^{w} \sigma_s^2 = \frac{1}{n} \sum_{s=1}^{w} (\xi_s, \xi_s) = \text{trace}(\nabla_0).$$

(v) The $w$ by $w$ *characteristic matrix*, of agent 1 is given by

$$C_1 = 2\beta(1 - 2\rho_1)\nabla_0 - I. \tag{6.9}$$

(vi) The *convergence coefficient of the model* $\mathbb{M}(\lambda, \beta)$ is

$$c \equiv c(\lambda, \beta) = 2\beta[1 - 2\rho_1]\sigma^2. \tag{6.10}$$

Observe that the $\beta$-parameter has dimension $L^{-2}$, so that $c$ is dimensionless. We can therefore use $c$ to compare different models.

Note also that agent 1 is by definition the agent with the lowest valence, and $\rho_1$, as defined above, is the probability that a generic voter will choose this agent when all agents are located at the origin. The estimate of the probability $\rho_1$ depends only on the comparison functions $\{f_{kj}\}$, as given above and these can be estimated in terms of the valence differences.

The following result is proved in Schofield (2007).

**The Mean Voter Valence Theorem**
(i) *The joint mean $z_0$ satisfies the first order condition to be a LNE for the model* $\mathbb{M}(\lambda, \beta)$.
(ii) *The necessary and sufficient second order condition for SLNE at* $z_0$ *is that $C_1$ has negative eigenvalues.*[1]
(iii) *A necessary condition for* $z_0$ *to be a SLNE for the model* $\mathbb{M}(\lambda, \beta)$ *is that $c(\lambda, \beta) < w$.*
(iv) *A sufficient condition for convergence to* $z_0$ *in the two dimensional case is that $c < 1$.*

Notice that (iii) follows from (ii) since the condition of negative eigenvalues means that

$$\text{trace}(C_1) = 2\beta[1 - 2\rho_1]\sigma^2 - w < 0.$$

---
[1] In the usual way, the condition for an LNE is that the eigenvalues are negative semi-definite.

In the case $c(\lambda, \beta) = w$, then trace $(C_1) = 0$, which means either that all eigenvalues are zero, or at least one is positive. This degenerate situation requires examination of $C_1$. The additional condition $c < 1$ is sufficient to guarantee that $\det(C_1) > 0$, which ensures that both eigenvalues are negative.

The expression for $C_1$ has a simple form because of the assumption of a single distance parameter $\beta$. It is possible to use a model with different coefficients $\beta = \{\beta_1, \beta_2, \ldots, \beta_w\}$ on each dimension. In this case the characteristic matrix can readily be shown to be

$$C_1 = 2(1 - 2\rho_1)\beta \nabla_0 \beta - \beta,$$

We require trace $(C_1) < 0$, or

$$2(1 - 2\rho_1)\,\text{trace}(\beta \nabla_0 \beta) < \beta_1 + \beta_2 + \cdots + \beta_w.$$

The convergence coefficient in this case is

$$c(\lambda, \beta) = \frac{2(1 - 2\rho_1)\,\text{trace}(\beta \nabla_0 \beta)}{\frac{1}{w}(\beta_1 + \beta_2 + \cdots + \beta_w)}$$

again giving the necessary condition of $c(\lambda, \beta) < w$.

Note that if $C_1$ has negative eigenvalues, then the Hessians of the vote shares for all agents are negative definite at the joint mean, $z_0$. When this is true, then the joint mean is a candidate for a PNE, and this property can be verified by simulation.

When the convergence condition $c(\lambda, \beta) < w$ is violated the joint origin cannot be a SPNE.

In the degenerate case $c(\lambda, \beta) = w$ it is again necessary to examine the characteristic matrix to determine whether the joint mean can be a PNE.

Model (1) in Table 6.4 shows the coefficients in 2008 for the pure spatial model, $\mathbb{M}(\lambda, \beta)$, to be

$$(\lambda_{Obama}, \lambda_{McCain}, \beta) = (0, -0.84, 0.85).$$

Table 6.4 indicates, the loglikelihood, Akaike information criterion (AIC) and Bayesian information criterion (BIC) are all quite acceptable, and all coefficients are significant with probability $< 0.001$.

Note that these parameters are estimated when the candidates are located at the estimated positions. Again, $\lambda_{McCain}$ is the relative negative exogenous valence of McCain, with respect to Obama, according to the pure spatial model $\mathbb{M}(\lambda, \beta)$. We assume that the parameters of the model remain close to these values as we modify the candidates positions in order to determine the equilibria of the model.

According to the model $\mathbb{M}(\lambda, \beta)$, the probability that a voter chooses McCain or Obama when both are positioned at the electoral mean, $z_0$, are

$$(\rho_{McCain}, \rho_{Obama}) = \left(\frac{e^0}{e^0 + e^{0.84}}, \frac{e^{0.84}}{e^0 + e^{0.84}}\right) = (0.30, 0.70).$$

## 6.3 Stochastic Choice

Now the covariance matrix can be estimated to be

$$\nabla_0 = \begin{bmatrix} 0.80 & -0.13 \\ -0.13 & 0.83 \end{bmatrix}.$$

Thus from Table 6.4, we obtain

$$C_{McCain} = [2\beta(1 - 2\rho_{McCain})\nabla_0 = [2 \times 0.85 \times 0.4 \times \nabla_0] - I$$

$$= (0.68)\nabla_0 - I$$

$$= (0.68)\begin{bmatrix} 0.80 & -0.13 \\ -0.13 & 0.83 \end{bmatrix} - I = \begin{bmatrix} 0.54 & -0.09 \\ -0.09 & 0.56 \end{bmatrix} - I$$

$$= \begin{bmatrix} -0.46 & -0.09 \\ -0.09 & -0.44 \end{bmatrix},$$

$$c = 2\beta(1 - 2\rho_{McCain}) \text{ trace } \nabla_0 = 2(0.85)(0.4)(1.63) = 1.1.$$

The determinant of $C_{McCain}$ is positive and the trace negative, so both eigenvalues are negative, showing that the mean is an LNE. The lower 95 % estimate for $\rho_{McCain}$ is 0.26, and the upper 95 % estimate for $\beta$ is 0.97, so a very conservative upper estimate for $\beta(1 - 2\rho_{McCain})$ is $0.97 \times 0.48 = 0.47$, so the upper estimate for $c$ is 1.53, giving an estimate for $C_{McCain}$ of

$$(0.94)\begin{bmatrix} 0.80 & -0.09 \\ -0.09 & 0.83 \end{bmatrix} - I$$

$$= \begin{bmatrix} 0.75 & -0.13 \\ -0.13 & 0.78 \end{bmatrix} - I$$

$$= \begin{bmatrix} -0.25 & -0.13 \\ -0.13 & -0.22 \end{bmatrix},$$

which still has negative eigenvalues.

We also considered a spatial model where the $x$ and $y$ axes had different coefficients, $\beta_1 = 0.8$, $\beta_2 = 0.92$.

Using

$$c(\lambda, \boldsymbol{\beta}) = \frac{2(1 - 2\rho_{lib}) \text{ trace}(\boldsymbol{\beta}\nabla_0\boldsymbol{\beta})}{\frac{1}{w}(\beta_1 + \beta_2 + \cdots + \beta_w)}$$

with $\frac{1}{2}(\beta_1 + \beta_2) = \frac{1}{2}(0.80 + 0.92) = 0.86$ and $\rho_{lib} = 0.25$, we find

$$c(\lambda, \boldsymbol{\beta}) = \frac{2(0.4)}{0.86} \text{ trace} \begin{bmatrix} (0.80)^2(0.80) & (0.80)(0.92)(-0.13) \\ (0.80)(0.92)(-0.13) & (0.92)^2(0.83) \end{bmatrix}$$

$$= (0.93) \text{ trace} \begin{bmatrix} 0.51 & -0.09 \\ -0.09 & 0.70 \end{bmatrix} = (0.93)(1.21) = 1.23.$$

For the characteristic matrix,

$$C_{McCain} = 2(1 - 2\rho_{McCain})\beta \nabla_0 \beta - \beta$$

$$= 2(0.4)\begin{bmatrix} 0.51 & -0.09 \\ -0.09 & 0.70 \end{bmatrix} - \begin{bmatrix} 0.80 & 0 \\ 0 & 0.92 \end{bmatrix}$$

$$= \begin{bmatrix} -0.41 & -0.07 \\ -0.07 & -0.56 \end{bmatrix} - \begin{bmatrix} 0.80 & 0 \\ 0 & 0.92 \end{bmatrix}$$

$$= \begin{bmatrix} -0.39 & -0.07 \\ -0.07 & -0.36 \end{bmatrix}.$$

The analysis showed the Hessian for this case had negative eigenvalues, so again $z_0$ is a LNE. This model is essentially the same as the model with a single $\beta$. Since these models imply the origin is an LNE but nether candidate is located there we infer that activists exert influence to move the candidate positions into opposite quadrants of the policy space. See Miller and Schofield (2003) for a devlopment of this observation.

# References

Banks, J. S. (1995). Singularity theory and core existence in the spatial model. *Journal of Mathematical Economics, 24*, 523–536.

Bergstrom, T. (1975). *The existence of maximal elements and equilibria in the absence of transitivity*. Typescript, University of Michigan.

Chichilnisky, G. (1995). Limited arbitrage is necessary and sufficient for the existence of a competitive equilibrium with or without short sales. *Economic Theory, 5*, 79–107.

Fan, K. (1961). A generalization of Tychonoff's fixed point theorem. *Mathematische Annalen, 42*, 305–310.

Hinich, M. J. (1977). Equilibrium in spatial voting: the median voter theorem is an artifact. *Journal of Economic Theory, 16*, 208–219.

Knaster, B., Kuratowski, K., & Mazurkiewicz, S. (1929). Ein Beweis des Fixpunktsatzes fur n-dimensionale Simplexe. *Fundamenta Mathematicae, 14*, 132–137.

Kramer, G. H. (1973). On a class of equilibrium conditions for majority rule. *Econometrica, 41*, 285–297.

Levine, D., & Palfrey, T. R. (2007). The paradox of voter participation. *American Political Science Review, 101*, 143–158.

Lin, T., Enelow, M. J., & Dorussen, H. (1999). Equilibrium in multicandidate probabilistic spatial voting. *Public Choice, 98*, 59–82.

McKelvey, R. D. (1976). Intransitivities in multidimensional voting models and some implications for agenda control. *Journal of Economic Theory, 12*, 472–482.

McKelvey, R. D. (1979). General conditions for global intransitivities in formal voting models. *Econometrica, 47*, 1085–1112.

McKelvey, R. D. (1986). Covering, dominance and institution free properties of social choice. *American Journal of Political Science, 30*, 283–314.

McKelvey, R. D., & Palfrey, T. R. (1995). Quantal response equilibria in normal form games. *Games and Economic Behavior, 10*, 6–38.

McKelvey, R. D., & Schofield, N. (1987). Generalized symmetry conditions at a core point. *Econometrica, 55*, 923–933.

# References

Miller, G., & Schofield, N. (2003). Activists and partisan realignment in the US. *American Political Science Review, 97*, 245–260.

Plott, C. R. (1967). A notion of equilibrium and its possibility under majority rule. *The American Economic Review, 57*, 787–806.

Saari, D. (1997). The generic existence of a core for $q$-rules. *Economic Theory, 9*, 219–260.

Schofield, N. (1978). Instability of simple dynamic games. *Review of Economic Studies, 45*, 575–594.

Schofield, N. (1983). Generic instability of majority rule. *Review of Economic Studies, 50*, 695–705.

Schofield, N. (1999). The heart and the uncovered set. *Journal of Economics. Supplementum, 8*, 79–113.

Schofield, N. (2006). Equilibria in the spatial stochastic model of voting with party activists. *Review of Economic Design, 10*, 183–203.

Schofield, N. (2007). The Mean Voter Theorem: necessary and sufficient conditions for convergent equilibrium. *Review of Economic Studies, 74*, 965–980.

Schofield, N. (2010). Social orders. *Social Choice and Welfare, 34*, 503–536.

Schofield, N. (2013). The probability of a fit choice. *Review of Economic Design, 17*, 129–150.

Schofield, N., Claassen, C., & Ozdemir, U. (2011). Empirical and formal models of the US presidential elections in 2004 and 2008. In N. Schofield & G. Caballero (Eds.), *The political economy of institutions, democracy and voting* (pp. 217–258). Berlin: Springer.

Walker, M. (1977). On the existence of maximal elements. *Journal of Economic Theory, 16*, 470–474.

# Review Exercises

## 7.1 Exercises to Chap. 1

**1.1.** Consider the relations:

$$P = \{(2,3), (1,4), (2,1), (3,2), (4,4)\} \quad \text{and} \quad Q = \{(1,3), (4,2), (2,4), (4,1)\}.$$

Compute $Q \circ P$, $P \circ Q$, $(P \circ Q)^{-1}$ and $(Q \circ P)^{-1}$. Let $\phi_Q$ and $\phi_P$ be the mappings associated with these two relations. Are either $\phi_Q$ and $\phi_P$ functions, and are they surjective and/or injective?

**1.2.** Suppose that each member $i$ of a society $M = \{1, \ldots, m\}$ has weak and strict preferences $(R_i, P_i)$ on a finite set $X$ of feasible states. Define the weak Pareto rule, $Q$, on $X$ by $xQy$ iff $xR_iy \forall i \in M$, and $xP_jy$ for some $j \in M$. Show that if each $R_i$, $i \in M$, is transitive, then $Q$ is transitive. Hence show that the Pareto choice set $C_Q(X)$ is non empty.

**1.3.** Show that the set $\Theta = \{e^{i\theta} : 0 \leq \theta \leq 2\pi\}$, of all $2 \times 2$ matrices representing rotations, is a subgroup of $(M^*(2), \circ)$, under matrix composition, $\circ$.

## 7.2 Exercises to Chap. 2

**2.1.** With respect to the usual basis for $\Re^3$, let $x_1 = (1, 1, 0)$, $x_2 = (0, 1, 1)$, $x_3 = (1, 0, 1)$. Show that $\{x_1, x_2, x_3\}$ are linearly independent.

**2.2.** Suppose $f : \Re^5 \to \Re^4$ is a linear transformation, with a 2-dimensional kernel. Show that there exists some vector $z \in \Re^4$, such that for any vector $y \in \Re^4$ there exists a vector $y_0 \in \text{Im}(f)$ with $y = y_0 + \lambda z$ for some $\lambda \in \Re$.

**2.3.** Find all solutions to the equations $A(x) = b_i$, for $i = 1, 2, 3$, where

$$A = \begin{pmatrix} 1 & 4 & 2 & 3 \\ 3 & 1 & -1 & 1 \\ 1 & -1 & 4 & 6 \end{pmatrix}$$

and

$$b_1 = \begin{pmatrix} 7 \\ 3 \\ 4 \end{pmatrix}, \quad b_2 = \begin{pmatrix} 1 \\ 1 \\ 1 \end{pmatrix} \quad \text{and} \quad b_3 = \begin{pmatrix} 3 \\ 2 \\ 1 \end{pmatrix}.$$

**2.4.** Find all solutions to the equation $A(x) = b$ where

$$A = \begin{pmatrix} 6 & -1 & 1 & 4 \\ 1 & 1 & 3 & -1 \\ 3 & 4 & 1 & 2 \end{pmatrix}$$

and

$$b = \begin{pmatrix} 4 \\ 3 \\ 7 \end{pmatrix}.$$

**2.5.** Let $F : \mathfrak{R}^4 \to \mathfrak{R}^2$ be the linear transformation represented by the matrix

$$\begin{pmatrix} 1 & 5 & -1 & 3 \\ -1 & 0 & -4 & 2 \end{pmatrix}.$$

Compute the set $F^{-1}(y)$, when $y = \begin{pmatrix} 4 \\ 1 \end{pmatrix}$.

**2.6.** Find the kernel and image of the linear transformation, $A$, represented by the matrix

$$\begin{pmatrix} 3 & 7 & 2 \\ 4 & 10 & 2 \\ 1 & -2 & 5 \end{pmatrix}.$$

Find new bases for the domain and codomain of $A$ so that $A$ can be represented as a matrix

$$\begin{pmatrix} I & 0 \\ 0 & 0 \end{pmatrix}$$

with respect to these bases.

## 7.3 Exercises to Chap. 3

**2.7.** Find the kernel of the linear transformation, $A$, represented by the matrix

$$\begin{pmatrix} 1 & 3 & 1 \\ 2 & -1 & -5 \\ -1 & 1 & 3 \end{pmatrix}.$$

Use the dimension theorem to compute the image of $A$. Does the equation $A(x) = b$ have a solution when

$$b = \begin{pmatrix} 1 \\ 1 \\ 1 \end{pmatrix}?$$

**2.8.** Find the eigenvalues and eigenvectors of

$$\begin{pmatrix} 2 & -1 \\ 1 & 4 \end{pmatrix}.$$

Is this matrix positive or negative definite or neither?

**2.9.** Diagonalize the matrix

$$\begin{pmatrix} 4 & 1 & 1 \\ 1 & 8 & 0 \\ 1 & 10 & 2 \end{pmatrix}.$$

**2.10.** Compute the eigenvalues and eigenvectors of

$$\begin{pmatrix} 1 & 0 & 0 \\ 0 & 0 & 1 \\ 0 & 1 & 0 \end{pmatrix}$$

and thus diagonalize the matrix.

## 7.3 Exercises to Chap. 3

**3.1.** Show that if $A$ is a set in a topological space $(X, \mathcal{T})$ then the interior, $\text{Int}(A)$, of $A$ is open and the closure, $\text{Clos}(A)$, is closed. Show that $\text{Int}(A) \subset A \subset \text{Clos}(A)$. What is the interior and what is the closure of the set $[a, b)$ in $\Re$, with the Euclidean topology? What is the boundary of $[a, b)$? Determine the limit points of $[a, b)$.

**3.2.** If two metrics $d_1, d_2$ on a space $X$ are equivalent write $d_1 \sim d_2$. Show that $\sim$ is an equivalence relation on the set of all metrics on $X$. Thus show that the Cartesian, Euclidean and city block topologies on $\Re^n$ are equivalent.

**3.3.** Show that the set, $L(\Re^n, \Re^m)$, of linear transformations from $\Re^n$ to $\Re^m$ is a normed vector space with norm

$$\|f\| = \sup_{x \in \Re^n} \left\{ \frac{\|f(x)\|}{\|x\|} : \|x\| \neq 0 \right\},$$

with respect to the Euclidean norms on $\Re^n$ and $\Re^m$. In particular verify that $\|\ \|_L$ satisfies the three norm properties. Describe an open neighbourhood of a member $f$ of $L(\Re^n, \Re^m)$ with respect to the induced topology on $L(\Re^n, \Re^m)$. Let $M(n, m)$ be the set of $n \times m$ matrices with the natural topology (see page 106), and let

$$M : L(\Re^n, \Re^m) \to M(n, m)$$

be the matrix representation with respect to bases for $\Re^n$ and $\Re^m$. Discuss the continuity of $M$ with respect to these topologies for $L(\Re^n, \Re^m)$ and $M(n, m)$.

**3.4.** Determine, from first principles, whether the following functions are continuous on their domain:
1. $\Re_+ \to \Re : x \to \log_e x$;
2. $\Re \to \Re_+ : x \to x^2$;
3. $\Re \to \Re_+ : x \to e^x$;
4. $\Re \to \Re : x \to \cos x$;
5. $\Re \to \Re : x \to \cos \frac{1}{x}$.

**3.5.** What is the image of the interval $[-1, 1]$ under the function $x \to \cos \frac{1}{x}$? Is the image compact?

**3.6.** Determine which of the following sets are convex:
1. $X_1 = \{(x_1, x_2) \in \Re^2 : 3x_1^2 + 2x_2^2 \leq 6\}$;
2. $X_2 = \{(x_1, x_2) \in \Re^2 : x_1 \leq 2, x_2 \leq 3\}$;
3. $X_3 = \{(x_1, x_2) \in \Re_+^2 : x_1 x_2 \leq 1\}$;
4. $X_4 = \{(x_1, x_2) \in \Re_+^2 : x_2 - 3 \geq -x_1^2\}$.

**3.7.** In $\Re^2$, let $B_C(x, r_1)$ be the *Cartesian* open ball of radius $r_1$ about $x$, and $B_E(y, r_2)$ the *Euclidean* ball of radius $r_2$ about $x$. Show that these two sets are convex. For fixed $x, y \in \Re^2$ obtain necessary and sufficient restrictions on $r_1, r_2$ so that these two open balls may be strongly separated by a hyperplane.

**3.8.** Determine whether the following functions are convex, quasi-concave, or concave:
1. $\Re \to \Re_+ : x \to e^x$;
2. $\Re \to \Re : x \to x^7$;
3. $\Re \to \Re : (x, y) \to xy$;
4. $\Re \to \Re : x \to \frac{1}{x}$;
5. $\Re^2 \to \Re : (x, y) \to x^2 - y$.

## 7.4 Exercises to Chap. 4

**4.1.** Suppose that $f: \Re^n \to \Re^m$ and $g: \Re^m \to \Re^k$ are both $C^r$-differentiable. Is $g \circ f: \Re^n \to \Re^k$, a $C^r$-differentiable function? If so, why?

**4.2.** Find and classify the critical points of the following functions:
1. $\Re^2 \to \Re: (x, y) \to x^2 + xy + 2y^2 + 3$;
2. $\Re^2 \to \Re: (x, y) \to -x^2 + xy - y^2 + 2x + y$;
3. $\Re^2 \to \Re: (x, y) \to e^{2x} - 2x + 2y$.

**4.3.** Determine the critical points, and the Hessian at these points, of the function $\Re^2 \to \Re: (x, y) \to x^2 y$.

Compute the eigenvalues and eigenvectors of the Hessian at critical points, and use this to determine the nature of the critical points.

**4.4.** Show that the origin is a critical point of the function:

$$\Re^3 \to \Re: (x, y, z) \to x^2 + 2y^2 + Zz^2 + xy + xz.$$

Determine the nature of this critical point by examining the Hessian.

**4.5.** Determine the *set* of critical points of the function

$$\Re^2 \to \Re: (x, y) \to -x^2 y^2 + 4xy - 4x^2.$$

**4.6.** Maximise the function $\Re^2 \to \Re: (x, y) \to x^2 y$ subject to the constraint $1 - x^2 - y^2 = 0$.

**4.7.** Maximise the function $\Re^2 \to \Re: (x, y) \to a \log x + b \log y$, subject to the constraint $px + qy \leq I$, where $p, q, I \in \Re_+$.

## 7.5 Exercises to Chap. 5

**5.1.** Show that if dimension $(X) > m$, then for almost every smooth profile $u = (u_1, \ldots, u_m): X \to \Re^m$ it is the case that Pareto optimal points in the interior of $X$ can be parametrised by at most $(m-1)$ strictly positive coefficients $\{\lambda_1, \ldots, \lambda_{m-1}\}$.

**5.2.** Consider a two agent, two good exchange economy, where the initial *endowment* of good $j$, by agent $i$ is $e_{ij}$. Suppose that each agent, $i$, has utility function $u_i: (x_{i1}, x_{i2}) \to a \log x_{i1} + b \log x_{i2}$. Compute the critical Pareto set $\Theta$, within the feasible set

$$Y = \{(x_{11}, x_{12}, x_{21}, x_{22}) \in \Re_+^4\},$$

**Fig. 7.1** The butterfly singularity

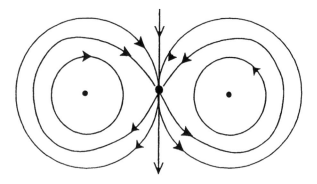

where the coordinates of $Y$ satisfy

$$x_{11} + x_{21} = e_{11} + e_{21} \quad \text{and} \quad x_{12} + x_{22} = e_{12} + e_{22}.$$

What is the dimension of $Y$ and what is the codimension of $\Theta$ in $Y$? Compute the market-clearing equilibrium.

**5.3.** Figure 7.1 shows a "butterfly singularity", $A$, in $\Re^2$. Compute the degree of this singularity. Show why such a singularity (though it is isolated) cannot be associated with a generic excess demand function on the two-dimensional price simplex.

# Subject Index

**A**
Abelian group, 20
Accumulation point, 83
Acyclic relation, 26
Additive inverse, 15
Additive relation, 22
Admissible set, 158
Antisymmetric relation, 24
Arrow's Impossibility Theorem, 33, 35
Associativity in a group, 15
Associativity of sets, 2
Asymmetric relation, 24
Attractor of a vector field, 214

**B**
Baire space, 212
Banach space, 123, 190
Base for a topology, 82
Basis of a vector space, 42
Bergstrom's theorem, 128
Bijective function, 11
Bilinear map, 67
Binary operation, 14
Binary relation, 24
Bliss point of a preference, 31
Boolean algebra, 2
Boundary of a set, 83
Boundary problem, 162
Bounded function, 29
Brouwer Fixed Point Theorem, 118, 121
Browder Fixed Point Theorem, 125
Budget set, 113
Butterfly dynamical system, 224

**C**
Calculation argument on economic information, 225
Canonical form of a matrix, 75
Cartesian metric, 85
Cartesian norm, 80
Cartesian open ball, 85
Cartesian product, 7
Cartesian topology, 85
Chain rule, 141
Change of basis, 55
Chaos, 221, 224
Characteristic equation of a matrix, 68
Choice correspondence, 27
City block metric, 87
City block norm, 80
City block topology, 86
Closed set, 83
Closure of a set, 83
Coalition feasibility, 179
Codomain of a relation, 7
Cofactor matrix, 53
Collegial rule, 35
Collegium, 35
Commutative group, 20
Commutativity of sets, 2
Compact set, 95
Competitive allocation, 178
Complement of a set, 2
Complete vector space, 123, 190
Composition of mappings, 8
Composition of matrices, 14
Concave function, 100
Connected relation, 24
Constrained optimisation, 155
Consumer optimisation, 113
Continuous function, 94
Contractible space, 119
Convergence coefficient, 244
Convex function, 99
Convex preference, 99
Convex set, 99
Corank, 196
Corank r singularity, 196
Core Theorem for an exchange economy, 179
Cover for a set, 6, 93
Critical Pareto set, 173
Critical point, 150

## D

Debreu projection, 204
Debreu-Smale Theorem, 183, 189, 203, 215, 221
Decisive coalition, 33
Deformation, 216
Deformation retract, 119
Degree of a singularity, 212
Demand, 114
Dense set, 83
Derivative of a function, 145
Determinant of a matrix, 14, 53
Diagonalisation of a matrix, 63, 64
Dictator, 35
Diffeomorphism, 140
Differentiable function, 135
Differential of a function, 138
Dimension of a vector space, 45
Dimension theorem, 50
Direction gradient, 139
Distributivity of a field, 21
Distributivity of sets, 2
Domain of a mapping, 8
Domain of a relation, 8

## E

Economic optimisation, 162
Edgeworth box, 182
Eigenvalue, 62
Eigenvector, 62
Endowment, 108
Equilibrium prices, 114
Equivalence relation, 28
Euclidean norm, 78
Euclidean scalar product, 77
Euclidean topology, 86
Euler characteristic of simplex, 214
Euler characteristic of sphere and torus, 199
Excess demand function, 207
Exchange theorem, 44
Existential quantifier, 7

## F

Fan theorem, 125, 225
Feasible input-output vector, 111
Field, 21
Filter, 34, 38
Fine topology, 87
Finite intersection property, 93
Finitely generated vector space, 44, 45
Fixed point property, 118
Frame, 41
Free-disposal equilibrium, 132

Function, 11
Function space, 92

## G

Game, 128
General linear group, 55
Generic existence of regular economies, 183, 203
Generic property, 183, 201, 212
Global maximum (minimum) of a function, 154
Global saddlepoint of the Lagrangian, 117
Graph of a mapping, 9
Group, 14

## H

Hairy Ball theorem, 215
Hausdorff space, 96
Heart, 237
Heine-Borel Theorem, 95
Hessian, 142, 144
Homeomorphism, 118, 211
Homomorphism, 18

## I

Identity mapping, 10
Identity matrix, 13, 52
Identity relation, 8
Image of a mapping, 8
Image of a transformation, 50
Immersion, 194
Implicit function theorem, 192, 195
Index of a critical point, 197
Index of a quadratic form, 70
Indifference, 24
Infimum of a function, 87
Injective function, 12
Interior of a set, 83
Intersection of sets, 2
Inverse element, 15
Inverse function, 11
Inverse function theorem, 190
Inverse matrix, 13, 54
Inverse relation, 8
Invisible dictator, 35
Irrational flow on torus, 218
Isomorphism, 18
Isomorphism theorem, 56

## J

Jacobian of a function, 140

## K

Kernel of a transformation, 50
Kernel rank, 50

Subject Index 259

Knaster-Kuratowski-Mazurkiewicz (KKM) Theorem, 123
Kuhn-Tucker theorems, 115–118

**L**
Lagrangian, 116
Lefschetz fixed point theorem, 216
Lefschetz obstruction, 217
Limit of a sequence, 94
Limit point, 83
Linear combination, 42
Linear dependence, 15
Linear transformation, 45
Linearly independent, 41
Local maximum (minimum), 154
Local Nash Equilibrium (LNE), 238
Locally non-satiated preference, 110
Lower demi-continuity, 118
Lower hemi-continuity, 128
Lyapunov function, 208, 209

**M**
Majority rule, 34
Manifold, 194
Mapping, 8
Marginal rate of technical substitution, 168
Market arbitrage, 236
Market equilibrium, 114
Matrix, 13, 46
Mean value theorem, 146
Mean Voter Theorem, 245
Measure zero, 198
Metric, 80
Metric topology, 82
Metrisable space, 80
Michael's Selection Theorem, 123
Monotonic rule, 36
Morphism, 18
Morse function, 154, 197
Morse lemma, 197
Morse Sard theorem, 202
Morse theorem, 202

**N**
Nakamura Lemma, 36
Nakamura number, 35
Nakamura Theorem, 127
Nash equilibrium, 128
Negation of a set, 2
Negative definite form, 70
Negative of an element, 21
Negatively transitive, 26
Neighbourhood, 82
Non-degenerate critical point, 150

Non-degenerate form, 70
Non-satiated preference, 110
Norm of a vector, 69, 78
Norm of a vector space, 80
Normal hyperplane, 106
Nowhere dense, 198
Null set, 2
Nullity of a quadratic form, 70

**O**
Oligarchy, 34, 38
Open ball, 81
Open cover, 93
Open set, 82
Optimum, 116
Orthogonal vectors, 69

**P**
Pareto correspondence, 171
Pareto set, 29, 171
Pareto theorem, 203, 216
Partial derivative, 139
Partition, 7
Peixoto-Smale theorem, 212
Permutation, 11
Phase portrait, 210
Poincaré-Hopf Theorem, 214
Positive definite form, 70
Preference manipulation, 180
Preference relation, 24
Prefilter, 38
Price adjustment process, 208
Price equilibrium, 170, 225
Price equilibrium existence, 126
Price vector, 108, 112
Producer optimisation, 111
Product rule, 141
Product topology, 84
Production set, 114
Profit function, 112
Propositional calculus, 4
Pseudo-concave function, 158

**Q**
$q$-Majority, 36
Quadratic form, 69
Quantal response, 238
Quasi-concave function, 100

**R**
Rank of a matrix, 51
Rank of a transformation, 50
Rationality, 25
Real vector space, 40
Reflexive relation, 24

Regular economy, 183, 206
Regular point, 192
Regular value, 192
Relation, 8
Relative topology, 6, 84
Repellor for a vector field, 214
Residual set, 83, 183
Resource manipulation, 181
Retract, 119
Retraction, 119
Rolle's Theorem, 145, 147
Rotations, 17, 74

**S**
Saddle, 70
Saddle point, 151
Sard's lemma, 198
Scalar, 22
Scalar product, 47, 77
Separating hyperplane, 108
Separation of convex sets, 107
Set theory, 1–3, 6
Shadow prices, 111
Shauder's fixed point theorem, 125
Similar matrices, 58
Singular matrix, 14
Singular point, 196
Singularity set of a function, 200
Singularity theorem, 202
Smooth function, 143
Social utility function, 28
Sonnenschein-Mantel-Debreu Theorem, 220
Stochastic choice, 238
Stratified manifold, 202
Strict Pareto rule, 28, 34
Strict partial order, 26, 33
Strict preference relation, 24
Strictly quasi-concave function, 156, 158
Structural stability of a vector field, 219, 222
Subgroup, 18
Submanifold, 202
Submanifold theorem, 202
Submersion, 194
Supremum of a function, 87

Surjective function, 11
Symmetric matrix, 68
Symmetric relation, 24

**T**
Tangent to a function, 138
Taylor's theorem, 148
Thom transversality theorem, 201
Topological space, 82
Topology, 6
Torus, 198
Trace of a matrix, 65
Transfer paradox, 181
Transitive relation, 24
Transversality, 200
Triangle inequality, 79
Truth table, 4
Two party competition, 130
Tychonoff's theorem, 97
Type I extreme value distribution, 238, 239

**U**
Union of sets, 2
Universal quantifier, 7
Universal set, 1
Utility function, 25

**V**
Valence, 238
Vector field, 211
Vector space, 40
Vector subspace, 40
Venn diagram, 2

**W**
Walras' Law, 218
Walras manifold, 203
Walrasian equilibrium, 183
Weak monotone function, 100
Weak order, 26
Weak Pareto rule, 28
Weierstrass theorem, 95
Welfare theorem, 177
Whitney topology, 182, 236

# Author Index

**A**
Aliprantis, C., 133
Aliprantis, D., 229
Arrow, K. J., 32, 35, 38, 187, 225
Aumann, R. J., 225

**B**
Balasko, Y., 181, 187, 206, 227
Banks, J. S., 236, 248
Bergstrom, T., 128, 131, 132, 133, 233
Bikhchandani, S., 226
Border, K., 132
Brouwer, L. E. J., 118, 121, 132
Browder, F. E., 125, 132
Brown, D., 133
Brown, R., 228
Burkenshaw, O., 229

**C**
Caballero, G., 239, 243, 244, 249
Calvin, W. H., 223
Chichilnisky, G., 236, 248
Chillingsworth, D. R. J., 195, 227
Claassen, C., 239, 243, 244, 249
Condorcet, M. J. A. N., 226

**D**
Debreu, G., 183, 203, 204, 206, 215, 220, 223, 227, 228
Dierker, E., 217, 221, 227
Dorussen, H., 248

**E**
Enelow, M. J., 248
Eldredge, N., 223

**F**
Fan, K., 125, 132, 225, 233

**G**
Gale, D., 181, 187
Gamble, A., 225
Gleick, J., 228
Golubitsky, M., 196, 227
Goroff, D., 223
Greenberg, J., 133
Guesnerie, R., 187
Guillemin, V., 196, 227

**H**
Hahn, F. H., 187
Hayek, F. A., 225
Heal, E. M., 132
Hewitt, F., 226
Hildenbrand, W., 108, 187
Hinich, M. J., 243, 248
Hirsch, M., 196, 227
Hirschleifer, D., 226
Hubbard, J. H., 228

**K**
Kauffman, S., 221
Keenan, D., 218, 228
Kepler, J., 222
Keynes, J. M., 226
Kirman, A. P., 38, 108, 187
Knaster, B., 123, 132, 233
Konishi, H., 133
Kramer, G. H., 234, 236, 248
Kuhn, H. W., 115, 132
Kuratowski, K., 123, 132

**L**
Laffont, J.-J., 187
Lange, O., 225
Laplace, P. S., 222
Levine, D., 239, 248
Lin, T., 248
Lorenz, E. N., 224, 229

## M
Mantel, R., 220, 221, 228
Mas-Colell, A., 220, 227
Mazerkiewicz, S., 132
McKelvey, R. D., 236, 237, 239, 248
McLean, I., 226
Michael, E., 123, 132
Miller, G., 248
Minsky, H., 226

## N
Nakamura, K., 35, 36, 38, 127, 236
Nash, J. F., 128, 132
Nenuefeind, W., 218, 228
Neuefeind, W., 128, 131, 132, 133
Newton, I., 222

## O
Ozdemir, U., 239, 243, 244, 249

## P
Palfrey, T. R., 239, 248
Peixoto, M., 203, 212, 219, 227, 228
Peterson, I., 223
Plott, C. R., 248
Poincaré, H., 214, 222, 223
Pontrjagin, L. S., 228
Prabhakar, N., 133

## R
Rader, T., 218, 228
Riezman, R., 128, 131, 132, 133, 218, 228

Rothman, N. J., 229

## S
Saari, D., 229, 236, 248
Safra, Z., 187
Scarf, H., 189, 207, 209, 217, 227
Schauder, J., 125, 132
Schofield, N., 133, 217, 228, 236, 237, 239, 241, 243–245, 248, 249
Schumpeter, J. A., 225
Shafer, W., 133
Skidelsky, R., 226
Smale, S., 183, 187, 203, 212, 215, 219, 221, 227, 228
Sondermann, D., 38
Sonnenschein, H., 133, 220, 221, 228
Strnad, J., 133

## T
Thom, R., 201, 228
Tucker, A. W., 115, 132

## V
von Mises, L., 225

## W
Welsh, I., 226
West, B. H., 228

## Y
Yannelis, N., 133

Printed by Printforce, the Netherlands